Advanced Magnetic Nanocomposites

Advanced Magnetic Nanocomposites: Structural, Physical Properties and Application

Editor

Raghvendra Singh Yadav

MDPI • Basel • Beijing • Wuhan • Barcelona • Belgrade • Manchester • Tokyo • Cluj • Tianjin

Editor
Raghvendra Singh Yadav
Centre of Polymer Systems
Tomas Bata University in Zlin
Zlín
Czech Republic

Editorial Office
MDPI
St. Alban-Anlage 66
4052 Basel, Switzerland

This is a reprint of articles from the Special Issue published online in the open access journal *Nanomaterials* (ISSN 2079-4991) (available at: www.mdpi.com/journal/nanomaterials/special_issues/advanced_magnetic).

For citation purposes, cite each article independently as indicated on the article page online and as indicated below:

LastName, A.A.; LastName, B.B.; LastName, C.C. Article Title. *Journal Name* **Year**, *Volume Number*, Page Range.

ISBN 978-3-0365-3143-4 (Hbk)
ISBN 978-3-0365-3142-7 (PDF)

Contents

About the Editor

Raghvendra Singh Yadav

Dr. Raghvendra Singh Yadav is working as a 'Senior Scientist/Researcher'at the Centre of Polymer Systems, Tomas Bata University in Zlin, Czech Republic.

He obtained his Ph.D. degree on the topic "Synthesis and Characterization of Nanophosphors (luminescent/optical nanoparticles)"from University of Allahabad, India. Dr.Yadav's current scientific activities are focused on "Lightweight, Flexible, Low-dimensional Electromagnetic Functional Nanocomposite Materials (MXene, MBene, Graphene, magnetic nanoparticles as nanofillers in a polymer matrix) and its Applications".

Article

Excellent, Lightweight and Flexible Electromagnetic Interference Shielding Nanocomposites Based on Polypropylene with MnFe$_2$O$_4$ Spinel Ferrite Nanoparticles and Reduced Graphene Oxide

Raghvendra Singh Yadav [1],[*], Anju [1], Thaiskang Jamatia [1], Ivo Kuřitka [1], Jarmila Vilčáková [1], David Škoda [1], Pavel Urbánek [1], Michal Machovský [1], Milan Masař [1], Michal Urbánek [1], Lukas Kalina [2] and Jaromir Havlica [2]

[1] Centre of Polymer Systems, University Institute, Tomas Bata University in Zlín, Trida Tomase Bati 5678, 760 01 Zlín, Czech Republic; deswal@utb.cz (A.); jamatia@utb.cz (T.J.); kuritka@utb.cz (I.K.); vilcakova@utb.cz (J.V.); dskoda@utb.cz (D.Š.); urbanek@utb.cz (P.U.); machovsky@utb.cz (M.M.); masar@utb.cz (M.M.); murbanek@utb.cz (M.U.)

[2] Materials Research Centre, Brno University of Technology, Purkyňova 464/118, 61200 Brno, Czech Republic; kalina@fch.vut.cz (L.K.); havlica@fch.vut.cz (J.H.)

[*] Correspondence: yadav@utb.cz; Tel.: +42-0576031725

Received: 18 November 2020; Accepted: 7 December 2020; Published: 10 December 2020

Abstract: In this work, various tunable sized spinel ferrite MnFe$_2$O$_4$ nanoparticles (namely MF20, MF40, MF60 and MF80) with reduced graphene oxide (RGO) were embedded in a polypropylene (PP) matrix. The particle size and structural feature of magnetic filler MnFe$_2$O$_4$ nanoparticles were controlled by sonochemical synthesis time 20 min, 40 min, 60 min and 80 min. As a result, the electromagnetic interference shielding characteristics of developed nanocomposites MF20-RGO-PP, MF40-RGO-PP, MF60-RGO-PP and MF80-RGO-PP were also controlled by tuning of magnetic/dielectric loss. The maximum value of total shielding effectiveness (SE$_T$) was 71.3 dB for the MF80-RGO-PP nanocomposite sample with a thickness of 0.5 mm in the frequency range (8.2–12.4 GHz). This lightweight, flexible and thin nanocomposite sheet based on the appropriate size of MnFe$_2$O$_4$ nanoparticles with reduced graphene oxide demonstrates a high-performance advanced nanocomposite for cutting-edge electromagnetic interference shielding application.

Keywords: spinel ferrite; nanocomposites; electromagnetic interference shielding; magnetic loss; dielectric loss

1. Introduction

Extensive practice of electronic and communication devices, liberating electromagnetic (EM) waves, generates EM radiation pollution [1]. Electromagnetic interference (EMI) does not only affect the working and life of electronic devices but also is harmful to human health [2]. This noble type of EM radiation pollution delivers a solid motivation to develop efficient EMI shielding materials [3]. Lightweight, thinness and cost efficiency are other additional necessities of high-performance EMI shielding materials for operational applications [4]. Polymer-based EMI shielding composite materials are lightweight, resistant to corrosion, flexible and simple in preparation [5]. The performance of polymer-based EMI shielding materials depends on the intrinsic electrical conductivity, aspect ratio, and concentration of the fillers [6]. Graphene has received considerable attention as nano-fillers due to their excellent electrical and thermal conductivities, and ultrahigh mechanical characteristics [7]. Additionally, spinel ferrite nanoparticles as nanofillers have been established as potential magnetic absorbers due to their outstanding magnetic loss, good stability and cost-effectiveness [8,9].

The particle shape and size of nanoparticles have a vital impact on the microwave absorption and electromagnetic interference shielding characteristics of nanoparticles and their nanocomposites [10]. In recent years, researchers have noticed the influence of particle size on microwave absorption and electromagnetic shielding performance [11]. Yi-Jun Liang et al. [12] noticed the size-dependent microwave absorption performance of Fe_3O_4 nanoparticles prepared by the rapid microwave-assisted thermal decomposition method. Niandu Wu et al. [13] observed particle size-dependent microwave absorption characteristics of carbon-coated nickel nanocapsules. A correlation of particle size with electromagnetic parameters can benefit us in better control of electromagnetic interference shielding performance. Our research group [14] also noticed that the particle size of $NiFe_2O_4$ nanoparticles correlates with the electromagnetic interference shielding performance of nanocomposites.

Efficient electromagnetic interference shielding nanocomposite material having a feature of lightweight, flexible and excellent shielding characteristics are highly essential. Here, lightweight, flexible and excellent EMI-shielding nanocomposites with control of magnetic loss/dielectric loss through the control of particle size of $MnFe_2O_4$ spinel ferrite embedded in polypropylene matric with reduced graphene oxide have been developed. Various sized $MnFe_2O_4$ spinel ferrite nanoparticles were synthesized by sonochemical synthesis at different sonication times.

2. Materials and Methods

2.1. Materials

The reagents manganese nitrate, iron nitrate and sodium hydroxide were procured from Alfa Aesar GmbH and Co KG (Karlsruhe, Germany). Potassium permanganate and graphite flakes were acquired from Sigma-Aldrich, (Munich, Germany). Sodium nitrate was obtained from Lach-Ner (Brno, Czech Republic). The utilized polypropylene (Vistamaxx 6202) was procured from Exxon Mobil (Machelen, Belgium). The reducing agent Vitamin C (Livsane) was obtained from Dr. Kleine Pharma GmbH, (Bielefeld, Germany).

2.2. Preparation of Nanoparticles

Various sized $MnFe_2O_4$ spinel ferrite nanoparticles were prepared by the sonochemical synthesis approach as reported in our previous report [15]. A schematic illustration of the preparation of $MnFe_2O_4$ spinel ferrite nanoparticles by the sonochemical synthesis approach is shown in Figure 1. Further, the synthesis condition for the preparation of these $MnFe_2O_4$ nanoparticles by the sonochemical method is tabulated in Table 1. For the preparation, manganese nitrate and iron nitrate was mixed with deionized water in a beaker. This solution was stirred on a magnetic stirrer for 5 min at room temperature. To this prepared mixed solution, sodium hydroxide aqueous solution was added and the whole mixed solution was placed under sonication (Ultrasonic homogenizer UZ SONOPULS HD 2070 (Berlin, Germany) (frequency: 20 kHz and power: 70 W)) for 20 min. The precipitate was collected and then washed with deionized water and ethanol and finally dried at 40 °C. Further, the increased particle size $MnFe_2O_4$ spinel ferrite nanoparticles were prepared for sonication time 40 min, 60 min and 80 min. The reaction temperature was 65 °C, 74 °C, 85 °C and 93 °C, after sonication time 20 min, 40 min, 60 min and 80 min, respectively. The synthesized $MnFe_2O_4$ nanoparticles were designated as MF20, MF40, MF60 and MF80 related to different sonication times 20 min, 40 min, 60 min and 80 min, respectively. Further, graphene oxide was prepared by the modified Hummer's method [16]. Furthermore, graphene oxide (GO) was converted into reduced graphene oxide (RGO) by utilizing vitamin C as a reducing agent.

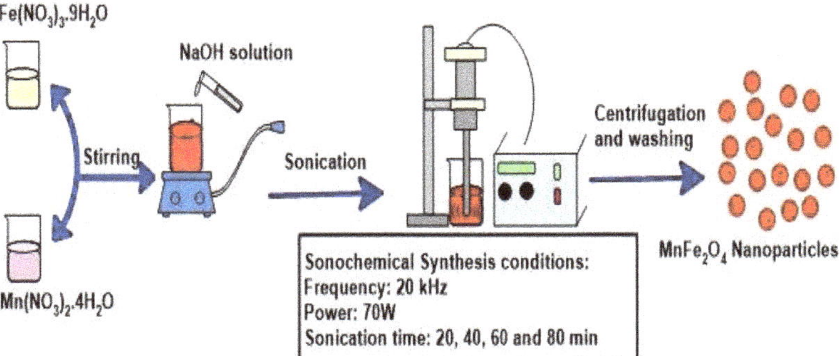

Figure 1. Schematic illustration of the preparation of MnFe$_2$O$_4$ nanoparticles by the sonochemical synthesis method.

Table 1. Synthesis condition for preparation of MnFe$_2$O$_4$ nanoparticles by the sonochemical method.

Sample	Concentration of Mn (NO$_3$)$_2$ 4H$_2$O	Concentration of Fe (NO$_3$)$_2$ 9H$_2$O	Concentration of NaOH	Sonication Time	Reaction Temperature
MF20	0.17 M	0.36 M	1.66 M	20 min	65 °C
MF40	0.17 M	0.36 M	1.66 M	40 min	74 °C
MF60	0.17 M	0.36 M	1.66 M	60 min	85 °C
MF80	0.17 M	0.36 M	1.66 M	80 min	93 °C

2.3. Preparation of Nanocomposites

A schematic illustration of the preparation of polypropylene (PP) based nanocomposites embedded with MnFe$_2$O$_4$ spinel ferrite nanoparticles and reduced graphene oxide (RGO) is shown in Figure 2. Nanocomposites of PP (50 wt %) with MnFe$_2$O$_4$ nanoparticles (40 wt %) and RGO (10 wt %) as nanofillers were developed by using the melt-mixing method. Four nanocomposite samples, namely (i) MF20-RGO-PP, (ii) MF40-RGO-PP, (iii) MF60-RGO-PP and (iv) MF80-RGO-PP were prepared. The rectangle-shaped sheet of a 22.86 × 10.16 × 0.5 mm^3 dimension of prepared nanocomposites was developed by the hot-press approach. A representative digital photograph of PP nanocomposite embedded with MnFe$_2$O$_4$ spinel ferrite nanoparticles and reduced graphene oxide (RGO) as nanofillers is shown in Figure 3.

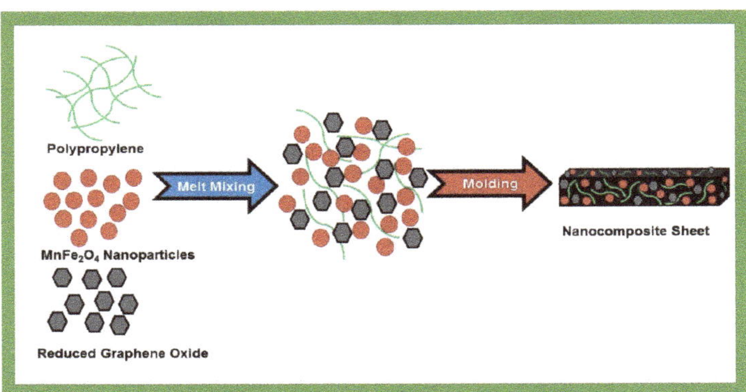

Figure 2. Schematic illustration of the preparation of polypropylene (PP) based nanocomposites embedded with MnFe$_2$O$_4$ spinel ferrite nanoparticles and reduced graphene oxide (RGO).

Figure 3. Digital photograph of PP nanocomposite embedded with $MnFe_2O_4$ spinel ferrite nanoparticles and reduced graphene oxide (RGO) as nanofillers.

2.4. Characterization Techniques

The EMI shielding effectiveness of prepared nanocomposite (MF20-RGO-PP, MF40-RGO-PP, MF60-RGO-PP and MF80-RGO-PP) sheets of dimension $22.86 \times 10.16 \times 0.5$ mm^3 was studied with a vector network analyzer (Agilent N5230A) at 8.2–12.4 GHz (the so-called X-band) frequency range using a waveguide sample holder. X-ray powder diffraction (Rigaku Corporation, Tokyo, Japan) characterization tool was employed to analyze the crystal structure of nanocomposites. A field emission scanning electron microscope (FEI NanoSEM450) was employed to observe the morphology and presence of $MnFe_2O_4$ nanoparticles and reduced graphene oxide in the polypropylene matrix. Raman spectrometer (Thermo Fisher Scientific, Waltham, MA, USA) was used for Raman spectra of prepared RGO, PP, and its nanocomposites. A vibrating sample magnetometer (VSM 7407, Lake Shore) was employed to study magnetic hysteresis curves of prepared nanocomposite (MF20-RGO-PP, MF40-RGO-PP, MF60-RGO-PP and MF80-RGO-PP). The FTIR spectrometer (Nicolet 6700, Thermo Scientific) was utilized to achieve the FTIR spectra of prepared nanocomposites. Thermogravimetric analyses of prepared nanocomposites were performed on a Setaram LabSys Evo with TG/DSC sensor in an atmosphere of air (heating ramp 5 °C min^{-1}, up to 1000 °C, and air flow 60 mL min^{-1}). Mechanical properties of prepared polypropylene based nanocomposites were measured on a Testometric universal-testing machine of type M 350–5CT (Testometric Co. Ltd., Rochdale, UK).

3. Results

3.1. XRD Study

XRD pattern of polypropylene (PP) and its prepared nanocomposites MF20-RGO-PP, MF40-RGO-PP, MF60-RGO-PP and MF80-RGO-PP is shown in Figure 4. The X-ray diffraction peaks indexed with (220), (311), (222), (400), (422), (511) and (440) confirm the presence of cubic spinel structure of $MnFe_2O_4$ nanoparticles in prepared nanocomposites [17]. It is noticeable in Figure 4 that the diffraction peak intensity of $MnFe_2O_4$ spinel ferrite nanoparticles was increased with the increase of sonication time, which signified an increase of crystallite size also [15]. The X-ray diffraction peaks at 14.2°, 16.8°, 18.2°, 21.1° and 21.9°, which is associated with (110), (040), (130), (111) and (131) + (041), respectively, crystal plane of the α-form of polypropylene [18]. Further, no diffraction peak associated with reduced graphene oxide was observed because of the low XRD intensity of RGO in prepared nanocomposites [19].

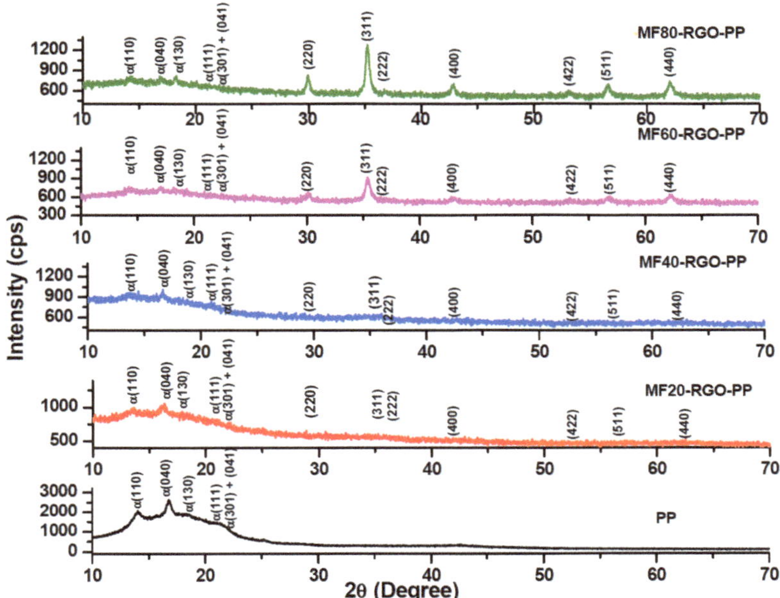

Figure 4. XRD pattern of polypropylene (PP) and prepared nanocomposites MF20-RGO-PP, MF40-RGO-PP, MF60-RGO-PP and MF80-RGO-PP.

3.2. FE-SEM Study

Field emission scanning electron microscopy (FE-SEM) was utilized to investigate morphology of prepared nanocomposites. FE-SEM image of cross-sections of prepared MF60-RGO-PP and MF80-RGO-PP nanocomposites is shown in Figure 5. Images display the existence of $MnFe_2O_4$ spinel ferrite nanoparticles and reduced graphene oxide in the polypropylene matrix system. Further, FE-SEM image of prepared MF20-RGO-PP and MF40-RGO-PP nanocomposites is shown in Figure 6a,c, respectively. The presence of $MnFe_2O_4$ nanoparticles and reduced graphene oxide can be noticed in the polypropylene matrix. In addition, energy dispersive X-ray spectrum (EDX) of the MF20-RGO-PP (Figure 6b) and MF40-RGO-PP (Figure 6d) showed the existence of C, O, Mn and Fe.

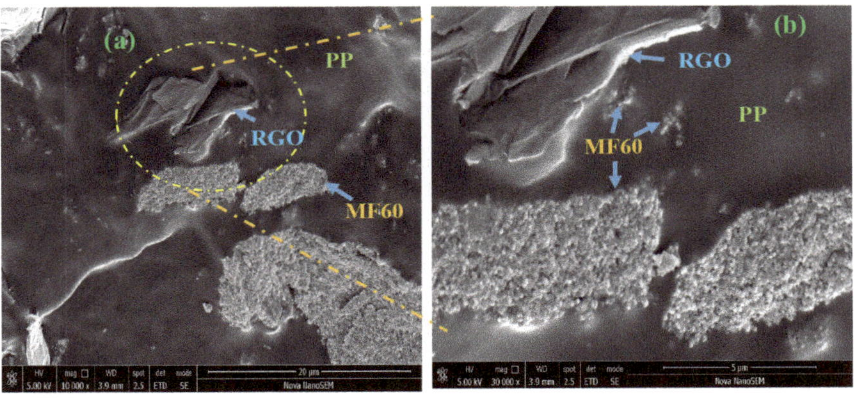

Figure 5. *Cont.*

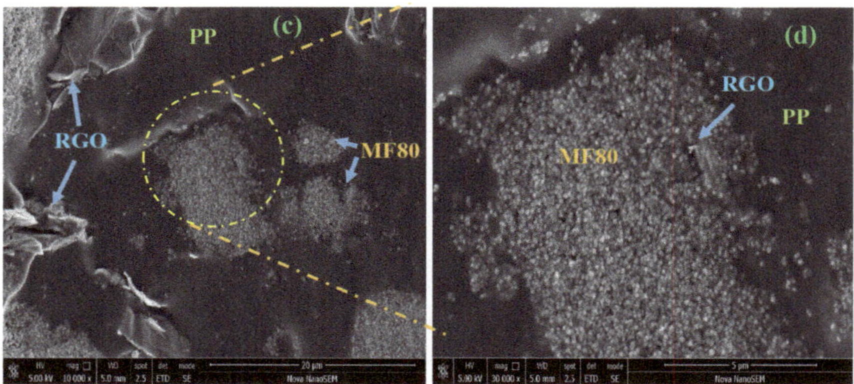

Figure 5. (**a,b**) FE-SEM image of cross-sections of the MF60-RGO-PP sample and (**c,d**) FE-SEM image of cross-sections of the MF80-RGO-PP sample.

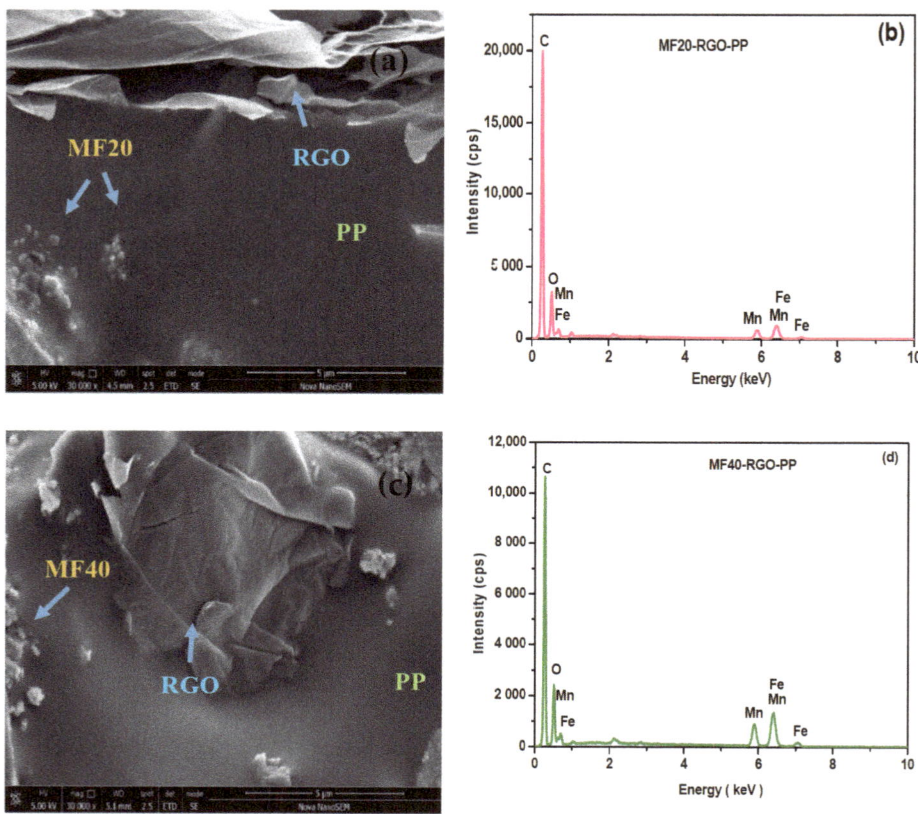

Figure 6. (**a**) FE-SEM image of cross-sections of MF20-RGO-PP, (**b**) EDX spectrum of MF20-RGO-PP, (**c**) FE-SEM image of cross-sections of MF40-RGO-PP and (**d**) EDX spectrum of MF40-RGO-PP.

3.3. Raman Spectroscopy

Figure 7 shows the Raman spectra of polypropylene (PP), reduced graphene oxide (RGO) and prepared nanocomposite MF20-RGO-PP, MF40-RGO-PP, MF60-RGO-PP and MF80-RGO-PP samples. The crystal structure and presence of $MnFe_2O_4$ in nanocomposites were confirmed through the measurement of A_{1g}, E_g and T_{2g} peak positions in the Raman spectrum. In Figure 7, the existence of characteristics Raman bands, i.e., E_g mode (296 cm^{-1}), T_{2g} mode (242 cm^{-1}, 355 cm^{-1} and 580 cm^{-1}) and A_{1g} mode (604 cm^{-1} and 657 cm^{-1}) of spinel ferrite can be noticed [20]. The appearance of two characteristics peaks of RGO at 1338 cm^{-1} and 1594 cm^{-1} corresponds to the D-band and G-band of RGO, respectively [21]. Additionally, the other Raman peaks in the nanocomposites are associated with the chemical group of polypropylene [22].

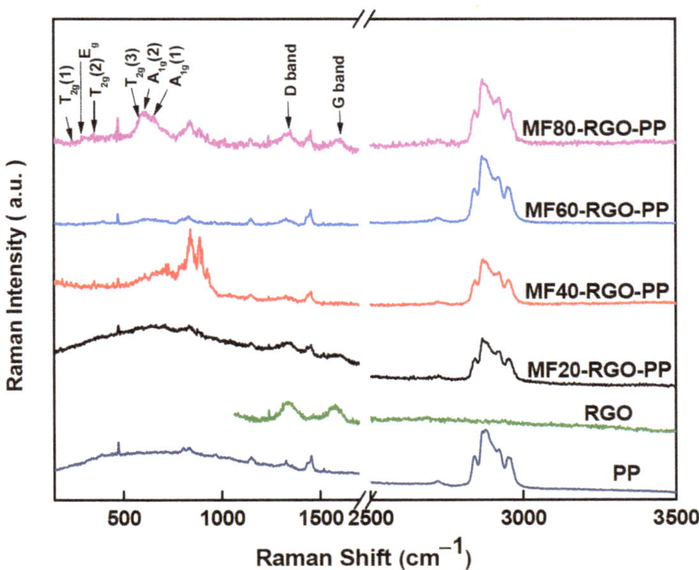

Figure 7. Raman spectrum of polypropylene (PP), reduced graphene oxide (RGO), MF20-RGO-PP, MF40-RGO-PP, MF60-RGO-PP and MF80-RGO-PP.

3.4. FTIR Spectroscopy

Figure 8 displays the FTIR spectra of polypropylene (PP) and developed nanocomposite samples MF20-RGO-PP, MF40-RGO-PP, MF60-RGO-PP and MF80-RGO-PP. The presence of characteristic FTIR peaks of $MnFe_2O_4$ spinel ferrite nanoparticles and polypropylene can be noticed in the prepared nanocomposites, as shown in Figure 8. In spinel ferrite, the infrared bands noticed between 100 and 600 cm^{-1} indicate the formation of single phase spinel ferrite material. The absorption band at 565 cm^{-1} was associated with the intrinsic stretching vibration of metals at tetrahedral sites in $MnFe_2O_4$ nanoparticles [23]. The absorption peak at 840 cm^{-1} was associated with C–CH$_3$ stretching vibration in PP. The peak 972 cm^{-1}, and 1165 cm^{-1} were associated with –CH$_3$ rocking vibration. The absorption peak at 1375 cm^{-1} and 2952 cm^{-1} were related to symmetric bending vibration of the –CH$_3$ group and –CH$_3$ asymmetric stretching vibration. The absorption peak at 1455 cm^{-1}, 2838 cm^{-1} and 2917 cm^{-1} were related to –CH$_2$-symmetric bending, –CH$_2$-symmetric stretching, and –CH$_2$-asymmetric stretching, respectively [24]. In amalgamation with Raman and FTIR spectroscopy results, the presence of $MnFe_2O_4$ spinel ferrite nanoparticles and reduced graphene oxide (RGO) in the polypropylene (PP) were confirmed.

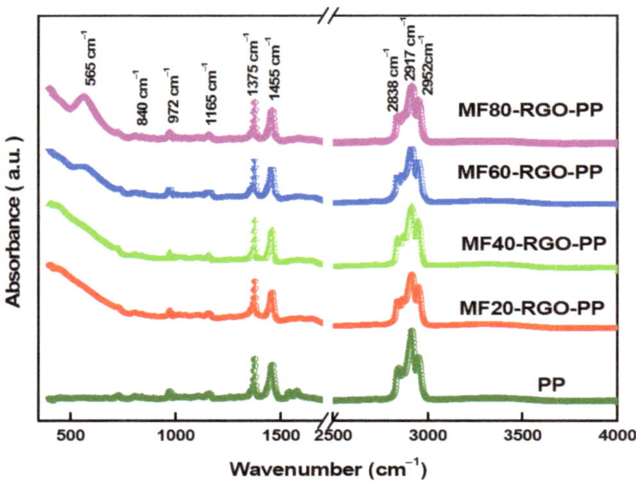

Figure 8. FTIR spectrum of polypropylene (PP), MF20-RGO-PP, MF40-RGO-PP, MF60-RGO-PP and MF80-RGO-PP.

3.5. Thermogravimetric Analysis (TGA)

Figure 9 depicts the TGA curves of polypropylene (PP) and its prepared MF20-RGO-PP, MF40-RGO-PP, MF60-RGO-PP and MF80-RGO-PP nanocomposites under air atmosphere. It can be noticed that the PP had lower degradation temperature in comparison with its prepared nanocomposites. Further, nanocomposites exhibited higher thermal stability as compared to PP, which is associated with the result of an interaction between PP, $MnFe_2O_4$ nanoparticles and RGO [25]. Furthermore, the oxidative residues at 1000 °C are 37%, 39%, 46% and 49.2% for MF20-RGO-PP, MF40-RGO-PP, MF60-RGO-PP and MF80-RGO-PP, respectively, with 50% nanofillers loading [26]. The slightly lower residue values especially for MF20-RGO-PP and MF40-RGO-PP sample than the corresponding actual residues (i.e., loaded nano-fillers) were mainly due to the evaporation of surface impurities/chemical functional group attached on surface of small sized nanoparticles MF20 and MF40 [27].

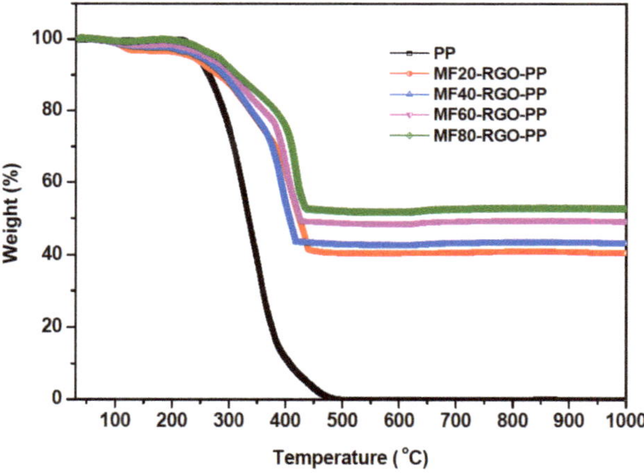

Figure 9. TGA curves of polypropylene (PP) and its prepared MF20-RGO-PP, MF40-RGO-PP, MF60-RGO-PP and MF80-RGO-PP nanocomposites under air atmosphere.

3.6. Magnetic Property

Magnetic properties of prepared MF20-RGO-PP, MF40-RGO-PP, MF60-RGO-PP and MF80-RGO-PP nanocomposites were investigated by using a vibrating sample magnetometer. The magnetic hysteresis curves of MF20-RGO-PP, MF40-RGO-PP, MF60-RGO-PP and MF80-RGO-PP nanocomposites are shown in Figure 10. Ferromagnetic behavior can be noticed in magnetic hysteresis curves as depicted in Figure 10 for MF20-RGO-PP (H_c = 33.2 Oe, M_r = 0.003 emu/g, M_s = 0.45 emu/g), MF40-RGO-PP (H_c = 43.57 Oe, M_r = 0.008 emu/g, M_s = 0.55 emu/g), MF60-RGO-PP (H_c = 61.0 Oe, M_r = 1.57 emu/g, M_s = 14.6 emu/g) and MF80-RGO-PP (H_c = 45.9 Oe, M_r = 2.03 emu/g, M_s = 24.8 emu/g) nanocomposites. The ferromagnetic behavior of nanoparticles MF20 (M_s = 1.9 emu/g, H_c = 45.0 Oe, M_r = 0.12 emu/g), MF40 (M_s = 2.5 emu/g, H_c = 42.0 Oe, M_r = 0.13 emu/g), MF60 (M_s = 30.2 emu/g, H_c = 34.0 Oe, M_r = 2.27 emu/g) and MF80 (M_s = 52.5 emu/g, H_c = 32.0, M_r = 4.50 emu/g) was noticed, as mentioned in our previous report [15]. The high-frequency resonance in terms of anisotropy constant (K), anisotropy energy (H_a) and resonance frequency (f_r) has the following interrelationship with coercivity (H_c) and saturation magnetization (M_s) [28]:

$$K = \frac{u_o M_s H_c}{2} \tag{1}$$

$$H_a = \frac{4|K|}{3u_o M_s} \tag{2}$$

$$2\pi f_r = r H_a \tag{3}$$

where μ_o is the universal value of permeability in free space ($4\pi \times 10^{-7}$ H/m) and r is the gyromagnetic ratio. The correlation of the above equations signifies that the value of H_c and M_s can influence the magnitude of K, H_a and f_r and consequently electromagnetic properties of nanocomposites [29].

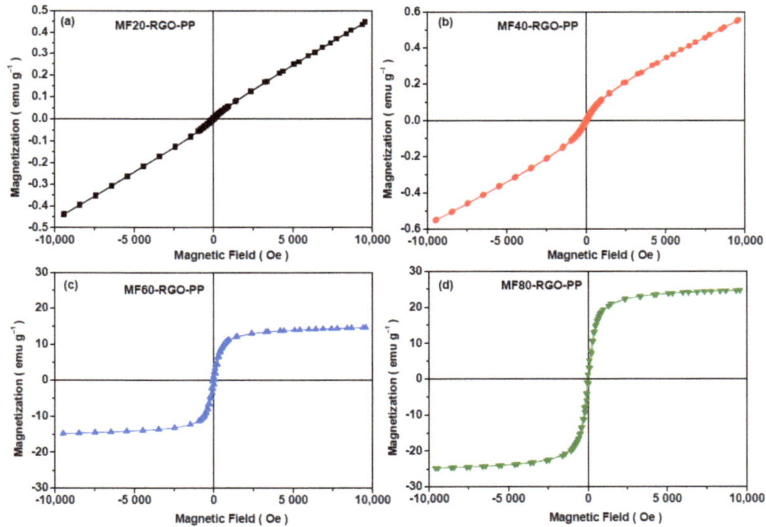

Figure 10. Magnetic hysteresis curves of prepared nanocomposites MF20-RGO-PP, MF40-RGO-PP, MF60-RGO-PP and MF80-RGO-PP samples.

3.7. Electromagnetic Interference Shielding Effectiveness

The electromagnetic interference (EMI) shielding effectiveness (SE) is the degree of the material's capability to block the electromagnetic waves. It is represented by the logarithm of the ratio of incident power (P_I) to transmitted power (P_T) in decibels

$$SE_T(\text{dB}) = 10 \log\left(\frac{P_I}{P_T}\right) \tag{4}$$

The attenuation of the electromagnetic waves involves generally three mechanisms: reflection (SE_R), absorption (SE_A) and multiple reflections (SE_M). When the shielding effectiveness associated with absorption has a higher value than 10 dB, i.e., approximately all the rereflected waves will be absorbed within the material, the contribution associated with multiple reflections can be neglected [30]. Then, the total shielding effectiveness (SE_T) can be expressed as

$$SE_T = SE_R + SE_A \tag{5}$$

A two-port network analyzer can be utilized to measure the scattering parameters (S_{11}, S_{12}, S_{21} and S_{22}), which correlates with reflection (R) and transmission coefficients (T) as [31]:

$$T = |S_{12}|^2 = |S_{21}|^2 \tag{6}$$

$$R = |S_{11}|^2 = |S_{22}|^2 \tag{7}$$

The shielding effectiveness due to absorption (SE_A) and reflection (SE_R) can be expressed in terms of the scattering parameters as

$$SE_R = 10 \log\left(\frac{1}{1-R}\right) = 10 \log\left(\frac{1}{1-|S_{11}|^2}\right) \tag{8}$$

$$SE_A = 10 \log\left(\frac{1-R}{T}\right) = 10 \log\left(\frac{1-|S_{11}|^2}{|S_{21}|^2}\right) \tag{9}$$

Therefore, total shielding effectiveness (SE_T) can be obtained from the above relations as

$$SE_T = 20 \log(S_{21}) \tag{10}$$

Figure 11 depicts the EMI shielding effectiveness of prepared nanocomposites MF20-RGO-PP, MF40-RGO-PP, MF60-RGO-PP and MF80-RGO-PP at a thickness of 0.5 mm. The maximum value of total shielding effectiveness (SE_T) was 58.6 dB, 66.4 dB, 69.4 dB and 71.3 dB for MF20-RGO-PP, MF40-RGO-PP, MF60-RGO-PP and MF80-RGO-PP samples, respectively, as shown in Figure 11a. Further, the maximum value of shielding effectiveness due to absorption (SE_A) was 35.3 dB, 41.3 dB, 44.3 dB and 45.8 dB for MF20-RGO-PP, MF40-RGO-PP, MF60-RGO-PP and MF80-RGO-PP samples, respectively, as shown in Figure 11b. Additionally, the maximum value of shielding effectiveness due to reflection (SE_R) was 23.4 dB, 25.2 dB, 25.1 dB and 25.6 dB for MF20-RGO-PP, MF40-RGO-PP, MF60-RGO-PP and MF80-RGO-PP samples, respectively, as shown in Figure 11c. The maximum value of total EMI SE (SE_T), absorption (SE_A) and reflection (SE_R) of developed nanocomposites were plotted in Figure 11d. The results imply an absorption dominant shielding mechanism instead of reflection in the designed nanocomposites.

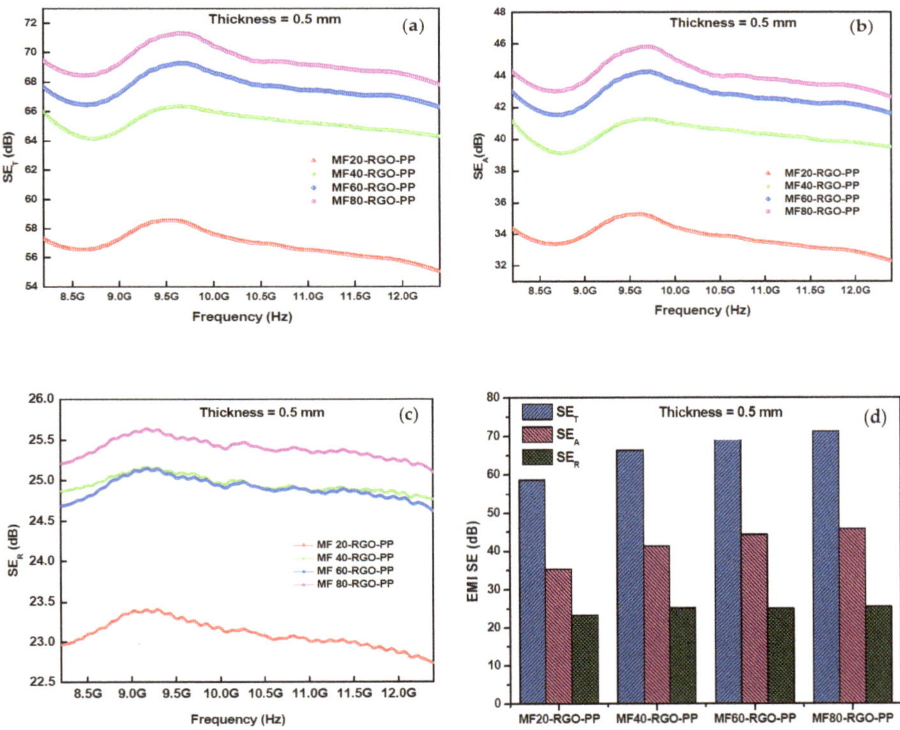

Figure 11. Electromagnetic interference (EMI) shielding effectiveness (**a**) SE_T, (**b**) SE_A, (**c**) SE_R and (**d**) comparison plot of SE_T, SE_A and SE_R for prepared nanocomposites MF20-RGO-PP, MF40-RGO-PP, MF60-RGO-PP and MF80-RGO-PP at a thickness of 0.5 mm.

A research group, X.-J. Zhang et al. [32] noticed the minimum reflection loss -29.0 dB at 9.2 GHz for RGO/MnFe$_2$O$_4$/PVDF composites, which contained 5 wt % filler content with a thickness of 3.0 mm. P. Yin et al. [33] observed the optimal microwave absorbing intensity -48.92 dB at 0.78 GHz at a 2.5 mm thickness for the Apium-derived biochar loaded with MnFe$_2$O$_4$@C. Another researcher, R. V. Lakshmi et al. [34] observed a total shielding effectiveness value 44 dB in the X band frequency range for PMMA modified MnFe$_2$O$_4$-polyaniline nanocomposites. R. K. Srivastava et al. [35] noticed the total shielding effectiveness of -38 dB filler 5 wt % RGO-MnFe$_2$O$_4$ and 3 wt % of MWCNTs in polyvinylidene fluoride (PVDF) matrix. Another research group, Y. Wang et al. [36] observed the maximum reflection loss of -32.8 dB at 8.2 GHz with the thickness of 3.5 mm for MnFe$_2$O$_4$/RGO composite. Further, P. Yin et al. [37] noticed the maximum reflection loss of -14.87 dB at 2.25 GHz with the thickness of 4 mm. Furthermore, Y. Wang et al. [38] showed the maximum absorption of -38 dB at 6 GHz with the thickness of 3.5 mm for a ternary composite of Ag/MnFe$_2$O$_4$/reduced graphene oxide (RGO). Additionally, a comparison of electromagnetic wave absorption performance between MnFe$_2$O$_4$ spinel ferrite nanoparticles based developed composites reported in recent years are tabulated in Table 2.

Table 2. Comparison of electromagnetic wave absorption performance between $MnFe_2O_4$ spinel ferrite nanoparticles based developed composites reported in recent years.

No.	Shielding Material	Frequency (GHz)	Specimens Thickness (mm)	Effect of Shielding	Ref.
1.	$RGO/MnFe_2O_4/PVDF$	2–18 GHz	3.0 mm	−29.0 dB	[32]
2.	$Biochar/MnFe_2O_4@C$	0.2–3 GHz	2.5 mm	−48.92 dB	[33]
3.	PMMA modified $MnFe_2O_4$-PANI	8–12 GHz		~44 dB	[34]
4.	$PVDF/RGO$-$MnFe_2O_4/MWCNTs$	8–18 GHz		−38dB	[35]
5.	$MnFe_2O_4/RGO$	2–18 GHz	3.5 mm	−32.8 dB	[36]
6.	SiO_2-$MnFe_2O_4$	0.2–3 GHz	4 mm	−14.87 dB	[37]
7.	$Ag/MnFe_2O_4/RGO$	2–18 GHz	3.5 mm	−38 dB	[38]
8.	$MnFe_2O_4$-RGO-PP	8.2–12.4 GHz	0.5 mm	~71.3 dB	This work

The influence of thickness on EMI shielding effectiveness of prepared nanocomposites was also investigated. Figure 12 depicts EMI shielding effectiveness of prepared nanocomposites MF20-RGO-PP, MF40-RGO-PP, MF60-RGO-PP and MF80-RGO-PP nanocomposite sheet at a thickness of 1 mm. The maximum value of total shielding effectiveness (SE_T) was 39.15 dB, 41.72 dB, 44.21 dB and 49.11 dB for MF20-RGO-PP, MF40-RGO-PP, MF60-RGO-PP and MF80-RGO-PP nanocomposite sheet, respectively, at a thickness of 1 mm, as depicted in Figure 12a. Further, the maximum value of shielding effectiveness due to absorption (SE_A) was 19.52 dB, 22.07 dB, 23.78 dB and 28.12 dB for MF20-RGO-PP, MF40-RGO-PP, MF60-RGO-PP and MF80-RGO-PP nanocomposite, respectively, at a thickness of 1 mm, as shown in Figure 12b. Furthermore, the maximum value of shielding effectiveness due to reflection (SE_R) was 19.64 dB, 19.67 dB, 20.44 dB and 21.00 dB for MF20-RGO-PP, MF40-RGO-PP, MF60-RGO-PP and MF80-RGO-PP nanocomposite, respectively, at a thickness of 1 mm, as represented in Figure 12c. A comparative value of SE_T, SE_A and SE_R for prepared nanocomposite sheet at a thickness of 1 mm is represented in Figure 12d. A slight decrease in shielding effectiveness was noticed with an increase in thickness from 0.5 to 1 mm [39,40], however, the observed value of EMI shielding effectiveness was higher enough than the limit (20 dB) needed for techno-commercial applications [41].

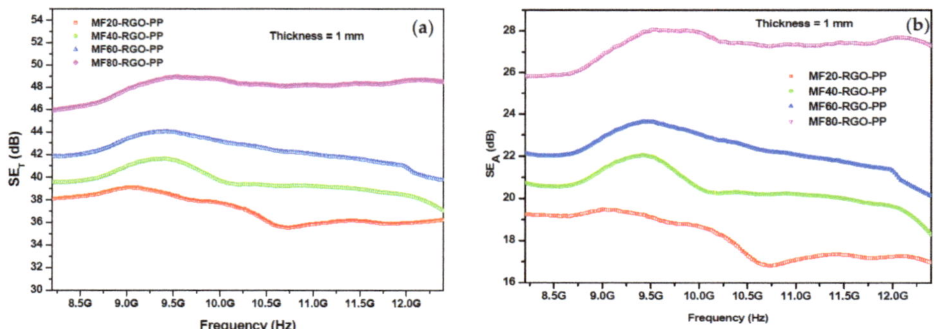

Figure 12. *Cont.*

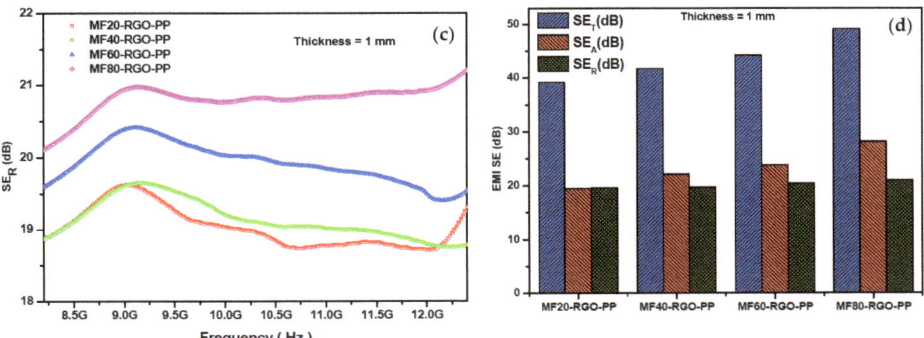

Figure 12. EMI shielding effectiveness (**a**) SE_T, (**b**) SE_A, (**c**) SE_R and (**d**) comparison plot of SE_T, SE_A and SE_R for MF20-RGO-PP, MF40-RGO-PP, MF60-RGO-PP and MF80-RGO-PP nanocomposite sheet at thickness of 1 mm.

3.8. Electromagnetic Properties and Parameters

It is well-known that the electromagnetic interference shielding performances of the nanocomposites are highly associated with complex permittivity ($\varepsilon_r = \varepsilon' + j\varepsilon''$) and complex permeability ($\mu_r = \mu' + j\mu''$). Figure 13a shows the real part of the complex permittivity (ε') of prepared nanocomposites. The real part of the complex permittivity (ε') corresponds to the storage of the electrical energy and can be controlled by polarization in the material. The existence of RGO and $MnFe_2O_4$ nanoparticles in the PP matrix created a heterogeneous medium that acted as interface accumulation in the developed nanocomposites. The value of the real part of the complex permittivity (ε') was in the range of 4.64–4.84, 4.62–4.76, 4.34–4.42 and 4.20–4.79 for MF20-RGO-PP, MF40-RGO-PP, MF60-RGO-PP and MF80-RGO-PP sample, respectively. The real part of the permittivity (ε') was associated with the polarization in the material, which consisted of dipolar polarization, interfacial/surface polarization, orientational polarization, ionic or electronic polarization. The higher value of the real part of permittivity (ε') of MF20-RGO-PP and MF40-RGO-PP was due to presence of a higher number of surface impurity bonds/residual bonds and cluster defects in MF20 and MF40 nanoparticles via a chemical synthesis route, the electrons were not evenly distributed, which led to orientation polarization and thereby further enhancement in the real part of permittivity [42]. The ionic or electronic polarization plays a dominant role in enhancing the real part of permittivity at a high frequency. The existence of this polarization enhanced slowly with increase of frequency. Therefore, the increase of real part of permittivity (ε') at higher frequency in MF80-RGO-PP sample could be associated with dominant role of electronic polarization [43]. The real part of the complex permittivity (ε') had no direct relation with the total shielding effectiveness (SE_T). The imaginary part of the permittivity (ε'') corresponded to the dielectric loss in the materials. Figure 13b depicts the imaginary part of the permittivity (ε'') of the developed nanocomposites. It can be observed that the value of the imaginary part of the complex permittivity (ε'') was in the range of 0.18–0.35, 0.17–0.36, 0.15–0.30 and −0.03–0.15 for MF20-RGO-PP, MF40-RGO-PP, MF60-RGO-PP and MF80-RGO-PP nanocomposite, respectively.

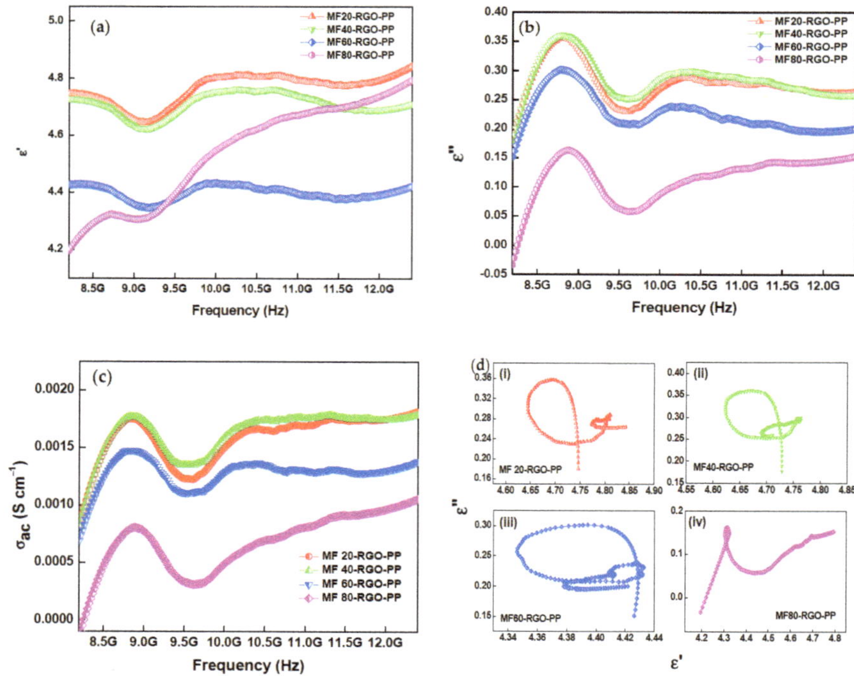

Figure 13. Frequency-dependent (**a**) real permittivity (ε'), (**b**) imaginary permittivity (ε''), (**c**) ac conductivity (σ_{ac}) and (**d**) Cole–Cole semicircles (ε' vs. ε'') of prepared MF20-RGO-PP, MF40-RGO-PP, MF60-RGO-PP and MF80-RGO-PP composite samples.

According to free-electron theory, the electrical conductivity (σ_{ac}) can be evaluated by the following relation [44]:

$$\sigma_{ac} = \varepsilon_0 \varepsilon'' \omega = \varepsilon_0 \varepsilon'' 2\pi f \tag{11}$$

where σ_{ac}, ε_0, ω and f are the electrical conductivity, the dielectric constant of the free space, the angular frequency and frequency of the electromagnetic waves, respectively. Figure 13c depicts the frequency dependence variation of the electrical conductivity (σ_{ac}) of the developed nanocomposites. The value of electrical conductivity was 9.04×10^{-4} to 1.84×10^{-3} S/cm, 8.41×10^{-4} to 1.80×10^{-3} S/cm, 7.25×10^{-4} to 1.47×10^{-3} S/cm and -8.55×10^{-5} to 1.05×10^{-3} S/cm for prepared MF20-RGO-PP, MF40-RGO-PP, MF60-RGO-PP and MF80-RGO-PP composite samples, respectively. The MF20 and MF40 nanoparticles-based nanocomposites had similar electrical conductivity and also higher than the other two nanoparticles MF60 and MF80 based nanocomposites. The enhanced electrical conductivity is associated with an increased induced microcurrent network and hopping phenomenon in prepared nanocomposites [45,46].

The relative complex permittivity can be expressed by the following relation [47]:

$$\varepsilon_r = \varepsilon_\infty + \frac{\varepsilon_s - \varepsilon_\infty}{1 + j2\pi f\tau} = \varepsilon' - j\varepsilon'' \tag{12}$$

where f, ε_∞, ε_s and τ corresponds to frequency, optical and stationary dielectric constant and polarization relaxation time, respectively. The dielectric parameters (ε', ε'') can be evaluated by the following relation [48,49]:

$$\varepsilon' = \varepsilon_\infty + \frac{\varepsilon_s - \varepsilon_\infty}{1 + \omega^2\tau^2} \tag{13}$$

$$\varepsilon'' = \varepsilon_r'' + \varepsilon_c'' = \frac{\varepsilon_s - \varepsilon_\infty}{1 + \omega^2 \tau^2} + \frac{\sigma}{\omega \varepsilon_o} \tag{14}$$

where σ, ε_r'' and ε_c'' corresponds to electrical conductivity, polarization loss and conductive loss, respectively.

The dielectric loss is generally associated with Debye polarization relaxation, which includes ionic polarization, electron polarization, dipole polarization and interfacial polarization [50]. The interfacial polarization originated because of the heterogeneous interfaces between $MnFe_2O_4$ nanoparticles and reduced graphene oxide in the polypropylene matrix. Based on Debye theory, the polarization characteristics can be confirmed by Cole–Cole semicircles, which is resulting from the following relation [51]:

$$\left(\varepsilon' - \frac{\varepsilon_s - \varepsilon_\infty}{2}\right)^2 + (\varepsilon'')^2 = \left(\frac{\varepsilon_s - \varepsilon_\infty}{2}\right)^2 \tag{15}$$

Figure 13d depicts the Cole–Cole semicircles (ε' vs. ε'') of prepared MF20-RGO-PP, MF40-RGO-PP, MF60-RGO-PP and MF80-RGO-PP composite samples. Several semicircles can be noticed for prepared nanocomposites, which indicate the coexistence of multiple polarization process. Further, since the prepared $MnFe_2O_4$-RGO-PP composites can simply form conductive networks and therefore conduction loss cannot be ignored also.

In general, the real permeability (μ') signifies the energy storage capacity of magnetic energy and the imaginary permeability (μ'') refers to the energy dissipation capacity of magnetic energy [52]. Figure 14a depicts the frequency dependence variation of the real part of permeability (μ') of designed nanocomposites. The value of the real part of permeability (μ') was 1.01–1.17, 1.06–1.31, 1.04–1.23 and 1.48–1.89 for the prepared MF20-RGO-PP, MF40-RGO-PP, MF60-RGO-PP and MF80-RGO-PP, respectively. Noticeably, the higher value of the real part of permeability (μ') of the MF80-RGO-PP sample suggesting the increased storage capacity of magnetic energy in comparison to other prepared nanocomposites. Figure 14b shows the frequency dependence variation of the imaginary part of permeability (μ'') of developed nanocomposite samples. The value of the imaginary part of permeability (μ'') was −0.015–0.057, −0.024–0.042, −0.033–0.023 and 0.52–1.62 for the prepared MF20-RGO-PP, MF40-RGO-PP, MF60-RGO-PP and MF80-RGO-PP, respectively. The negative value of imaginary permeability in the frequency range 9–11.4 GHz for MF60-RGO-PP, 10–11.2 GHz for MF40-RGO-PP and 10.2–10.7 GHz for MF20-RGO-PP can be associated with the eddy current caused by an extra magnetic field, which nullifies the inherent magnetic field [53]. Based on the Maxwell equations, the negative values of the imaginary part of permeability signify that the magnetic energy is radiated out and converted into electric energy [54].

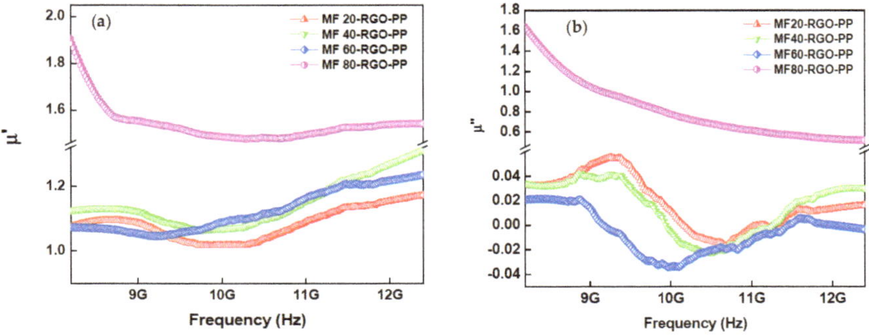

Figure 14. *Cont.*

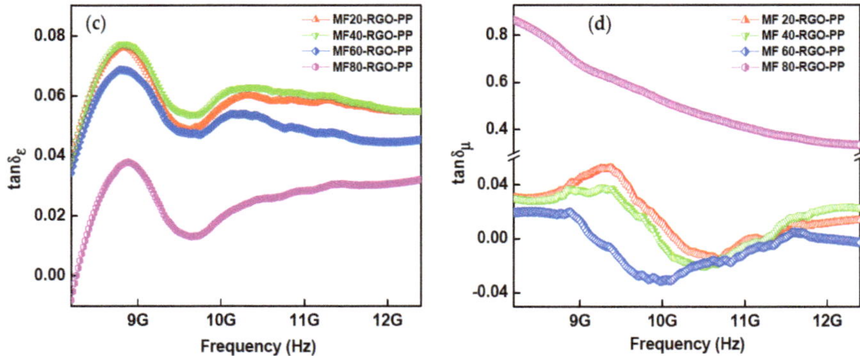

Figure 14. Frequency-dependent (**a**) real part of permeability (μ'), (**b**) imaginary part of permeability (μ''), (**c**) dielectric loss (tanδ_ε) and (**d**) magnetic loss (tanδ_μ) of prepared MF20-RGO-PP, MF40-RGO-PP, MF60-RGO-PP and MF80-RGO-PP composite samples.

Additionally, the dielectric loss tangent (tan$\delta_\varepsilon = \varepsilon''/\varepsilon'$) was utilized to calculate the loss capability against the stored capacity for electric energy. Figure 14c depicts the frequency dependence variation of dielectric loss (tanδ_ε) of developed nanocomposites. Noteworthy, the trend of value of dielectric loss (tanδ_ε) of developed nanocomposites was similar to the trend of the imaginary part of permittivity (ε''). The value of dielectric loss (tanδ_ε) was 0.042–0.075, 0.037–0.077, 0.035–0.068 and −0.008–0.038 for prepared MF20-RGO-PP, MF40-RGO-PP, MF60-RGO-PP and MF80-RGO-PP composite samples, respectively.

Moreover, the magnetic loss ability of the prepared nanocomposites can be evaluated by the magnetic tangent loss (tan$\delta_\mu = \mu''/\mu'$). Figure 14d depicts the frequency dependence changes in magnetic loss (tanδ_μ) of prepared nanocomposites. The value of magnetic loss (tanδ_μ) was −0.014–0.053, −0.021–0.037, −0.031–0.021 and 0.331–0.853 for prepared MF20-RGO-PP, MF40-RGO-PP, MF60-RGO-PP and MF80-RGO-PP, respectively.

Generally, the magnetic loss is attributed to the magnetic resonance (natural resonance and exchange resonance), eddy current loss, magnetic hysteresis loss and domain wall resonance [55]. The magnetic hysteresis loss had no appearance in the weak electromagnetic field, whereas the domain wall resonance had occurrence only at 1–100 MHz. The magnetic resonance and the eddy current effect induced the magnetic loss in the range of GHz frequency. When the magnetic loss was associated with the eddy current loss, the value of $\mu''(\mu')^{-2}f^{-1}$ should be constant with the variation of the frequency [56].

The eddy current can be calculated by using the following relation [57]:

$$C_o = \mu''(\mu')^{-2}f^{-1} = 2\pi\sigma\mu_o d^2/3 \tag{16}$$

where μ_o is the permeability of the vacuum, σ is the electric conductivity and d is the thickness of the material. As shown in Figure 15a, the value of C_o was constant at a lower frequency range from 8.2 to 8.8 GHz for MF20-RGO-PP, MF40-RGO-PP and MF60-RGO-PP nanocomposites, which implies that the magnetic loss in this frequency range was eddy current loss. Further, for these nanocomposites, the value of $\mu''(\mu')^{-2}f^{-1}$ varied at a higher frequency from 8.8 to 12.4 GHz, which suggests that the magnetic loss was not only induced by eddy current effect but also natural ferromagnetic resonance. Furthermore, for MF80-RGO-PP composite sample, the value of $\mu''(\mu')^{-2}f^{-1}$ was not constant throughout the whole frequency range.

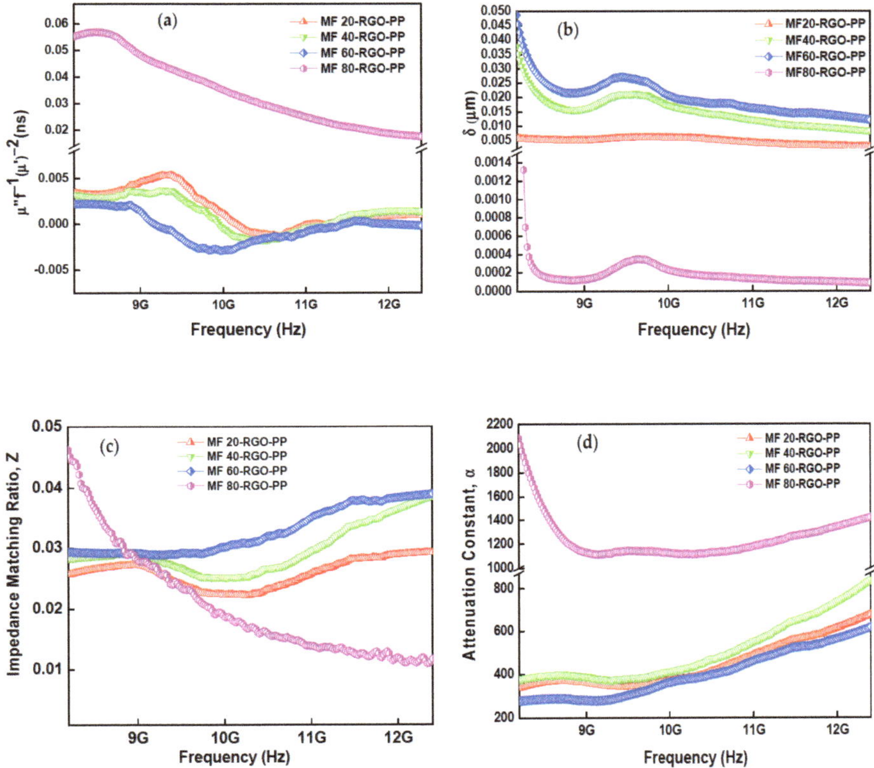

Figure 15. Frequency-dependent (**a**) eddy current loss, (**b**) skin depth, (**c**) impedance matching ratio and (**d**) attenuation constant of prepared MF20-RGO-PP, MF40-RGO-PP, MF60-RGO-PP and MF80-RGO-PP composite samples.

The superior value of EMISE was associated with the low skin depth of the prepared nanocomposites. The skin depth (δ) is the depth where the incident power of the EM waves fell to 1/e of its value at the surface. It can be given by the following relation [58]:

$$\delta = (\pi f \mu \sigma)^{-1/2} \tag{17}$$

where f is the frequency, μ is the permeability of the material and σ is the electrical conductivity. Figure 15b depicts the variation of skin depth (δ) of prepared MF20-RGO-PP, MF40-RGO-PP, MF60-RGO-PP and MF80-RGO-PP composite samples. The skin depth varied from 0.003 to 0.007 µm, 0.008 to 0.036 µm, 0.012 to 0.048 µm and 0.0012 to 0.0013 µm for MF20-RGO-PP, MF40-RGO-PP, MF60-RGO-PP and MF80-RGO-PP nanocomposites, respectively. It was noticed that the value of the skin depth of nanocomposites was much lower than their thickness, which leads to a high EMI SE [59]. Further, in general, the material with the shallowest skin depth exhibits high absorption loss [60].

In general, to achieve a large role of electromagnetic absorption, the shielding material should exhibit a large impedance matching ratio (Z) to free space [61]. The impedance matching ratio (Z) can be evaluated from the following relation [62]:

$$Z = Z_1/Z_0 = (\mu_r/\varepsilon_r)^{1/2} \tag{18}$$

where, Z_1 is the impedance matching of the electromagnetic wave absorber material, and Z_o is the impedance in free space. As shown in Figure 15c, the impedance matching ratio was increased with the increase of nanoparticle size of $MnFe_2O_4$ spinel ferrite in developed MF20-RGO-PP, MF40-RGO-PP and MF60-RGO-PP nanocomposites, whereas it was increased more at a lower frequency and decreased more at a higher frequency in case of the MF80-RGO-PP nanocomposite sample.

The other important electromagnetic parameter for electromagnetic interference shielding nanocomposites is the electromagnetic wave attenuation, and the attenuation constant (α) can be evaluated by the following relation [63]:

$$\alpha = \frac{\sqrt{2}\pi f}{c}\left[(\mu''\varepsilon'' - \mu'\varepsilon') + \sqrt{(\mu''\varepsilon'' - \mu'\varepsilon')^2 + (\mu'\varepsilon'' + \mu''\varepsilon')^2}\right]^{1/2} \tag{19}$$

Figure 15d depicts the attenuation constant of prepared MF20-RGO-PP, MF40-RGO-PP, MF60-RGO-PP and MF80-RGO-PP composite samples. In general, a large value of attenuation constant (α) indicates a good attenuation ability, which reveals the great dissipation characteristics of materials [64]. It can be observed that the nanocomposites MF20-RGO-PP, MF40-RGO-PP and MF60-RGO-PP had a very similar attenuation constant (α) value in the frequency range 8.2–12.4 GHz, whereas there was a noticeable gap in the attenuation constant (α) value of MF80-RGO-PP composite samples, especially in the low-frequency range. These results indicate that the total loss ability of MF80 spinel ferrite nanoparticles based nanocomposites displayed high magnetic loss in comparison with other samples. Moreover, a clear design of the electromagnetic wave shielding mechanism as reflected above is illustrated in Figure 16.

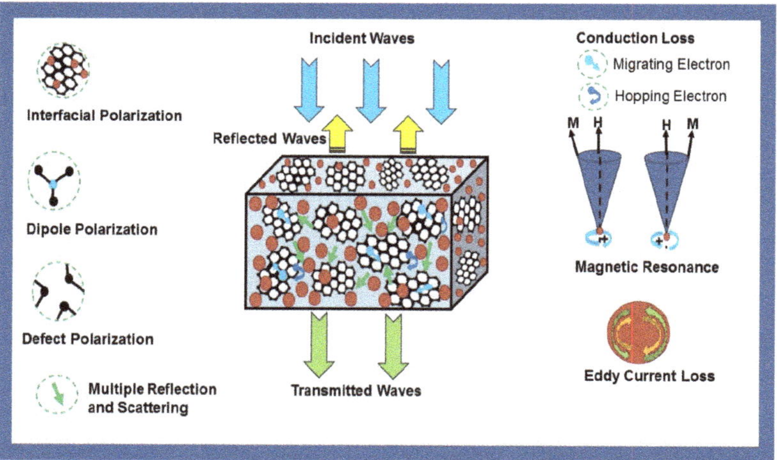

Figure 16. Schematic illustration of the electromagnetic interference shielding mechanism in prepared nanocomposites.

3.9. Mechanical Properties

In general, the variation in mechanical properties is associated with particle size, morphology and loading amount of fillers in polymer matrix [65,66]. Figure 17a depicts representative strain-stress curves of prepared MF20-RGO-PP, MF40-RGO-PP, MF60-RGO-PP and MF80-RGO-PP nanocomposites. The extracted mechanical parameter tensile strength of prepared nanocomposites is depicted in Figure 17b. The value of tensile strength was 4.84 MPa, 4.32 MPa, 5.56 MPa and 6.42 MPa for MF20-RGO-PP, MF40-RGO-PP, MF60-RGO-PP and MF80-RGO-PP, respectively. Further, Figure 17c depicts the extracted mechanical parameter elongation at break for prepared MF20-RGO-PP,

MF40-RGO-PP, MF60-RGO-PP and MF80-RGO-PP nanocomposites. The value of elongation at break was 711%, 486%, 598% and 699% for MF20-RGO-PP, MF40-RGO-PP, MF60-RGO-PP and MF80-RGO-PP, respectively. Furthermore, the extracted mechanical parameters Young's modulus is shown in Figure 17d. The value of Young's modulus was 15.2 MPa, 17.2 MPa, 17.8 MPa and 8.7 MPa for MF20-RGO-PP, MF40-RGO-PP, MF60-RGO-PP and MF80-RGO-PP, respectively.

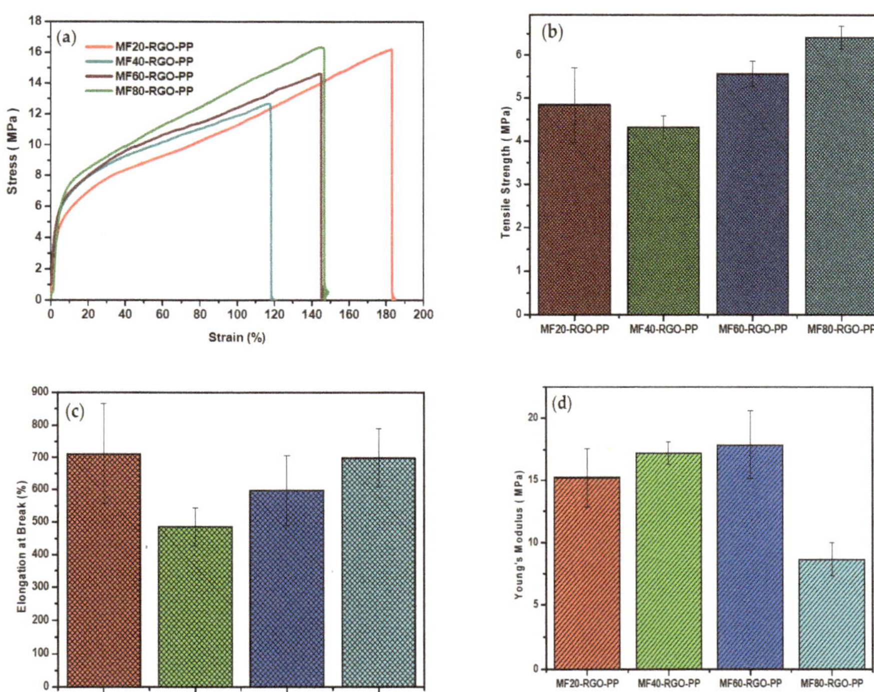

Figure 17. Mechanical behavior of prepared MF20-RGO-PP, MF40-RGO-PP, MF60-RGO-PP and MF80-RGO-PP nanocomposites: (**a**) representative strain–stress curves, (**b**) the tensile strength, (**c**) elongation at break and (**d**) Young's modulus.

4. Conclusions

We developed electromagnetic interference shielding nanocomposites based on polypropylene (PP) matrix with reduced graphene oxide (RGO) and $MnFe_2O_4$ spinel ferrite nanoparticles as nanofillers. Different sized magnetic filler $MnFe_2O_4$ (namely MF20, MF40, MF60 and MF80 samples) nanoparticles were prepared by the sonochemical approach at sonication synthesis time 20, 40, 60 and 80 min. It was noticed that the electromagnetic interference shielding performances of designed nanocomposites MF20-RGO-PP, MF40-RGO-PP, MF60-RGO-PP and MF80-RGO-PP were also controlled with the tuning of dielectric/magnetic loss. The maximum value of total shielding effectiveness (SE_T) was 71.3 dB for MF80-RGO-PP nanocomposite with a thickness of 0.5 mm in the frequency range (8.2–12.4 GHz). The excellent electromagnetic interference shielding properties with a lightweight, flexible and thinness sheet of developed nanocomposites was realized.

Author Contributions: A. and T.J. performed the experiments; D.Š., P.U., M.M. (Michal Machovský), M.M. (Milan Masar), M.U., and L.K. performed the characterizations; R.S.Y., I.K., J.V. and J.H. analyzed the data and wrote the manuscript. All authors have read and agreed to the published version of the manuscript.

Funding: We thank the financial support by the Czech Science Foundation (GA19–23647S) project at the Centre of Polymer Systems, Tomas Bata University in Zlin, Czech Republic.

Conflicts of Interest: The authors declare no conflict of interest.

References

1. Biswas, S.; Arief, I.; Panja, S.S.; Bose, S. Absorption-Dominated Electromagnetic Wave Suppressor Derived from Ferrite-Doped Cross-Linked Graphene Framework and Conducting Carbon. *ACS Appl. Mater. Interfaces* **2017**, *9*, 3030–3039. [CrossRef]
2. Cao, W.-T.; Chen, F.-F.; Zhu, Y.-J.; Zhang, Y.-G.; Jiang, Y.-Y.; Ma, M.-G.; Chen, F. Binary Strengthening and Toughening of MXene/Cellulose Nanofiber Composite Paper with Nacre-Inspired Structure and Superior Electromagnetic Interference Shielding Properties. *ACS Nano* **2018**, *12*, 4583–4593. [CrossRef]
3. Zhang, Y.; Qiu, M.; Yu, Y.; Wen, B.; Cheng, L. A Novel Polyaniline-Coated Bagasse Fiber Composite with Core-Shell Heterostructure Provides Effective Electromagnetic Shielding Performance. *ACS Appl. Mater. Interfaces* **2017**, *9*, 809–818. [CrossRef]
4. Shen, B.; Li, Y.; Zhai, W.; Zheng, W. Compressible Graphene-Coated Polymer Foams with Ultralow Density for Adjustable Electromagnetic Interference (EMI) Shielding. *ACS Appl. Mater. Interfaces* **2016**, *8*, 8050–8057. [CrossRef]
5. Hsiao, S.-T.; Ma, C.-C.M.; Tien, H.-W.; Liao, W.-H.; Wang, Y.-S.; Li, S.-M.; Yang, C.-Y.; Lin, S.-C.; Yang, R.-B. Effect of Covalent Modification of Graphene Nanosheets on the Electrical Property and Electromagnetic Interference Shielding Performance of a Water-Borne Polyurethane Composite. *ACS Appl. Mater. Interfaces* **2015**, *7*, 2817–2826. [CrossRef]
6. Wang, H.; Zhu, D.; Zhou, W.; Luo, F. Effect of Multiwalled Carbon Nanotubes on the Electromagnetic Interference Shielding Properties of Polyimide/Carbonyl Iron Composites. *Ind. Eng. Chem. Res.* **2015**, *54*, 6589–6595. [CrossRef]
7. Kim, S.; Oh, J.-S.; Kim, M.-G.; Jang, W.; Wang, M.; Kim, Y.; Seo, H.-W.; Kim, Y.-C.; Lee, J.-H.; Lee, Y.; et al. Electromagnetic Interference (EMI) Transparent Shielding of Reduced Graphene Oxide (RGO) Interleaved Structure Fabricated by Electrophoretic Deposition. *ACS Appl. Mater. Interfaces* **2014**, *6*, 17647–17653. [CrossRef]
8. Liu, P.; Yao, Z.; Zhou, J. Fabrication and microwave absorption of reduced graphene oxide/$Ni_{0.4}Zn_{0.4}Co_{0.2}Fe_2O_4$ nanocomposites. *Ceram. Int.* **2016**, *42*, 9241–9249. [CrossRef]
9. Dippong, T.; Toloman, D.; Levei, E.-A.; Cadar, O.; Mesaros, A. A possible formation mechanism and photocatalytic properties of $CoFe_2O_4$/ $PVA-SiO_2$ nanocomposites. *Thermochim. Acta* **2018**, *666*, 103–115. [CrossRef]
10. Jazirehpour, M.; Ebrahimi, S.S. Synthesis of magnetite nanostructures with complex morphologies and effect of these morphologies on magnetic and electromagnetic properties. *Ceram. Int.* **2016**, *42*, 16512–16520. [CrossRef]
11. Yang, Y.; Li, M.; Wu, Y.; Zong, B.; Ding, J. Size-dependent microwave absorption properties of Fe_3O_4 nanodiscs. *RSC Adv.* **2016**, *6*, 25444–25448. [CrossRef]
12. Liang, Y.-J.; Fan, F.; Ma, M.; Sun, J.; Chen, J.; Zhang, Y.; Gu, N. Size-dependent electromagnetic properties and the related simulations of Fe_3O_4 nanoparticles made by microwave-assisted thermal decomposition. *Colloids Surf. A* **2017**, *530*, 191–199. [CrossRef]
13. Wu, N.; Liu, X.; Zhao, C.; Cui, C.; Xia, A. Effects of particle size on the magnetic and microwave absorption properties of carbon-coated nickel nanocapsules. *J. Alloy. Compd.* **2016**, *656*, 628–634. [CrossRef]
14. Yadav, R.S.; Kuřitka, I.; Vilcakova, J.; Machovsky, M.; Skoda, D.; Urbánek, P.; Masař, M.; Jurča, M.; Urbánek, M.; Kalina, L.; et al. $NiFe_2O_4$ Nanoparticles Synthesized by Dextrin from Corn-Mediated Sol-Gel Combustion Method and Its Polypropylene Nanocomposites Engineered with Reduced Graphene Oxide for the Reduction of Electromagnetic Pollution. *ACS Omega* **2019**, *4*, 22069–22081. [CrossRef] [PubMed]
15. Yadav, R.S.; Kuřitka, I.; Vilcakova, J.; Jamatia, T.; Machovsky, M.; Skoda, D.; Urbánek, P.; Masař, M.; Urbánek, M.; Kalina, L.; et al. Impact of sonochemical synthesis condition on the structural and physical properties of $MnFe_2O_4$ spinel ferrite nanoparticles. *Ultrason. Sonochem.* **2020**, *61*, 104839. [CrossRef] [PubMed]

16. Bai, Y.; Rakhi, R.B.; Chen, W.; Alshareef, H.N. Effect of pH induced chemical modification of hydrothermally reduced graphene oxide on supercapacitor performance. *J. Power Sources* **2013**, *233*, 313–319. [CrossRef]
17. Patade, S.R.; Andhare, D.D.; Somvanshi, S.B.; Jadhav, S.A.; Khedkar, M.V.; Jadhav, K.M. Self-heating evaluation of superparamagnetic $MnFe_2O_4$ nanoparticles for magnetic fluid hyperthermia application towards cancer treatment. *Ceram. Int.* **2020**, *46*, 25576–25583. [CrossRef]
18. Hsiao, M.-C.; Liao, S.-H.; Lin, Y.-F.; Wang, C.-A.; Pu, N.-W.; Tsai, H.-M.; Ma, C.-C.M. Preparation and characterization of polypropylene-graft-thermally reduced graphite oxide with an improved compatibility with polypropylene-based nanocomposite. *Nanoscale* **2011**, *3*, 1516. [CrossRef]
19. Yadav, R.S.; Kuritka, I.; Vilcáková, J.; Machovský, M.; Škoda, D.; Urbánek, P.; Masar, M.; Goralik, M.; Urbánek, M.; Kalina, L.; et al. Polypropylene Nanocomposite Filled with Spinel Ferrite $NiFe_2O_4$ Nanoparticles and In-Situ Thermally-Reduced Graphene Oxide for Electromagnetic Interference Shielding Application. *Nanomaterials* **2019**, *9*, 621. [CrossRef]
20. Varshney, D.; Verma, K.; Kumar, A. Structural and vibrational properties of $Zn_xMn_{1-x}Fe_2O_4$ (x = 0.0, 0.25, 0.50, 0.75, 1.0) mixed ferrites. *Mater. Chem. Phys.* **2011**, *131*, 413–419. [CrossRef]
21. Gupta, A.; Jamatia, R.; Patil, R.A.; Ma, Y.-R.; Pal, A.K. Copper Oxide/Reduced Graphene Oxide Nanocomposite-Catalyzed Synthesis of Flavanones and Flavanones with Triazole Hybrid Molecules in One Pot: A Green and Sustainable Approach. *ACS Omega* **2018**, *3*, 7288–7299. [CrossRef]
22. Wadi, V.S.; Jena, K.K.; Halique, K.; Alhassan, S.M. Enhanced Mechanical Toughness of Isotactic Polypropylene Using Bulk Molybdenum Disulfide. *ACS Omega* **2020**, *5*, 11394–11401. [CrossRef]
23. Thakur, A.; Kumar, P.; Thakur, P.; Rana, K.; Chevalier, A.; Mattei, J.-L.; Quéffélec, P. Enhancement of magnetic properties of $Ni_{0.5}Zn_{0.5}Fe_2O_4$ nanoparticles prepared by the co-precipitation method. *Ceram. Int.* **2016**, *42*, 10664–10670. [CrossRef]
24. Gopanna, A.; Mandapati, R.N.; Thomas, S.P.; Rajan, K.; Chavali, M. Fourier transform infrared spectroscopy (FTIR), Raman spectroscopy and wide-angle X-ray scattering (WAXS) of polypropylene (PP)/cyclic olefin copolymer (COC) blends for qualitative and quantitative analysis. *Polym. Bull.* **2019**, *76*, 4259–4274. [CrossRef]
25. Hassan, M.M.; Koyama, K. Enhanced thermal, mechanical and fire retarding properties of polystyrene sulphonate-grafted- nanosilica/polypropylene composites. *RSC Adv.* **2015**, *5*, 16950–16959. [CrossRef]
26. He, Q.; Yuan, T.; Zhang, X.; Luo, Z.; Haldolaarachchige, N.; Sun, L.; Young, D.P.; Wei, S.; Guo, Z. Magnetically Soft and Hard Polypropylene/Cobalt Nanocomposites: Role of Maleic Anhydride Grafted Polypropylene. *Macromolecules* **2013**, *46*, 2357–2368. [CrossRef]
27. Zhu, J.; Wei, S.; Li, Y.; Sun, L.; Haldolaarachchige, N.; Young, D.P.; Southworth, C.; Khasanov, A.; Luo, Z.; Guo, Z. Surfactant-Free Synthesized Magnetic Polypropylene Nanocomposites: Rheological, Electrical, Magnetic, and Thermal Properties. *Macromolecules* **2011**, *44*, 4382–4391. [CrossRef]
28. Lv, H.; Liang, X.; Ji, G.; Zhang, H.; Du, Y. Porous Three- Dimensional Flower-like Co/CoO and Its Excellent Electromagnetic Absorption Properties. *ACS Appl. Mater. Interfaces* **2015**, *7*, 9776–9783. [CrossRef]
29. Manna, K.; Srivastava, S.K. Fe_3O_4@Carbon@Polyaniline Trilaminar Core–Shell Composites as Superior Microwave Absorber in Shielding of Electromagnetic Pollution. *ACS Sustain. Chem. Eng.* **2017**, *5*, 10710–10721. [CrossRef]
30. Shahzad, F.; Kumar, P.; Kim, Y.-H.; Hong, S.M.; Koo, C.M. Biomass-Derived Thermally Annealed Interconnected Sulfur-Doped Graphene as a Shield against Electromagnetic Interference. *ACS Appl. Mater. Interfaces* **2016**, *8*, 9361–9369. [CrossRef]
31. Song, W.-L.; Gong, C.; Li, H.; Cheng, X.-D.; Chen, M.; Yuan, X.; Chen, H.; Yang, Y.; Fang, D. Graphene-Based Sandwich Structures for Frequency Selectable Electromagnetic Shielding. *ACS Appl. Mater. Interfaces* **2017**, *9*, 36119–36129. [CrossRef] [PubMed]
32. Zhang, X.-J.; Wang, G.-S.; Cao, W.-Q.; Wei, Y.-Z.; Liang, J.-F.; Guo, L.; Cao, M.-S. Enhanced Microwave Absorption Property of Reduced Graphene Oxide (RGO)-$MnFe_2O_4$ Nanocomposites and Polyvinylidene Fluoride. *ACS Appl. Mater. Interfaces* **2014**, *6*, 7471–7478. [CrossRef] [PubMed]
33. Yin, P.; Zhang, L.; Sun, P.; Wang, J.; Feng, X.; Zhang, Y.; Dai, J.; Tang, Y. Apium-derived biochar loaded with $MnFe_2O_4$@C for excellent low frequency electromagnetic wave absorption. *Ceram. Int.* **2020**, *46*, 13641–13650. [CrossRef]

34. Lakshmi, R.V.; Bera, P.; Chakradhar, R.P.S.; Choudhury, B.; Pawar, S.P.; Bose, S.; Nair, R.U.; Barshilia, H.C. Enhanced microwave absorption properties of PMMA modified $MnFe_2O_4$-polyaniline nanocomposites. *Phys. Chem. Chem. Phys.* **2019**, *21*, 5068–5077. [CrossRef]

35. Srivastava, R.K.; Xavier, P.; Gupta, S.N.; Kar, G.N.; Bose, S.; Sood, A.K. Excellent Electromagnetic Interference Shielding by Graphene- $MnFe_2O_4$ -Multiwalled Carbon Nanotube Hybrids at Very Low Weight Percentage in Polymer Matrix. *ChemistrySelect* **2016**, *1*, 5995–6003. [CrossRef]

36. Wang, Y.; Wu, X.; Zhang, W.; Huang, S. One-pot synthesis of $MnFe_2O_4$ nanoparticles-decorated reduced graphene oxide for enhanced microwave absorption properties. *Mater. Technol.* **2017**, *32*, 32–37. [CrossRef]

37. Yin, P.; Zhang, L.; Wang, J.; Feng, X.; Zhao, L.; Rao, H.; Wang, Y.; Dai, J. Preparation of SiO_2- $MnFe_2O_4$ Composites via One-Pot Hydrothermal Synthesis Method and Microwave Absorption Investigation in S-Band. *Molecules* **2019**, *24*, 2605. [CrossRef]

38. Wang, Y.; Wu, X.; Zhang, W.; Huang, S. Synthesis and electromagnetic absorption properties of Ag-coated reduced graphene oxide with $MnFe_2O_4$ particles. *J. Magn. Magn. Mater.* **2016**, *404*, 58–63. [CrossRef]

39. Kashi, S.; Gupta, R.K.; Bhattacharya, S.N.; Varley, R.J. Experimental and simulation study of effect of thickness on performance of (butylene adipate-co-terephthalate) and poly lactide nanocomposites incorporated with graphene as stand-alone electromagnetic interference shielding and metal-backed microwave absorbers. *Compos. Sci. Technol.* **2020**, *195*, 108186.

40. Sui, M.; Fu, T.; Sun, X.; Cui, G.; Lv, X.; Gu, G. Unary and binary doping effect of M^{2+} (M=Mn, Co, Ni, Zn) substituted hollow Fe_3O_4 approach for enhancing microwave attenuation. *Ceram. Int.* **2018**, *44*, 17138–17146. [CrossRef]

41. Sankaran, S.; Deshmukh, K.; Ahamed, M.B.; Pasha, S.K.K. Recent advances in electromagnetic interference shielding properties of metal and carbon filler reinforced flexible polymer composites: A review. *Compos. Part A* **2018**, *114*, 49–71. [CrossRef]

42. Mishra, M.; Singh, A.P.; Singh, B.P.; Singh, V.N.; Dhawan, S.K. Conducting Ferrofluid: A High-performance Microwave Shielding Material. *J. Mater. Chem. A* **2014**, *2*, 13159–13168. [CrossRef]

43. Behera, C.; Choudhary, R.N.P.; Das, P.R. Size dependent electrical and magnetic properties of mechanically-activated $MnFe_2O_4$ nanoferrite. *Ceram. Int.* **2015**, *41*, 13042–13054. [CrossRef]

44. Lyu, L.; Wang, F.; Zhang, X.; Qiao, J.; Liu, C.; Liu, J. CuNi alloy/carbon foam nanohybrids as high-performance electromagnetic wave absorbers. *Carbon* **2021**, *172*, 488–496. [CrossRef]

45. Wang, Y.; Guan, H.; Dong, C.; Xiao, X.; Du, S.; Wang, Y. Reduced graphene oxide(RGO)/Mn_3O_4 nanocomposites for dielectric loss properties and electromagnetic interference shielding effectiveness at high frequency. *Ceram. Int.* **2016**, *42*, 936–942. [CrossRef]

46. Yin, Y.; Zeng, M.; Liu, J.; Tang, W.; Dong, H.; Xia, R.; Yu, R. Enhanced high-frequency absorption of anisotropic Fe_3O_4/graphene nanocomposites. *Sci. Rep.* **2016**, *6*, 25075. [CrossRef]

47. Zhang, H.; Wang, B.; Feng, A.; Zhang, N.; Jia, Z.; Huang, Z.; Liu, X.; Wu, G. Mesoporous carbon hollow microspheres with tunable pore size and shell thickness as efficient electromagnetic wave absorbers. *Compos. Part B* **2019**, *167*, 167,690–699. [CrossRef]

48. Guanglei Wu, G.; Jia, Z.; Zhou, X.; Nie, G.; Lv, H. Interlayer controllable of hierarchical MWCNTs@C@Fe_xO_y cross-linked composite with wideband electromagnetic absorption performance. *Compos. Part A* **2020**, *128*, 105687.

49. Jia, Z.; Gao, Z.; Feng, A.; Zhang, Y.; Zhang, C.; Nie, G.; Wang, K.; Wu, G. Laminated microwave absorbers of A-site cation deficiency perovskite $La_{0.8}FeO_3$ doped at hybrid RGO carbon. *Compos. Part B* **2019**, *176*, 107246. [CrossRef]

50. Meng, X.M.; Zhang, X.J.; Lu, C.; Pan, Y.F.; Wang, G.-S. Enhanced absorbing properties of three-phase composites based on a thermoplastic-ceramic matrix ($BaTiO_3$+PVDF) and carbon black nanoparticles. *J. Mater. Chem.* **2014**, *2*, 18725–18730. [CrossRef]

51. Zhao, Z.; Kou, K.; Wu, H. 2-Methylimidazole-mediated hierarchical Co_3O_4/N-doped carbon/short-carbon-fiber composite as high-performance electromagnetic wave absorber. *J. Colloid Interface Sci.* **2020**, *574*, 1–10. [CrossRef] [PubMed]

52. Dong, S.; Hu, P.; Li, X.; Hong, C.; Zhang, X.; Han, J. $NiCo_2S_4$ nanosheets on 3D wood-derived carbon for microwave absorption. *Chem. Eng. J.* **2020**, *398*, 125588. [CrossRef]

53. Wang, F.; Li, X.; Chen, Z.; Yu, W.; Loh, K.P.; Zhong, B.; Shi, Y.; Xu, Q.-H. Efficient low-frequency microwave absorption and solar evaporation properties of γ-Fe$_2$O$_3$ nanocubes/graphene composites. *Chem. Eng. J.* **2021**, *405*, 126676. [CrossRef]

54. Shi, X.-L.; Cao, M.-S.; Yuan, J.; Fang, X.-Y. Dual nonlinear dielectric resonance and nesting microwave absorption peaks of hollow cobalt nanochains composites with negative permeability. *Appl. Phys. Lett.* **2009**, *95*, 163108. [CrossRef]

55. Sun, X.; He, J.; Li, G.; Tang, J.; Wang, T.; Guo, Y.; Xue, H. Laminated magnetic graphene with enhanced electromagnetic wave absorption properties. *J. Mater. Chem. C* **2013**, *1*, 765. [CrossRef]

56. Luo, J.; Shen, P.; Yao, W.; Jiang, C.; Xu, J. Synthesis, Characterization, and Microwave Absorption Properties of Reduced Graphene Oxide/Strontium Ferrite/Polyaniline Nanocomposites. *Nanoscale Res. Lett.* **2016**, *11*, 141. [CrossRef]

57. Ibrahim, I.R.; Matori, K.A.; Ismail, I.; Awang, Z.; Rusly, S.N.A.; Nazlan, R.; Idris, F.M.; Zulkimi, M.M.M.; Abdullah, N.H.; Mustaffa, M.S.; et al. A Study on Microwave Absorption Properties of Carbon Black and Ni$_{0.6}$Zn$_{0.4}$Fe$_2$O$_4$ Nanocomposites by Tuning the Matching-Absorbing Layer Structures. *Sci. Rep.* **2020**, *10*, 3135. [CrossRef]

58. Hou, Y.; Cheng, L.; Zhang, Y.; Du, X.; Zhao, Y.; Yang, Z. High temperature electromagnetic interference shielding of lightweight and flexible ZrC/SiC nanofiber mats. *Chem. Eng. J.* **2021**, *404*, 126521. [CrossRef]

59. Lai, H.; Li, W.; Xu, L.; Wang, X.; Jiao, H.; Fan, Z.; Lei, Z.; Yuan, Y. Scalable fabrication of highly crosslinked conductive nanofibrous films and their applications in energy storage and electromagnetic interference shielding. *Chem. Eng. J.* **2020**, *400*, 125322. [CrossRef]

60. Gupta, T.K.; Singh, B.P.; Mathur, R.B.; Dhakate, S.R. Multi-walled carbon nanotube-graphene-polyaniline multiphase nanocomposite with superior electromagnetic shielding effectiveness. *Nanoscale* **2014**, *6*, 842. [CrossRef]

61. Lv, H.; Zhang, H.; Zhao, J.; Ji, G.; Du, Y. Achieving excellent bandwidth absorption by a mirror growth process of magnetic porous polyhedron structures. *Nano Res.* **2016**, *9*, 1813–1822. [CrossRef]

62. Deng, Y.D.; Zheng, Y.; Zhang, D.; Han, C.; Cheng, A.; Shen, J.; Zeng, G.; Zhang, H. A novel and facile-to-synthesize three-dimensional honeycomb-like nano-Fe$_3$O$_4$@C composite: Electromagnetic wave absorption with wide bandwidth. *Carbon* **2020**, *169*, 118–128. [CrossRef]

63. Xu, Z.; Du, Y.; Liu, D.; Wang, Y.; Ma, W.; Wang, Y.; Xu, P.; Han, X. Pea-like Fe/Fe$_3$C Nanoparticles Embedded in Nitrogen-Doped Carbon Nanotubes with Tunable Dielectric/Magnetic Loss and Efficient Electromagnetic Absorption. *ACS Appl. Mater. Interfaces* **2019**, *11*, 4268–4277. [CrossRef] [PubMed]

64. Dong, S.; Lyu, Y.; Li, X.; Chen, J.; Zhang, X.; Han, J.; Hu, P. Construction of MnO nanoparticles anchored on SiC whiskers for superior electromagnetic wave absorption. *J. Colloid Interface Sci.* **2020**, *559*, 186–196. [CrossRef]

65. Zhang, H.-B.; Yan, Q.; Zheng, W.-G.; He, Z.; Yu, Z.-Z. Tough Graphene-Polymer Microcellular Foams for Electromagnetic Interference Shielding. *ACS Appl. Mater. Interfaces* **2011**, *3*, 918–924. [CrossRef]

66. Zou, H.; Li, S.; Zhang, L.; Yan, S.; Wu, H.; Zhang, S.; Tian, M. Determining factors for high performance silicone rubber microwave absorbing materials. *J. Magn. Magn. Mater.* **2011**, *323*, 1643–1651. [CrossRef]

Publisher's Note: MDPI stays neutral with regard to jurisdictional claims in published maps and institutional affiliations.

Review

Magnetic Nanomaterials for Arterial Embolization and Hyperthermia of Parenchymal Organs Tumors: A Review

Natalia E. Kazantseva [1,2,*], Ilona S. Smolkova [1], Vladimir Babayan [1], Jarmila Vilčáková [1,2], Petr Smolka [1] and Petr Saha [1,2]

1. Centre of Polymer Systems, Tomas Bata University in Zlín, Třída Tomáše Bati 5678, 760 01 Zlín, Czech Republic; smolkova@utb.cz (I.S.S.); babayan@utb.cz (V.B.); vilcakova@utb.cz (J.V.); smolka@utb.cz (P.S.); saha@utb.cz (P.S.)
2. Polymer Centre, Faculty of Technology, Tomas Bata University in Zlín, Vavrečkova 275, 760 01 Zlín, Czech Republic
* Correspondence: kazantseva@utb.cz; Tel.: +420-608607035

Abstract: Magnetic hyperthermia (MH), proposed by R. K. Gilchrist in the middle of the last century as local hyperthermia, has nowadays become a recognized method for minimally invasive treatment of oncological diseases in combination with chemotherapy (ChT) and radiotherapy (RT). One type of MH is arterial embolization hyperthermia (AEH), intended for the presurgical treatment of primary inoperable and metastasized solid tumors of parenchymal organs. This method is based on hyperthermia after transcatheter arterial embolization of the tumor's vascular system with a mixture of magnetic particles and embolic agents. An important advantage of AEH lies in the double effect of embolotherapy, which blocks blood flow in the tumor, and MH, which eradicates cancer cells. Consequently, only the tumor undergoes thermal destruction. This review introduces the progress in the development of polymeric magnetic materials for application in AEH.

Keywords: magnetic hyperthermia; arterial embolization hyperthermia; magnetic nanoparticles; embolic agents; animal model; clinical application (results)

Citation: Kazantseva, N.E.; Smolkova, I.S.; Babayan, V.; Vilčáková, J.; Smolka, P.; Saha, P. Magnetic Nanomaterials for Arterial Embolization and Hyperthermia of Parenchymal Organs Tumors: A Review. *Nanomaterials* **2021**, *11*, 3402. https://doi.org/10.3390/nano11123402

Academic Editor: Joachim Clement

Received: 12 November 2021
Accepted: 8 December 2021
Published: 15 December 2021

Publisher's Note: MDPI stays neutral with regard to jurisdictional claims in published maps and institutional affiliations.

1. Introduction

Cancer is the second leading cause of death around the world. According to the World Health Organization statistics report, 19.3 million new cancer cases were diagnosed worldwide, with almost 10 million deaths from cancer in 2020 [1]. It is also notable from this report that the number of deaths due to cancer does not change from year to year, notwithstanding the application of new drugs and combination treatments. Moreover, the number of cancer patients is expected to rise to almost 30 million annually by 2040 (Figure 1).

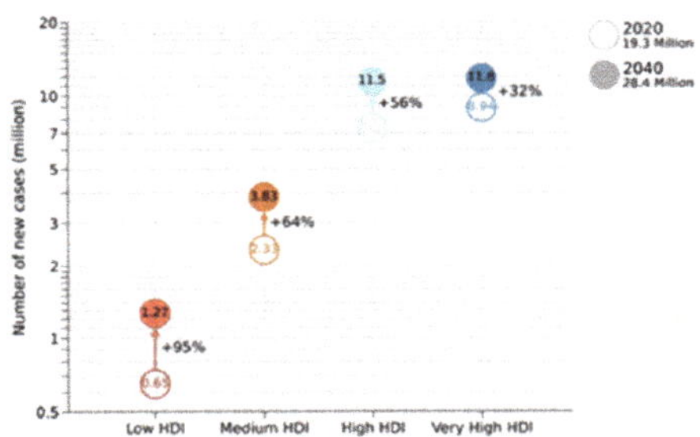

Figure 1. Projected number of new cancer cases in 2040 according to the 4-Tier Human Development Index. Source: GOBOCAN 2020. Reprinted with permission [1]. Copyright 2021, John Wiley, and Sons.

New approaches to cancer detection and treatment are needed. The current tendency in oncotherapy is focused on the complex palliative therapy in treating cancer patients. One of such methods is Hyperthermia (HT) combined with RT and ChT, standardized by different organizations such as Radiation Oncology Group, European Society for Hyperthermic Oncology, and others [2–5]. Hyperthermia in oncology refers to the treatment of malignant diseases by controlled heating between 39–45 °C for a period of time with minimal unwanted side effects. Depending on tumor location and tissue volume, conventional HT is subdivided into three categories: local, locoregional, and whole-body [5,6]. Local hyperthermia aims to increase the temperature of near-surface primary malignant tumors before the metastases stage by ultrasound, electroporation, and more often by converting electromagnetic energy into heat. Locoregional hyperthermia is useful for large inoperable deep-seated tumors and is based on perfusion of organ and body with heated fluids or electromagnetic energy. Whole-body hyperthermia is used for patients with solid metastatic tumors. This type of hyperthermia is based on heating the blood in extracorporeal circulation using infrared radiation, hot water blankets, or thermal chambers. The medical hyperthermia devices using electromagnetic waves are called applicators. According to the heating principle, applicators are principally divided into dielectric applicator systems (capacitive heating applicator) and inductive heating systems (inductive heating applicators). A detailed description of HT technology currently used in clinical practice is presented in a book by Andre Vander Vorst [7] and review articles by H. Petra Kok et al. [8] and H. Dobšíček Trefna et al. [9].

Hyperthermia in combination with RT and ChT is widely used in Europe, the United States, Japan, China, Russia, and other countries for the treatment of different tumor types and sites: mammary gland, prostate gland, lung, liver, intestinal tract, bonny tissue, glioblastoma, etc. [9–21]. Preclinical in vitro and in vivo studies have shown two aspects of cancer inhibition by combining HT with RT and ChT: the death of individual cells from hyperthermia and enhancing the effects of RT and ChT [15]. The additional benefit of HT in combination with RT and ChT has been shown in a number of randomized clinical trials. Thus, for example, the overall response rate increased from 38% to 60% for patients with breast cancer who received HT + RT, while the treatment of cervical cancer at stage IIB-III-IVA with a combination of RT, ChT, and HT showed improved complete response rates and significantly increased overall survival [16,17]. Combining these methods has also proven to be effective in the preoperative treatment of stage III lung cancer [19]. According to the results obtained, the overall response to treatment was about 94%, including a complete response of about 22% and a partial response of about 72%. The significant regression of the tumor achieved in the preoperative period makes it possible to reduce the volume of surgical treatment and thereby facilitate the course of the postoperative period.

Currently, hyperthermia's cellular and molecular basis and its effect on cancer treatment in combination with RT and ChT are better understood due to the substantial technical improvement in the sources used to supply heat and measure its output. Depending on the applied temperature and duration of treatment, various biological effects of hyperthermia on macroscopic and microscopic levels have been revealed [18,22,23].

The dominant mechanisms of cancer cell death caused by heating tissues to a temperature within the range of 41–45 °C are necrosis, apoptosis, and modes related to mitotic catastrophe [24–28]. The macroscopic effect of hyperthermia manifests itself in the tumor's vascular system, i.e., heat increases the blood flow, which increases the vessel permeability and tissue oxygenation, which, in turn, causes a temporarily increased radiosensitivity. Concerning the microscopic effect, hyperthermia causes changes in the cellular components of the tumor and thus leads to a loss of cellular homeostasis. The mechanisms involved in heat-induced cell damage are protein denaturation, lipid peroxidation, and DNA damage. Moreover, hyperthermia modulates the immune system due to the production of heat shock proteins (HSPs), which, in turn, stimulates macrophages by acting in damage-associated molecular patterns. On the other hand, HSPs protect cells from apoptosis, which reduces the effect of hyperthermia.

The chemo-sensitizing property of hyperthermia is determined by (1) an increase in the cellular membrane permeability so that chemotherapeutic drugs can more easily pass the cell barrier, (2) a drug-induced DNA adduct formation, and (3) inhibition of DNA repair, which all enhance the cytotoxic activity of drugs [14,27]. Moreover, intratumoral heat affects DNA damage pathways by deactivating specific repair proteins. These processes are highly dependent on several factors: the degree of temperature elevation, the duration of heat, and the cell type and microenvironmental conditions, such as the acidity and oxygenation status of the tumor. In addition, hyperthermic chemosensitization depends on the type and concentration of drugs due to different mechanisms by which heat affects drug activity, i.e., transport, intracellular cytotoxicity, and metabolism. In vivo and in vitro studies, as well as clinical results, have demonstrated that most chemotherapeutic drugs are effective when delivered just before or during an HT session [28].

Critical problems of conventional hyperthermia in clinical practice are insufficient heat localization in the tumor, especially in deep-seated tumors, heterogeneity in temperature distribution within the tumor, and overheating of healthy tissues due to deficiency in temperature monitoring. For these reasons, the use of hyperthermia alone gives an overall response of only about 15% [5]. The main challenge of hyperthermia is to achieve a precise energy delivery and controlled heating of primary and metastasized tumors while avoiding heating of normal tissues and overcoming the thermotolerance [26]. In particular, MH proposed in 1957 by R.K. Gilchrist as local hyperthermia [29] is still undergoing preclinical and clinical trials as an independent method for cancer therapy and as a multimodal treatment in combination with RT and ChT [30–38]. The general methodology of MH comprises the introduction of magnetic material into the tumor followed by exposure to an alternating magnetic field (AMF) at moderate frequencies and amplitudes (f = 0.05–1.5 MHz, H $\leq$ 15 kA·m^{-1}) to limit peripheral nerve stimulation due to induced eddy currents occurring in the body [39–41]. The free current loss also needs to be considered since it may lead to nonspecific induction heating. The heating ability of magnetic material in AMF is usually estimated by specific power loss (SLP) or specific absorption rate (SAR), which are defined as heating power (P [W]) generated per unit mass of magnetic nanoparticles (m_{MNP} [g]): SAR = P/m_{MNP} [42,43]. The heating power produced by the mediator depends on nanoparticles (NPs) concentration, core size, and magnetic properties (saturation magnetization, magnetic anisotropy energy), the viscosity and heat capacity of dispersion media, as well as on the extrinsic factors, i.e., frequency and amplitude of AMF [7]. To eliminate these extrinsic factors, intrinsic loss power (ILP) as a system-independent parameter was introduced to compare results obtained in different field conditions: ILP = SAR/f $\times$ H^2 [44].

The key requirement in MH is maximizing heat generation within medically safe limits of the AMF. Therefore, particle size and particle size distribution must be taken under control. Experimental determination of the heating effect of magnetic materials is usually conducted by the nonadiabatic calorimetric method [45] and rarely by adiabatic calorimetry [46]. It is also possible to predict the size-dependent heating efficiency of MNPs by stochastic Neel–Brown Langevin equation and Monte Carlo (MC) simulations [47]. U.M. Endelmann et al. used this method to calculate the heating efficiency of MNPs with a size of 10–30 nm and various values of effective anisotropy constant (K = 4000 J/m^3–11 J/m^3) and damping parameter (α = 0.5–1). The magnetic parameters of MNPs at the same time were obtained using VSM. Experimental and simulated results have shown that the maximum SLP value demonstrated particles in the 22–28 nm range. Moreover, MC simulation revealed a strong dependence of SLP on K [48]. Besides, various empirical and analytical methods are used to evaluate the SLP from an experimental setup, such as the initial slope, corrected slope, Box–Lucas, and steady-state methods [49]. Although recently developed bioheat models for MH are used to understand heat transfer phenomena in living tissue. These methods are fully considered in newly published articles by I. Raouf et al. [50] and M. Suleman et al. [51].

To optimize the experimental conditions, magnetic fluids of different compositions and volumes were analyzed in several laboratories. As a result, the necessary AMF parameters and sample volume were determined as f = 300 kHz; H = 10.6 kA·m^{-1}–15 kA·m^{-1}, volume −1 mL [52].

For MH, the concentration, distribution, and retention of magnetic material within tumor volume are critical parameters. Currently, there are two main directions in MH dependent on the magnetic heating agent used and the manner of its intratumoral administration. Those are «Magnetic Fluid Hyperthermia» (MFH) [41–44,53,54] and «Arterial Embolization Hyperthermia» (AEH) [55–61]. These methods are based on the use of a liquid carrier medium typically containing magnetic iron oxide nanoparticles due to their good biocompatibility. The carrier medium is usually water or saline in MFH, while, in AEH, it is oily contrast media (Lipiodol Ultra Fluid, France, and its analogs) [56–63], and in-situ gelling materials [64] and low-viscosity polymers [65–69].

In clinical practice, the main problem of MFH is the way mediator administration which is realized either by direct intratumoral injection or intravenous medication that do not provide a uniform distribution of magnetic phase in the target tissue due to the typical heterogeneous structure of malignant tumors [54]. Therefore, heat distribution within tumors is not uniform, and there is a risk of proliferation of survived cancer cells. This method is also unacceptable for treating patients with hepatocellular carcinoma (HCC) due to the bleeding tendency in such highly vascularized tumors. Another problem of MFH is the low accumulation of particles in the cancerous area due to metabolism. Based on the clinical trials, the dose of magnetic phase for effective MFH is reported to be 40 mgFe/mL$_{tissue}$ per site, which is difficult to achieve in practice [31]. Indeed, as S. Wilhelm et al. reported, only 0.7% of nanoparticle dose can be delivered to a solid tumor [59]. However, despite these challenges, MFH has now received approval for clinical testing in humans by the German Federal Institute for Drugs and Medical Devices and the United States Food and Drug Administration to treat glioblastoma and prostate gland by colloidal suspension of aminosilane-coated iron oxide nanoparticles (NanoTherm®, MagForce company, Berlin, Germany) [70–72]. This is most likely due to the development of a unique inductive heating applicator operating at an AMF frequency of 100 kHz with a field strength of up 18 kA/m (MFH 300F NanoActivator®; MagForce Nanotechnologies AG, Berlin, Germany) [30].

In contrast to MFH, the challenges discussed above can be eliminated in AEH, developed to treat parenchymal organs. The concept of AEH is based on selective capacitive or inductive hyperthermia after transcatheter embolization of a tumor's arterial supply with a mixture of magnetic particles and an embolic agent [65–69]. As a result of embolization, the size of the tumor decreases due to a decrease in the blood supply, leading to partial necrosis of the tumor. The technique for transarterial embolization dictates the choice of materials for AEH. It depends on the patient's structural features and the treated lesion. However, the technique should meet the requirements of nontoxicity, nonantigenicity, stability to lysis, and radio-opacity. Moreover, at the delivery stage, the material should be of low viscosity to pass through angiographic catheters and fill up not only the main artery but also peripheral arteries and small blood vessels, i.e., ensure both proximal and distal embolization [66,67]. Then, the material should prevent blood flow, for example, due to the rapid increase of the material's viscosity.

The combined effect of embolization and hyperthermia on the tumor leads to ischemic necrosis of the tumor and programmed cell death, apoptosis. The first clinical trials of AEH were conducted 20 years ago at the Russian Research Centre for Radiology and Surgical Technologies (St. Petersburg, Russia) with the permission of the Russian Ministry of Health [67]. At the first stage of treatment, X-ray endovascular embolization with silicone composition containing microsized carbonyl iron particles (Ferrocomposite®, Linorm, Saint Petersburg, Russia) was performed for 46 patients with stage III and IV renal cell carcinoma of the kidney [66]. The occlusion of the vascular system of the kidney was controlled angiographically. Then, 7–10 days after the post-embolization period, capacitive

RF hyperthermia was performed at a frequency of 27.12 MHz with an input power of 80 W. The temperature at the treatment area was monitored by an invasive method: needle-shaped temperature sensors were inserted into the peripheral parts of the kidney under ultrasound control. The treatment time of 30–45 min was necessary to heat a tumor to 43–45 °C. Morphological and histological analysis of kidneys after palliative nephrectomy showed the complete occlusion of the renal tumor blood supply and massive necrosis of the tumor tissue (Figure 2) [66]. As a result, 3–5-year survival of inoperable patients after embolization and hyperthermia was about 18% and 5%, respectively. Nevertheless, AEH based on the dielectric heating principle remains an experimental method in medical practice due to the difficulties associated with the overheating of healthy tissue. Contrariwise, the AEH method involving inductive heating principle (induction hyperthermia) has the advantage to heat selectively a tumor filled with ferromagnetic material without heat generation in the fat layers [68,69,73–75].

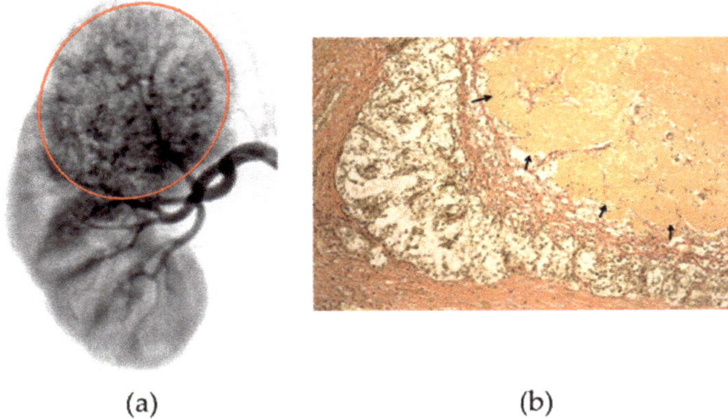

(a)   (b)

Figure 2. (a) Roentgenograph of the kidney after transarterial embolization with Ferrocomposite®; (b) Results of histological analysis indicating massive necrosis of tumor tissue of patient's kidney after ferromagnetic embolization and capacitive RF hyperthermia [66].

The purpose of this review is to demonstrate the potential of AEH for the treatment of deep-sited tumors of the vascular organs, kidneys, liver, and pancreas gland. Considering the strict limitations on the frequency and amplitude of AMF in MH, the heating ability of the mediator should be maximized considering the physical mechanisms responsible for the losses in magnetic materials. Therefore, the mechanisms of magnetic losses in nanomaterials are discussed to determine the relationship between the heating efficiency and magnetostructural properties of NPs. Besides, the role of interparticle magnetic interactions and the properties of the carrier medium on the heating efficiency are considered. The review also presents the results of in vitro and in vivo preclinical trials of magnetic nanomaterials in treating several oncological diseases using AEH.

2. Properties of Magnetic Materials for Their Application in Magnetic Hyperthermia: Nanomagnetism over Micromagnetism

The primary tenet of micromagnetism is that ferro and ferrimagnetic materials are mesoscopic continuous media where atomic-scale structure can be ignored since the magnetization (M), and the demagnetizing field (H_d) are nonuniform but continuously varying functions of distance (r) [76]. The main characteristic of macroscopic samples is the irreversible nonlinear response of M when exposed to an external magnetic field (H) (Figure 3).

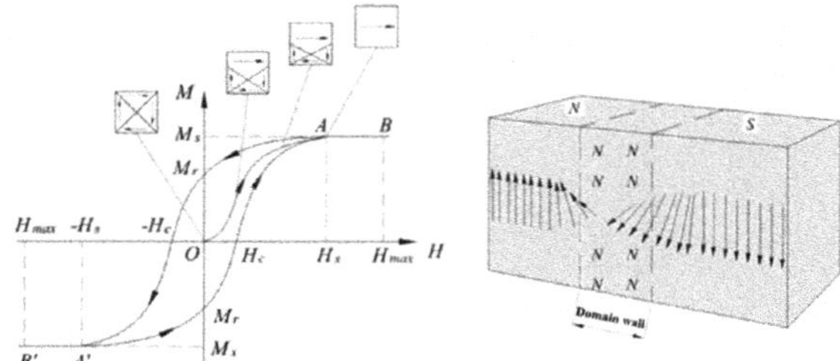

Figure 3. Schematic representation of initial magnetization curve, hysteresis loop, and domain wall structure for a typical ferromagnetic material, where M_S is the saturation magnetization, M_r is the remanent magnetization at H = 0, and H_C is the coercivity.

In irreversible processes, energy is dissipated in the crystal lattice in the form of heat, known as hysteresis loss. The actual physical processes by which energy is dissipated during a quasistatic traversal of the hysteresis loop are identical to those responsible for the dynamic losses. In most materials with multidomain structures, the hysteresis of magnetization arises from domain wall motion or domain nucleation and growth.

There are several contributions to the free energy of magnetic samples with multidomain structure [77]:

Magnetostatic energy E_m, resulting from the interaction of atomic magnetic moments with local internal magnetic field H_i:

$$E_m = - \int_{vol} M{\cdot}H_i \, dV \tag{1}$$

Magnetic free energy, determined by an interaction between atomic magnetic moments and crystalline lattice expressed by magneto-crystalline energy E_k and magnetostrictive energy E_λ:

$$E_k = - \int_{vol} f_k \, dV \tag{2}$$

where f_k is magnetocrystalline anisotropy density

$$E_\lambda = - \int_{vol} f_\lambda \, dV \tag{3}$$

where f_λ is magnetostriction anisotropy density.

Free energy, which is related to the magnetic-exchange interaction.

The magnetostatic energy with dipole–dipole nature is inversely proportional to the volume of the particle, while the domain-wall energy is proportional to the area of the wall (Figure 4) [78]. Considering the balance between the magnetostatic energy and the domain-walls energy, the formation of a multidomain structure is energetically unfavorable when the particle size is less than the width of the domain walls.

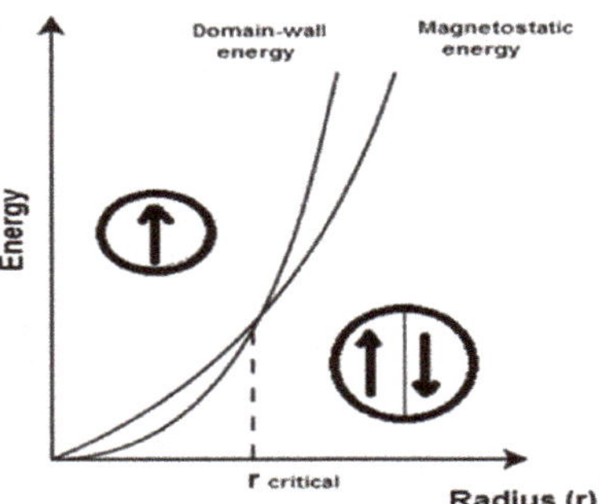

Figure 4. The relative stability of multidomain and single-domain particles. Reprinted with permission [78]. Copyright 2021, Cambridge University Press.

Modern methods for studying the micromagnetic structure of materials, such as transmission microscopy and off-axis electron holography, as well as numerical micromagnetic simulation, revealed three typical magnetic configurations (states) in magnetic materials, depending on the particle size: single-domain (SD) with a uniform arrangement of magnetic moments, pseudo-single-domain (PSD) with a vortex spin arrangement, and multidomain (MD) structure where magnetic structure breaks up into discrete regions separated by domain walls (Figure 5) [79–82].

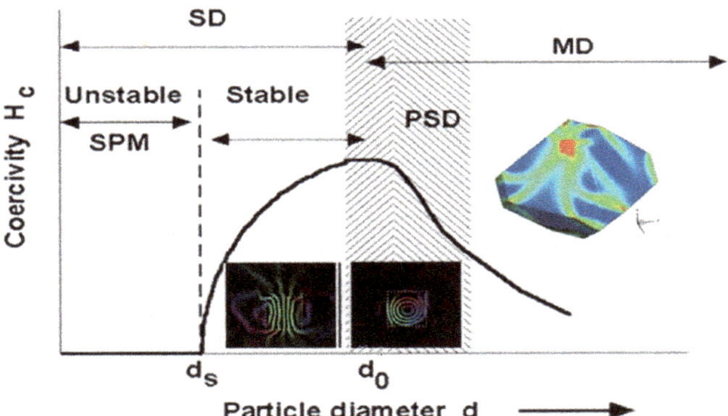

Figure 5. Schematic representation of the transitional state in magnets that spans the particle size range between SD and MD states. The inset collar pictures indicate the direction of the magnetic induction in magnetite particles with 25 nm, 200 nm, and microns sizes, modified from. Reprinted with permission [82]. Copyright 2021, American Chemical Society.

In the case of SD particles, these configurations may exhibit two states: (1) superparamagnetic (SPM) with unstable behavior due to the thermal instability of the magnetization if the thermal energy k_BT ($k_B = 1.38 \times 10^{-23}$ J·K^{-1} is Boltzmann constant, and T is temperature sufficient to change the orientation of the magnetic moment of particle, and, (2) stable SD with ferromagnetic-like behavior when the magnetic moment is pinned along the magnetic anisotropy axis as a result of effective magnetic anisotropy [83].

For many practical applications, such as magnetic storage media and MH, the suitable particle size is within the range of a stable SD state, namely, in the vicinity of SD to PSD transition where the coercivity approaches maximum (Figure 5). The critical size for an SD magnetic state depends on several parameters, including M_S and K. For magnetite and maghemite approved for biomedical applications, the particle size range for the stable SD state is about 20–80 nm for spherical particles and about 200 nm for elongated particles with 2:1 axial ratio [83,84].

The magnetization reversal mechanism in nanosized magnetic materials differs from that for MD ferromagnets. In SD particles smaller than 100 nm, magnetization occurs only by coherent rotation of all atomic magnetic moments within the sample against an energy barrier (ΔE) given mainly by the shape and the crystalline anisotropy fields [85–90]:

$$\Delta E = K - HM_S|\sin\varphi| \pm HM_S\cos\varphi \tag{4}$$

where K is the anisotropy energy density, and φ is the angle between the easy axis and the magnetic field.

The shape anisotropy comes from the demagnetizing field:

$$H_d = -N_dM_S \tag{5}$$

where N_d is the demagnetizing shape factor of a magnetized unit.

A dominant effect of the size and shape anisotropy on H_C and M_S and the heating efficiency has been observed in anisotropic magnetite NPs, such as wire, ring, rod, cube, octahedron, etc. (Figure 6) [87–90].

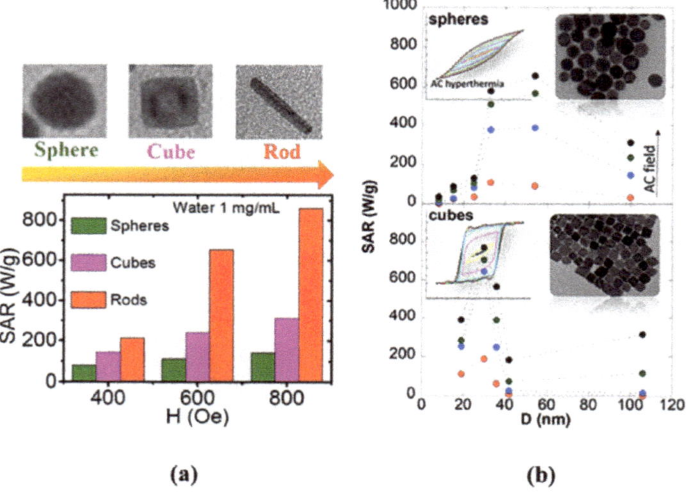

Figure 6. (a) SAR vs. H for the Fe_3O_4 NPs (spheres, cubes, nanorods) with the volume of about 2000 nm^3, (b) SAR vs. H for the sphere and cube-shaped Fe_3O_4 NPs dispersed in water under AMF (300 kHz, from tens to 800 Oe). Reprinted with permission [87,88]. Copyright 2021, American Chemical Society.

It is known that MNPs with particle sizes larger than 20 nm are in a stable-domain state with ferromagnetic-like behavior when the magnetic moment is pinned along the magnetic anisotropy axis as a result of effective magnetic anisotropy. Such NPs exhibit much higher heat loss in AMF. This class of magnetic nanomaterials also includes the novel octahedral monocrystalline magnetite NPs obtained by thermal decomposition [91]. Owing to the octahedral morphology, these NPs show one of the largest SARs rates reported to date for a colloidal suspension of magnetite: 1000 W/gFe_3O_4 at 40 mT and 300 kHz. Such behavior has been explained by the shape of NPs that imprints a biaxial or bi-stable character to the magnetic anisotropy.

Besides the size and shape, other properties of magnetic NPs should be considered for modulating the heat generation, such as particle size distribution and the presence of interparticle interactions.

The polydisperse sample represents an ensemble (mixture) of particles with various magnetization states corresponding to the distributions in magnetic properties, especially the magnetic anisotropy, which governs the height of the energy barrier. Many works have been focused on how polydispersity influences the hyperthermia performance of magnetic NPs. The detrimental influence of size polydispersity (σ) on the heat outcome is usually reported as follows: heat generation can drop between 30% to 50% for σ varying between 0.2 and 0.4 [92–94]. However, the decrease in the heating efficiency in polydisperse materials can be associated not only with polydispersity per se but also with low (unsaturated) magnetic field strength [95]. The use of low amplitudes in the experiments results from a number of unsuccessful attempts in clinical trials to apply the amplitudes of AMF beyond 15 kA/m. For example, M. Johannsen et al. reported that patients undergoing thermotherapy treatment of prostate cancer with exposure to AMF of 100 kHz felt discomfort at amplitudes higher than 5 kA/m [30], whereas for the treatment of brain tumor, field strength up to 13.5 kA/m was reported to be well tolerated [71]. Therefore, in each treatment case, the frequency and amplitude of AMF must be correctly selected regardless of the polydispersity of particles.

As mentioned above, along with polydispersity, the interparticle interactions can significantly affect the magnetization dynamics of NPs, since they lead to aggregation, especially when particles are without surface coating.

In an ensemble of noninteracting SD NPs, the energy losses are associated with the Neel–Brown relaxation process [95]:

$$\tau_N = \tau_0 \exp\left\{ \frac{KV_C}{k_BT} \left(1 - \frac{H}{H_A} \right)^2 \right\}; \quad \tau_B = \frac{3V_h\eta}{k_BT} \tag{6}$$

where τ_N and τ_B are time scales of Neel and Brownian relaxation, τ_0 is a pre-exponential factor ($10^{-9} \div 10^{-11}$ s), H_A is the anisotropy field equal to 2 K/μ_0M_S, V is the particle volume, η is the viscosity of the carrier medium, V_h is the hydrodynamic volume of the particle, and H is the field amplitude.

For a material in which both Neel and Brown relaxation takes place with an effective relaxation time τ_{eff}, the mechanism with shortest τ_{eff} dominates. However, due to the exponential dependence of τ_N on the particle volume, while τ_B linear grows with hydrodynamic volume, different Neel and Brownian contributions can be realized for the same magnetic material with different particle sizes (Figure 7a) [95,96]. Both relaxation processes strongly depend on the amplitude of AMF. It is established that the Neel mechanism manifests itself predominantly at high field amplitudes, while at low field amplitudes, the Brownian mechanism prevails [97]. In addition, the relaxation time of a Brownian process is proportional to the viscosity of the carrier medium; thus, Brownian relaxation is largely suppressed when the particles are immobilized in a viscous medium such as cancerous tissue (Figure 7b) [98].

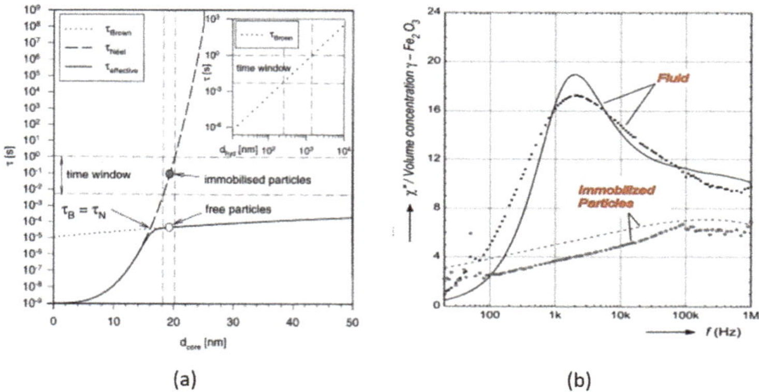

(a)   (b)

Figure 7. (**a**) Néel and Brown relaxation times calculated over a range of particle sizes for a water-based magnetite ferrofluid [96]; (**b**) Imaginary part of susceptibility of maghemite based aqueous suspension in comparison to the identical particles immobilized in the gel. Reprinted with permission [96,98]. Copyright 2021, Elsevier.

The assembly process of magnetic NPs in liquid media is driven by the attractive–repulsive interactions between NPs, van der Waals (vdW), magnetic, electrostatic, and solvophobic forces [98–100]. The former two are core–core interactions that dominate the interaction potential and hold NPs together. The van der Waals interactions scale linearly with the particle's radius, while the magnetic interaction scales with its volume. Magnetic interactions always coexist with vdW forces, which becomes increasingly important with decreasing particle size. For example, the formation of aggregates from NPs in the absence of an external magnetic field already takes place at the beginning of coprecipitation reaction; thus, the contribution of vdW forces can be notable (Figure 8) [93]. However, the formation of dense aggregates that are stable against segregation into individual nanoparticles is possible only under the influence of interparticle magnetic interactions [100]. Estimating the threshold sizes for the agglomeration of magnetite NPs has shown that they are relatively stable against agglomeration up to 20–25 nm in diameter [101,102].

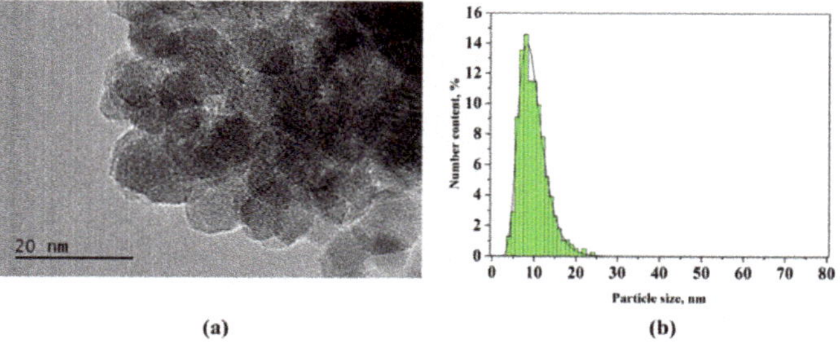

(a)   (b)

Figure 8. (**a**) TEM image of the five-minute reaction product for magnetite synthesized by coprecipitation method, and (**b**) particle size distribution histogram. Reprinted with permission [93]. Copyright 2021, American Chemical Society.

The interactions between magnetic NPs are interpreted in terms of magnetodipole and exchange interactions. Exchange interaction can be neglected when interparticle spacing is of the order of 2 nm, which approximately corresponds to the distance between two NPs with a dead layer of thickness of about 1 nm [93]. In most cases, the dominant contribution to interparticle energy is a magnetodipole coupling, which increases with the volume of NPs and depends on the mutual distance between particles [103–105].

Dipole–dipole interactions can be either attractive (in-line dipoles) or repulsive (antiparallel aligned dipoles). The predominant type of configuration of dipoles is the antiparallel orientation of the magnetic moments of a pair of particles [93,106]. Subsequently, the pair of dipoles stick together to form larger aggregates, and, in the absence of an external field, these aggregates have closed magnetic flux with random orientation of magnetic moments of individual NPs (Figure 9).

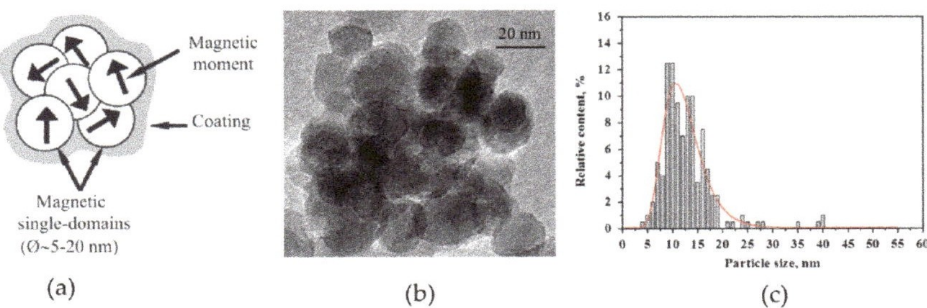

Figure 9. (**a**) Schematic picture of dense aggregate (multicore particle) with surface coating [106], (**b**) TEM image of magnetite NPs clustered into dense aggregate, and (**c**) particle size distribution in aggregate. Reprinted with permission [93,106]. Copyright 2021, Elsevier, and American Chemical Society.

Even though NPs in the aggregate are in the SPM state, in some cases, the material itself may demonstrate ferromagnetic-like behavior, which is evidenced by distinct sextets on the Mössbauer spectrum and blocking temperature well above room temperature on the FC/ZFC curves (Figure 10) [107].

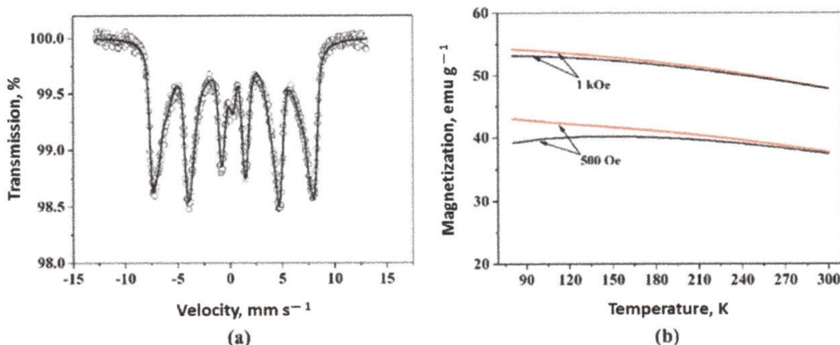

Figure 10. (**a**) Mössbauer spectroscopy and (**b**) FC/ZFC magnetization curves of as-prepared magnetite NPs in a powder form. Reprinted with permission [107], Copyright 2021, Elsevier.

The ferromagnetic-like behavior of such materials can be explained by the internal structure of the aggregate formed, namely, when SPM NPs are combined into a dense 3D cluster, the so-called multicore particles [105,107,108] and nanoflowers [109]. These materials can provide efficient and rapid heating in AMF at low-field regimes [110].

The experimental results and numerical simulations have shown the different effects of inter and intra-aggregate magnetodipole interactions on heat generation [93,107–113]. It is found that an increase in the concentration of NPs leads to a nonmonotonic behavior of SAR (SLP) with its reduction at a specific aggregate size (Figure 11).

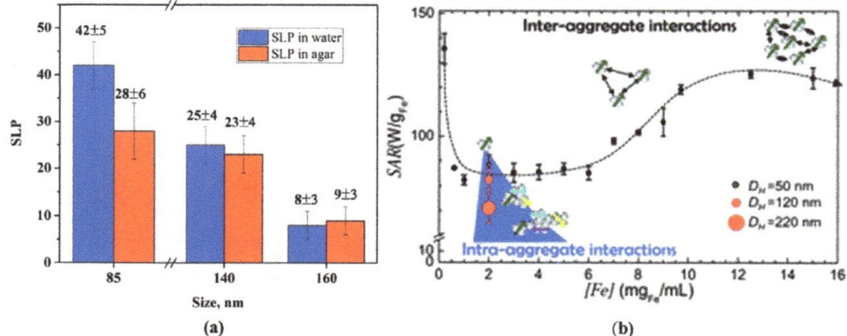

Figure 11. (a) SLP for water and agar dispersions of magnetite-based multicore particles as a function of hydrodynamic size in AMF of 1048 kHz and 5.8 kA/m [93]; (b) Effect of intra and intercluster magnet–dipole interactions in the dispersion of magnetite NPs on SAR as a function of concentration and hydrodynamic cluster size (D_H) obtained under the given AMF conditions of 105 kHz and 13.1 kA/m [111]. Reprinted with permission [93,111]. Copyright 2021, Elsevier and Royal Society of Chemistry.

A theoretical study of the effect of magnetodipole interactions on the heating ability of an ensemble of particles, and especially multicore particles, is challenging since it is a multiparameter task that must account for the magnetic characteristics of primary particles, as well as morphostructural properties, i.e., polydispersity, shape anisotropy, packing density of NPs in a cluster, etc. [114–117].

An increase in the heating ability is usually explained either by a change in the characteristic height of the energy barrier related to the thermal energy [92] or by a change in the magnetic state due to the collective behavior of closely spaced NPs [84,85,113,114]. In turn, a decrease in the heating ability in the ensemble of interacting NPs is explained by the disorienting effect of a random magnetic field, causing a deviation of the magnetic moments of NPs from the direction of the AMF [116].

The nanoparticle size is one of the most important parameters that affects the magnetic properties of multicores. The study of the heating ability of multicore particles of approximately the same hydrodynamic diameter (100 nm) but formed by magnetite NPs of different sizes (7.1 and 11.5 nm) showed different results [118]. Both types of multicores improve their heating efficiency compared with individual NPs when exposure to AMF at f = 302 kHz and H = 15 kA/m, but multicores composed of larger NPs show two times higher values of SAR. The crucial role of core particle size in a cluster has also been established by C.H. Jonasson et al. They investigated the heating efficiency for SD particles of different sizes and multicore particles both experimentally and theoretically using dynamic Monte-Carlo simulations [119]. It was found that for a given AMF (1 MHz, 3–10 kA/m), core–core interactions can lead to a different character of ILP dependence on the core diameter (D_c) when D_c is higher or lower than the size maximizing the ILP value, i.e., D_c = 20 nm (Figure 12).

Apart from magnetite and maghemite single phase-based systems, exchange-coupled magnetic NPs have been proposed so far as possible candidates for efficient MH. These particles have a core–shell structure with different combinations of magnetically soft and hard materials, for example, $CoFe_2O_4@MnFe_2O_4$, $CoFe_2O_4@Fe_3O_4$, $Fe_3O_4@CoFe_2O_4$, $CoFe_2O_4@\gamma Fe_2O_3$, $FeO@Fe_3O_4$, etc. [120–123]. The main idea underlying exchange-coupled magnetic NPs in MH is to increase the hysteresis losses by controlling the anisotropy constant (K) while maintaining superparamagnetism, thereby preventing aggregation and the formation of large clusters. The magnetic properties of exchange-coupled core–shell particles and their effect on SLP are dependent on the composition, as well as on the core and shell size (Figure 13). As can be seen from this figure, the SLP of core–shell NPs exhibits

SLP values significantly higher than the SLP of single-phase magnetic NPs; however, the amplitude of AMF is twice the value allowed for medical application.

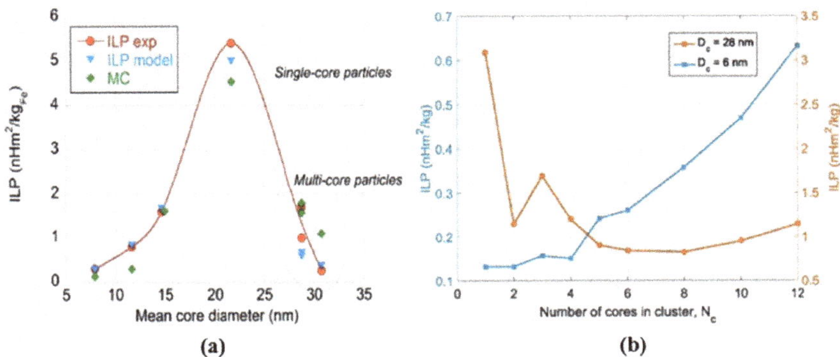

Figure 12. (**a**) Experimental (red) and simulated ILP data analysis for noninteracting (blue) and interacting single-core particles (green); (**b**) Simulated ILP versus the number of cores in a core cluster for two core sizes, below (6 nm) and above (28 nm) the core size maximizing the ILP value. Reprinted with permission [119]. Copyright 2021, Elsevier.

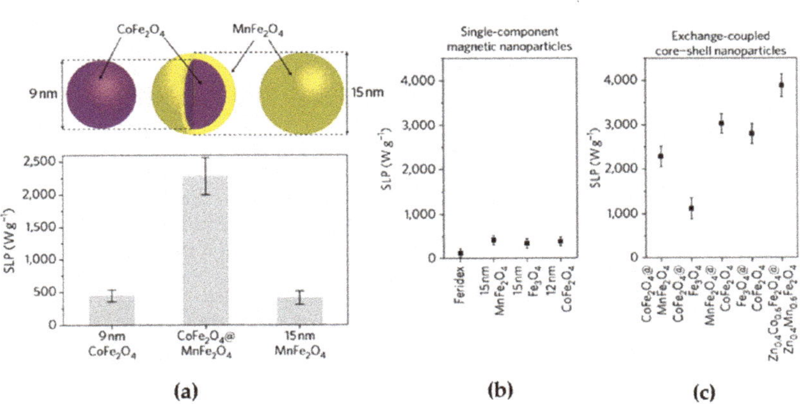

Figure 13. (**a**) Schematic representation of an exchange-coupled core–shell $CoFe_2O_4@MnFe_2O_4$ nanoparticle and its SLP value compared with SLP values of single-phase $CoFe_2O_4$ and $MnFe_2O_4$; (**b**,**c**) SLP values of various combinations of core–shell NPs and single-component NPs measured at f = 500 kHz and H = 37.3 kA/m. Reprinted with permission [120]. Copyright 2021, Springer Nature.

To sum up, dynamic magnetic properties and an increase in the heating efficiency of the mediator (for given amplitudes and frequencies of AMF) are determined by the following factors: (1) material composition and degree of crystallinity; (2) average particle size within the range of stable SD state, which is between 16–20 nm in diameter for ferrimagnetic iron oxides; (3) particle size distribution (preference to monodispersity over polydispersity), and (4) magnetodipole interaction. The first affects K and M_S, the next two control the interaction strength of NPs, thus regulating the hydrodynamic size and internal structure of multicore particles, and the last one determines collective magnetic behavior and thus modifies the amount of heat generation during the hyperthermia session.

3. Magnetic Materials for Application in AEH

As already mentioned in the introduction, AEH is a multiple treatment modality involving transarterial embolization of tumors with magnetic material followed by exposure to AMF at clinically relevant frequencies and amplitudes. The embolotherapy itself is widely used in clinical practice for diagnostic (coronary angiography), preoperative management of malignant renal tumors, chemoembolization of malignant hypervascular tumors such as hepatocellular carcinoma, as well as in the treatment of aneurysms, hemorrhage, angiomyolipoma, and other conditions [63,67,124].

At the first stage of the treatment by AEH, transcatheter injection of the embolic agent is administered under an angiographic control. This procedure usually lasts about 20–25 min. During this time, the embolic agent should maintain low viscosity ($\eta < 0.5\,\mathrm{Pa/s}$) for transportation and filling of the tumor vascular system. After this induction period, the viscosity should increase rapidly with forming a soft and stable embolus, which can occlude the tumor blood vessels. The selective MH can be carried out after the passing of the post-embolization period of patients (fever, elevated white blood count, etc.), which usually takes from one to two weeks [66,67].

Embolic agents are generally classified into mechanical [56] and flow-directed agents, but only the latter is used in AEH as a carrier of magnetic particles. The list of such materials includes Lipiodol (Lipiodol Ultra Fluid, Guerbet, France) and its analogs [57,60,63,125], in-situ gelling materials (e.g., Onyx® (DMSO, Acros Organics, Basel, Switzerland): ethylene-vinyl alcohol copolymer dissolved in dimethylsulfoxide, etc.) [64,126], and low-viscosity polymers (polyorganosiloxanes, radiopaque degradable polyurethane) [65–69,127–129].

Most studies on the effectiveness of AEH have been conducted in vivo in mouse and rabbit models with liver cancer disease after hepatic intra-arterial injection of magnetic NPs suspended in Lipiodol [59–63,130–133]. The use of Lipiodol as a carrier medium is determined by a set of its properties: radio-opacity (~48% of iodine), ability to induce plastic and transient embolization of tumor, microcirculations causing ischemic necrosis of tumor, limiting the ingress of viable cancer cells as well as their debris in the bloodstream. With the right choice of the magnetic particle size and concentration in Lipiodol, it is possible to achieve a homogeneous intratumoral distribution of the magnetic phase and, thus, to significantly increase the specific heating of the tumor, but only at field amplitudes of about 20 kA/m and higher, which is beyond the permissible limit in MH (Figure 14) [62].

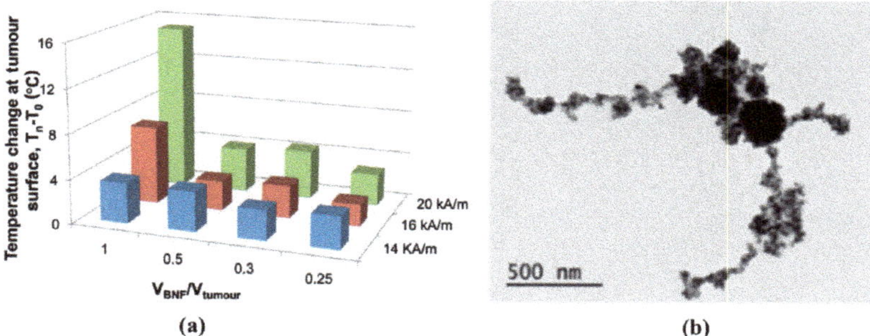

(a) (b)

Figure 14. (a) Temperature change at the tumors of mice with hepatocellular carcinoma after injection of magnetite NPs emulsion in Lipiodol (cBNf-lip) followed by MH at 155 kHz and amplitudes within 14–28 kA/m. (b) TEM image of cBNf-lip. Reprinted with permission [62]. Copyright 2021, Taylor & Francis.

Similar results were obtained earlier by Moroz et al., who reported the superiority of AEH compared with direct injection hyperthermia, studied on a model of a rabbit liver tumor [58,60,133]. It was found that after hepatic arterial embolization by maghemite NPs (100–200 nm) suspended in Lipiodol and subsequent MH (53 kHz, 30–45 kA/m), the tumor volume can be reduced by almost 94%, depending on the particle concentration and the uniformity of its distribution in the tumor.

Along with Lipiodol, other materials have been investigated as embolic agents in AEH. Exemplarily, the embolic properties and heating efficiency of organogel (Onyx®) containing silica microbeads filled with magnetic iron oxide NPs have been studied in nude mice carrying subcutaneous human carcinomas [64,126]. Thus, the intratumoral injection of nanocomposite followed by 20 min MH at 141 kHz resulted in extensive tumor necrosis (78%) but only for a group of animals exposed to AMF of high intensity (~12 mT) (Figure 15). At such an intensity of AMF, the tumor heats up to 44–45 °C, thus, undergoing thermal ablation. A survival study using magnetic resonance imaging has shown that 45% of the 12 mT-treated groups survived one year without any tumor recurrence.

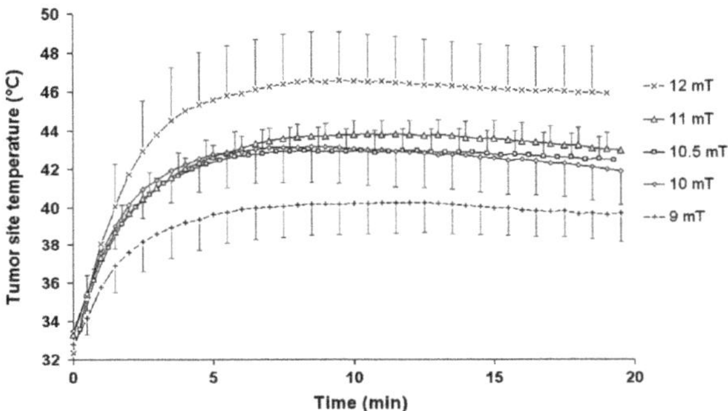

Figure 15. Thermogram of mean intratumoral tumor temperature as a function of treatment time and magnetic field strength. Reprinted with permission [64]. Copyright 2021, Taylor & Francis.

Despite the demonstrated efficacy of AEH in vivo, some obstacles remain that limit the use of this method in clinical practice, namely, difficulty in assessing the actual temperature of tumors. It is generally clear that, for successful performance of AEH, two significant factors should be accomplished simultaneously: achieving the total embolization of the tumor vascular system and attaining hyperthermia temperatures causing ischemic necrosis of the tumor. Thus, a balance must be achieved between the mechanical properties of embolic agents comprising low initial viscosity, rapid solidification, and robust embolus stiffness. From that perspective, the biocompatible polyorganosiloxanes are promising embolic agents: (1) their viscosity and the curing rate can be regulated by their composition, (2) silicone elastomers do not display adhesion to living tissues, (3) they are soft materials and do not injure blood vessels [134]. Moreover, silicones closely adjoin the walls of blood vessels, reducing the probability of blood flow recovery. The low specific heat of silicon rubber of 1.05–1.30 J/g·K is also advantageous because it favors the distribution of the heat generated by incorporated magnetic particles [73]. The advantage of using these polymers for transcatheter embolization in patients with renal cell carcinoma was demonstrated with Ferrocomposite® (Saint Petersburg, Linorm, Russia) [65–69]. The X-ray image (Figure 2a) shows a patient's kidney with a large tumor uniformly filled with Ferrocomposite®. Complete occlusion of renal tumor blood supply results in necrosis of tumor tissue, and in combination with RF capacitive hyperthermia, it provided in massive necrosis (Figure 2b). Considering the positive results of clinical trials of Ferrocomposite®, further research aimed to improve its properties: reducing the viscosity of the embolic agent and increasing its heating ability in AMF permitted for medical application. As a result, maghemite-based silicone composition (Nanoembosil®) (Saint Petersburg, Linorm, Russia) was developed, which possesses a set of properties required for its use as a mediator in AEH: the ability for secure embolization of the tumor blood vessels, the high heating rate in AMF at moderate frequencies, and amplitudes, and radiopacity [68,69]. Nanoembosil®

has been tested in vitro at the Tomas Bata University in Zlin (CZ) and in vivo at Moscow's Blokhin Russian Cancer Research Centre.

4. Nanoembosil®: Synthesis, Characterization, In Vitro and In Vivo Study

The magnetic phase of Nanoembosil® is maghemite NPs prepared by annealing (6 h at 300 °C) of magnetite NPs synthesized by coprecipitation method under certain reaction conditions that guarantee the formation of monodisperse NPs with a high degree of crystallinity [107,135]. The annealing does not change the morphology of NPs but predictably decreases the M_S value. Notably, the original (as-prepared) and annealed samples demonstrate almost identical heating efficiency (Figure 16), which can be explained by increased effective anisotropy due to magnetodipole interactions resulting in the formation of stable multicore particles (Figures 8 and 9).

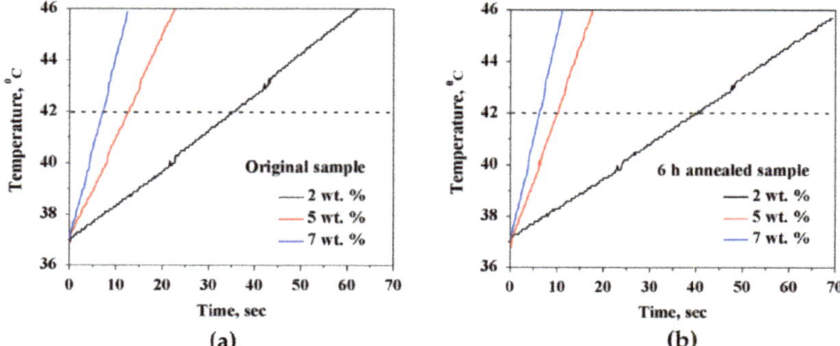

Figure 16. Inductive heating of magnetite (**a**) and maghemite (**b**) glycerol dispersions in AMF (525 kHz, 7.6 kA/m). Reprinted with permission [107]. Copyright 2021, Elsevier.

The characteristics of raw materials used for the preparation of Nanoembosil® are presented in Tables 1 and 2.

Table 1. Magnetic phase: Iron oxide NPs Structural and magnetic properties.

Property	Magnetite	Maghemite
d_{TEM} (nm)	13	13
σ_{TEM}	0.3	0.3
d_{XRD} (nm)	12	12
ε (%)	0.3	0.6
Magnetite content determined from XRD (%)	72	8
Magnetite content determined from MS (%)	60	0
M_s (emu g^{-1})	56 ± 2	48 ± 1
M_r (emu g^{-1})	0.8 ± 0.2	0.8 ± 0.2
H_c (Oe)	11 ± 4	10 ± 3
SLP (W/g$_{Fe}$)	23.0 ± 0.6	20.3 ± 1.5

d_{TEM}, d_{XRD}—average particle size determined by TEM and XRD, respectively; σ_{TEM}—polydispersity index; ε—crystal lattice strain; M_S—Mössbauer spectra.

Table 2. Polymer phase: Characteristics of raw polymer for Nanoembosile® preparation.

Reagent	Molar Weight (g·mol^{-1})	Polydispersity	Viscosity @ 25 °C (Pa·s)	Concentration of Substitutions (wt. %)
PVS	100,126	1.62	2.6	0.1–0.4, vinyl groups
PHS	17,202	1.83	0.65	0.55, hydrosubstitutions
PDMS	73,78	1.20	0.03	-
CTS	345	-	3.9	-
Speier's catalyst	-	-	-	Hexachloroplatinic acid [H$_2$PtCl$_2$]·H$_2$O dissolved in PDMS
Karstedt's catalyst	-	-	-	Platinum [0] complex containing vinyl–siloxane ligands

PVS [poly(dimethylsiloxane-co-methylvinylsiloxane)]; PHS [poly(dimethylsiloxane-co-methylhydrosiloxane)]; PDMS [poly(dimethylsiloxane)]; CTS/Cyclotetrasiloxan [1,3,5,7-tetravinyl-1,3,5,7-tetramethylcyclotetrasiloxane].

The Nanoembosil® is supplied in two compositions in separate containers: the first contains PVS, catalyst, maghemite NPs, and radiopaque potassium iodide (KI) (Container 1), whereas the second contains PHS, CTS, and PDMS (Container 2). Once the content of Container 2 is added into Container 1, the hydrosilylation reaction of hydro-and vinyl-functional silicone polymers starts. To monitor the polymerization process, the optimal concentration of reagents was chosen to provide low initial viscosity (0.25–0.3 Pa/s) during the induction period (20–25 min), followed by an abrupt increase in viscosity and the formation of a soft embolus (Table 3). To this end, the impact of each component on the kinetics of Nanoembosil® formation was investigated through measurements of the rheological properties on a Rheometer with parallel plate geometry. Variation of rheological properties of the composite during its formation is presented in Figure 17.

Table 3. Nanoembosile® composition.

Composite Type	Concentration of the Initial Components, wt. %							η^*_{in}	η^*_{fin}	t_{in}
	PVS	CAT	PHS	PDMS	CTS	NPs	KI	(Pa/s)		(min)
I	40	2	11	33	-	7	7	0.3	3000	20
II	36	2	10	32	6	7	7	0.25	3000	25

η^*_{in} is initial viscosity, η^*_{fin} is final viscosity, t_{in} is the duration of the induction period.

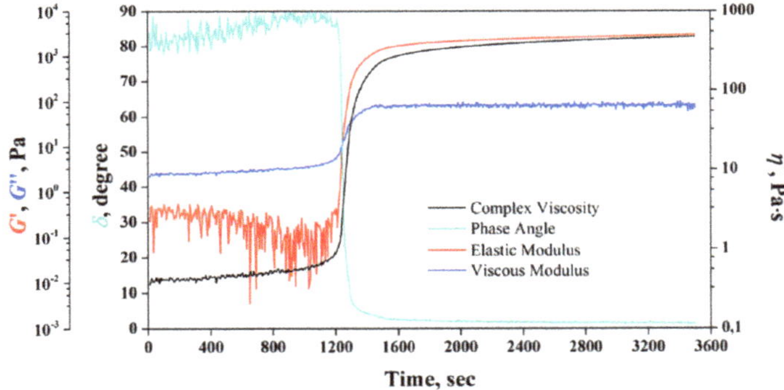

Figure 17. Kinetics of Nanoembosil® formation. Reprinted with permission [68]. Copyright 2021, Elsevier.

As it can be seen from this figure, no changes in the elastic modulus (G′) and viscous modulus (G″) of complex shear modulus G* are observed at the beginning of the reaction, and the system's viscosity stays constant, which corresponds to the induction period of the reaction. The complex shear modulus is defined as [136]:

$$G^* = \frac{\tau^*}{\gamma_M} = \sqrt[2]{\left(G\prime^2 + G\prime\prime^2\right)},$$

(7)

where τ^* is the complex stress, and γ_M is the maximum value of fixed strain.

During the induction period, the composition is fluid, evidenced by a greater value of G″ over G′. As the reaction between vinyl- and hydro-groups starts, both G′ and G″ increase, but G′ increases faster than G″. At the end of the induction period (~20 min), there is an intersection between G′ and G″, indicating reaching the vulcanization point. Above the vulcanization point, the elastic component dominates over the viscous one, G′ > G″; the viscosity of the composition rapidly increases till it reaches a maximum and then stays constant after all the hydro-groups have reacted. Thus, the dominant influence on the kinetics of composite formation is exerted by PHS, PVS, PDMS, and CTS, where the last two components play the role of plasticizers to adjust composite viscosity. As to the influence of the magnetic filler concentration on the rheological properties of the composite, it is negligible up to 14 wt.%.

The heating efficiency of Nanoembosil® was estimated in the AMFs at frequencies and amplitudes (f = 0.05–1.5 MHz, H $\leq$ 15 kA·m^{-1}) standardized for medicine. The heating rate of the composites, as well as SLP, depends on the AMF parameters since the magnetization process is determined solely by the Neel relaxation (Figure 18, Table 4). Nevertheless, according to the results obtained, a high heating rate can be achieved in the entire frequency range even at sufficiently low field amplitudes. Therefore, an increase in SLP in composites of this type is possible with increased field amplitude to the allowed power level.

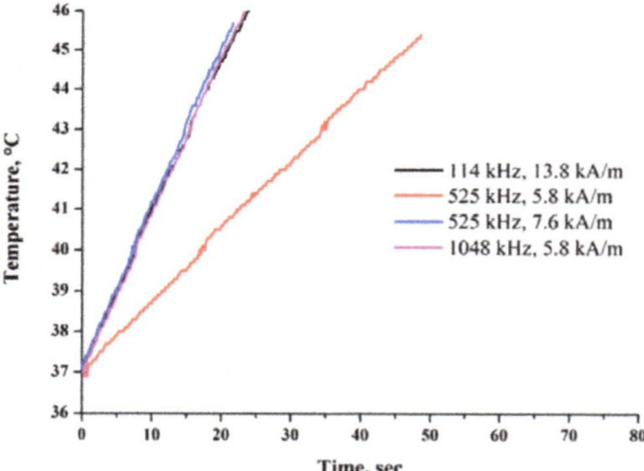

Figure 18. Inductive heating of composites with 7 wt.% of NPs at different alternating magnetic field parameters. Reprinted with permission [68]. Copyright 2021, Elsevier.

Table 4. SLP values (W·g$_{Fe}$$^{-1}$) and heating rates (°C/min) of Nanoembosil® in AMF of various frequencies and amplitudes.

AC Magnetic	f (kHz)	114	525	1048	
Field	H (kA·m^{-1})	13.8	5.8	7.6	5.8
SLP		8.6 ± 0.1	4.3 ± 0.4	8.6 ± 1.0	9.0 ± 1.5
Heating rate	(°C·min^{-1})	13.8	5.8	7.6	5.8

Considering that the embolic agent should also possess thermal expansion similar to or higher than blood to prevent the blood flow recovery during hyperthermia session, the thermomechanical properties of Nanoembosil® were studied by dynamic mechanical analysis. The study revealed that the material possesses the rubber-elastic properties: shear modulus, G', is almost independent of the applied frequency, and the loss tangent, $\tan \delta$, is slight (0.1–0.2) (Figure 19). The value of G' within the range of hyperthermia temperatures slightly increases from 9.6 to 9.9 kPa at 38 °C and 10 Hz shear rate. The obtained shear modulus values are smaller than those reported for the artery, 30–3000 kPa [136,137]; thus, the composite will easily deform with the artery. Furthermore, the thermal expansion coefficient of the composite is 760 ppm °C^{-1} at 37 °C and slightly decreases to 710 ppm °C^{-1} with the temperature rise to 45 °C. The high value of α for the composite ensures the prevention of blood flow recovery during heating.

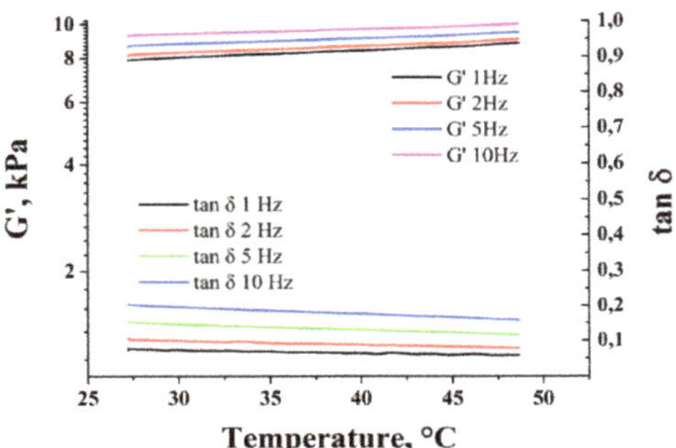

Figure 19. Temperature dependence of shear modulus and loss tangent at different shear rates for Nanoembosil®. Reprinted with permission [68]. Copyright 2021, Elsevier.

In vivo study of Nanoembosil® was done at the N.N. Blokhin Russian Cancer Research Centre in Moscow. The study aimed to determine the embolic agent dose for intra-arterial administration and filling the tumor vascular system and estimate its antitumor effect [138]. All experiments were conducted according to the National and European guidelines on the ethical use of animals [139,140]. Throughout the procedures, animals were anesthetized with Zoletil 100 (Virbac, Carros, France). The experimental study was performed using 25 rats (8-week-old males of body weight 200 g) and 16 male rabbits weighing 2.0–3.0 kg. Some animals were kept healthy to estimate tolerability of embolization, and others were intramuscularly implanted with hepatocellular carcinoma PC-1 (rats) and VX2 (rabbits). It was established that intravenous tissue tolerance of embolic agents for rats is 0.1 mL for the composition I with η^*_{in} = 0.3 Pa/s, and 0.2 mL for composition II with η^*_{in} = 0.25 Pa/s, while for rabbits, 1.5 mL, which is well below the guidelines for the maximum intravenous injection volumes of experimental compounds in rats and rabbits [141]. Thus, the study of the effect of embolization on tumor growth was conducted using the established volumes of compositions. To this end, four independent animal groups with the same number of rats (n = 5) were used. Animals in group 1 were exposed to embolization by composition I (0.1 mL) and in groups 2 and 3, respectively, by composition II with volumes of 0.1 mL and 0.2 mL. To monitor the treatment of tumor growth, the test group received only a physiological solution.

The embolization was carried out on the 20th day after transplantation when a full regional blood flow in the tumor node was formed. The embolic agents were prepared extempore and administered once by transarterial infusion into the femoral artery using an intravenous catheter.

The efficacy of embolization was estimated according to standard criteria: inhibition of tumor growth, tumor growth dynamic (3, 7, 10, and 14 days after embolization), tumor doubling time (τ_2), and therapeutic response ($\tau_{trial}/\tau_{control}$). The tumor growth was estimated by the change in the tumor volume (V_t/V_0), where V_0 is the mean volume of the tumor before, and V_t is the mean volume of the tumor after treatment. The results of in vivo study are presented in Table 5 and Figure 20.

Table 5. The effect of embolization by silicone-based magnetic composites on the dynamics of PC1 tumor growth in rats.

Group of Animals	Dose (mL)	V_t/V_0 n Days after Embolization		τ_2 (Days)	$\tau_{trial}/\tau_{control}$ (%)
		3	7		
Test group saline infusion	0.2	2.3	5.3	3.0	-
Group 1 Composite I	0.1	1.4	2.5	6.0	2.0
Group 2 Composite II	0.1	1.4	2.5	6.0	2.0
Group 3 Composite II	0.2	1.2	1.5	>11	4.0

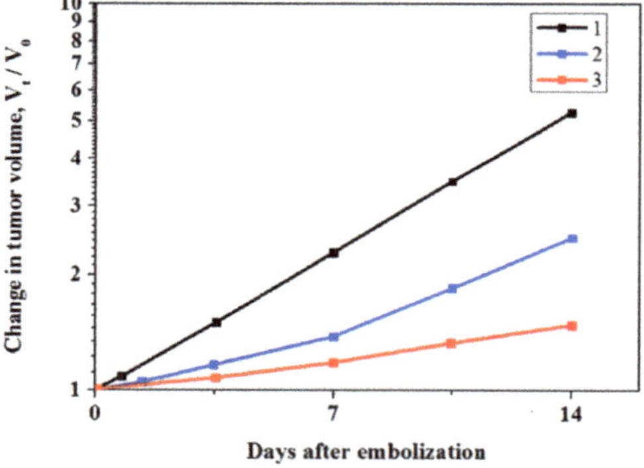

Figure 20. Inhibition of rat liver tumor growth after embolization of tumor vascular system by silicone-based magnetic composites: 1: test group; 2: embolization with composite I (administrative dose 0.1 mL); 3: embolization with composite II (administrative dose 0.2 mL); standard deviation is ±0.5.

Accordingly, in the case of the test group of rats, the tumor growth rate increased by a factor of 1/5. The embolization of the tumor vascular system by both types of compositions significantly inhibits the tumor growth, and this effect is more pronounced in the case of embolization by Nanoembosil® (composition II). Indeed, over a period of seven days after embolization, the rate of tumor growth was two times lower in group 3 compared with the test group. This is due to the low viscosity of composition II (0.25 Pa/s), which allowed one to increase the administered dose without causing side effects.

Similar results were obtained in rabbits with intramuscularly transplanted VX-2 liver cancer. Intra-arterial injection of Nanoembosil® at a dose of 1.5 mL on the 20th day of tumor growth led to a significant decrease in growth rate within two weeks. Stabilization of tumor growth within 14 days is associated with tumor cytoreduction by more than 50%.

The efficacy of silicone-based magnetic composites as heat mediators was studied in vitro on the human Hepatocellular carcinoma cell line (HepG2). The treatment was conducted in the following scheme. Prior to in vitro testing, a certain amount of Nanoembosil® was mixed with cells in the ratios of 1:1 and 2:1, which correspond to iron concentrations of 3.5 and 5 g/L in the tested value. All samples were preheated to 37 °C in a hot water bath and further heated up to 44 °C via a homemade inductive heating applicator (Figure 21).

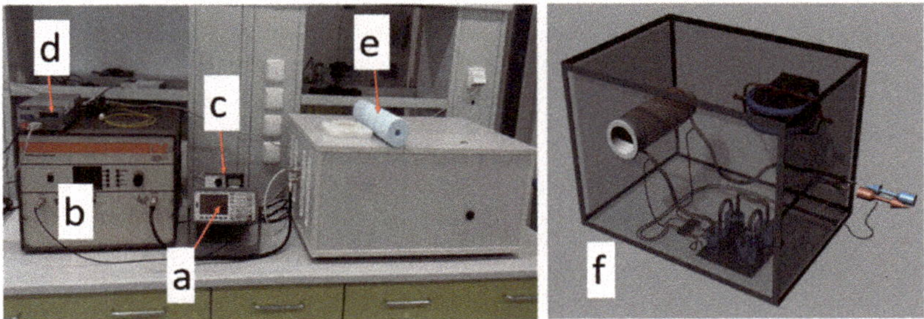

Figure 21. Laboratory device for measuring the heating ability of materials and conducting in vitro and in vivo studies: (**a**) signal generator, (**b**) signal amplifier, (**c**) AMF amplitude meter, (**d**) temperature recorder with fiber optic thermocouples, (**e**) sample holder, (**f**) schematic illustration of a device with a horizontal position of an impedance coil for in vivo study.

The measurements were carried out in AMFs with frequency 525 kHz and amplitude of about $9\ \mathrm{kA \cdot m^{-1}}$. After reaching 44 °C, this temperature was maintained for 30 min, which is a common treatment time for hyperthermia sessions [5]. To determine cell viability, the MTT (3-(4,5-dimethylthiazol-2-yl)-2.5-diphenyltetrazolium) test was performed. The results obtained indicate the cytotoxic effect of the hyperthermia treatment with Nanoembosil®. Moreover, the observed effect increases with the concentration of the mediator (Figure 22).

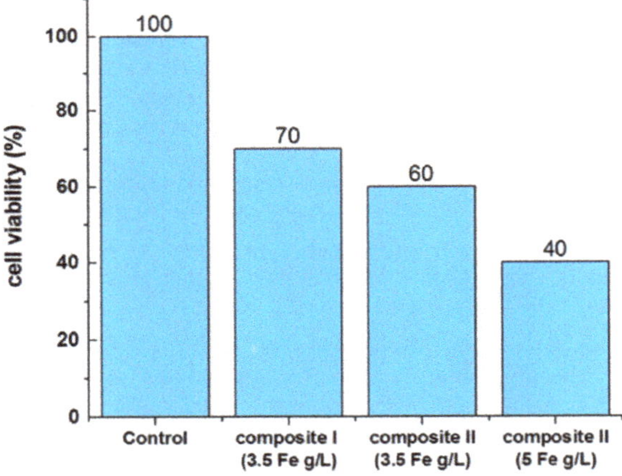

Figure 22. Results of in vitro test for magnetic hyperthermia-treated HepG2 cells. The AFM parameters are f = 525 kHz, H = 9 kA·m.

5. Concluding Remarks and Future Perspectives

Of considerable interest is the development of magnetic polymeric composites to treat malignant tumors of parenchymal organs by AEH. In this method, it is possible to achieve significant benefits from the dual effect of embolization of the tumor vascular system and subsequent MH. Embolization blocks the tumor due to the stagnation of its blood supply and leads to the shrinkage of the tumor. In turn, magnetic hyperthermia promotes the heating of the tumor to temperatures of 43–44 °C and can be repeated several times. The combined effect of embolization and hyperthermia on a tumor leads not only to ischemic necrosis but also to programmed cell death (apoptosis). Moreover, AEH combined with RT and ChT can significantly improve the survival of patients with various types of cancer.

The clinical purposes dictate the choice of materials for AEH. Thus the development of silicone-based magnetic composites that possess the set of properties necessary for conducting AEH, namely, the ability for secure embolization of the tumor blood vessels, radiopacity, and the high heating rate at moderate frequencies and amplitudes of AMFs, allowed for medical applications. The possibility of securing embolization was achieved by the optimization of the composition of the embolic agent to provide low initial viscosity (0.2–$0.3\,\text{Pa·s}^{-1}$), allowing the delivery and distribution of the material uniformly in tumor's blood vessels, and 20–25 min induction period after which the viscosity of the composite rapidly increases forming soft embolus. Such embolus displays rubber–elastic properties with shear modulus lower than arteries, within the range of HT temperatures, and higher thermal expansion coefficient than blood. Therefore, silicone-based magnetic elastomer can deform with the blood vessels and prevent blood flow recovery during heating.

The in vivo study has shown that intra-arterial administration of Nanoembosil® to animals with intramuscular developed tumors (PCI/mouse and VX2/rabbits) significantly inhibits the tumor growth rate. In vivo study of Nanoembosil® should be continued towards using a combined method including embolization followed by MH. Moreover, when studying the dynamics of tumor growth, the possible influence of colloidal platinum, which is formed during the hydrosilylation reaction of hydro- and vinyl-functional silicone polymers, must be considered. In vitro and in vivo study have shown that Pt-NPs can lead to high toxicity due to their large surface area [142].

To increase the effectiveness of the MH, in addition to mastering the shape and size of magnetite and maghemite NPs, other ferro- and ferri-magnetic materials can be used to increase the heating efficiency in AMFs at moderate frequencies and amplitudes, such as exchange-coupled MNPs with a core–shell structure [120], as well as hybrid systems, that is, combinations of soft and hard magnets or iron oxide NPs doped with Co [143]., The transarterial embolization must be controlled by enhancing the X-ray contrast with the addition of radiopaque materials or binding radionuclides to coated MNPs [144].

The future challenge lies in developing mathematical heat transfer models in tumors and their surrounding biological tissue to preserve healthy tissue [145].

Author Contributions: All authors have made a substantial and direct intellectual contribution to this article. All authors have read and agreed to the published version of the manuscript.

Funding: This work was supported by the Ministry of Education, Youth and Sports of the Czech Republic–DKRVO (RP/CPC/2020/005), Program Multilateral Scientific and Technological Cooperation in the Danube Region (8X20041).

Acknowledgments: The research of the development of magnetic polymeric materials for AEH was carried out under the direct guidance and with the active participation of Kira N. Makovetskaya. The research resulted in the development of ideas about the magnetic polymeric materials usage in locoregional methods applied for cancer patients' treatment. We, her colleagues, dedicate this work to the blessed memory of our teacher.

Conflicts of Interest: The authors declare no conflict of interest.

References

1. Sung, H.; Ferlay, J.; Siegel, R.L.; Laversanne, M. Global Cancer Statistics 2020: GLOBOCAN Estimates of Incidence and Mortality Worldwide for 36 Cancers in 185 Countries. *CA Cancer J. Clin.* **2021**, *71*, 209–249. [CrossRef]
2. Myerson, R.J.; Moros, E.G.; Diederich, C.J.; Haemmerich, D.; Hurwitz, M.D.; Hsu, I.J.; McGough, R.J.; Nau, W.H.; Straube, W.L.; Turner, P.F.; et al. Components of a hyperthermia clinic: Recommendations for staffing, equipment, and treatment monitoring. *Int. J. Hyperth.* **2014**, *30*, 1–5. [CrossRef]
3. Datta, N.R.; Ordonez, S.G.; Gaipl, U.S.; Paulides, M.M.; Crezee, H.; Gellermann, J.; Marder, D.; Puric, E.; Bodis, S. Local hyperthermia combined with radiotherapy and-/or chemotherapy: Recent advances and promises for the future. *Cancer Treat. Rev.* **2015**, *41*, 742–753. [CrossRef]
4. Kroesen, M.; Mulder, H.T.; van Holthe, J.M.L.; Aangeenbrug, A.A.; Mens, J.W.M.; van Doorn, H.C.; Paulides, M.M.; Oomen-de Hoop, E.; Vernhout, R.M.; Lutgens, L.C.; et al. Confirmation of thermal dose as a predictor of local control in cervical carcinoma patients treated with state-of-the-art radiation therapy and hyperthermia. *Radiother. Oncol.* **2019**, *140*, 150–158. [CrossRef]
5. Van der Zee, J. Heating the patient: A promising approach? *Ann. Oncol* **2002**, *13*, 1173–1184. [CrossRef] [PubMed]

6. Van der Zee, J.; Vujaskovic, Z.; Kondo, M.; Sugahara, T. Part I. Clinical hyperthermia. The Kadota Fund International Forum 2004—Clinical group consensus. *Int. J. Hyperth.* **2008**, *24*, 111–122. [CrossRef]

7. Vorst, A.V.; Rosen, A.; Kotsuka, Y. *RF/Microwave Interaction with Biological Tissues*, 1st ed.; John Wiley &Sons, Inc.: Hoboken, NJ, USA, 2006; pp. 1–330.

8. Kok, H.P.; Cressman, E.N.K.; Ceelen, W.; Brace, C.L.; Ivkov, R.; Grüll, H.; Ter Haar, G.; Wust, P.; Crezee, J. Heating technology for malignant tumors: A review. *Int. J. Hyperth.* **2020**, *37*, 711–741. [CrossRef] [PubMed]

9. Dobšíček Trefná, H.; Crezee, H.; Schmidt, M.; Marder, D.; Lamprecht, U.; Ehmann, M.; Hartmann, J.; Nadobny, J.; Gellermann, J.; Van Holthe, N.; et al. Quality assurance guidelines for superficial hyperthermia clinical trials: I. Clinical requirements. *Int. J. Hyperth.* **2017**, *33*, 471–482. [CrossRef]

10. Cihoric, N.; Tsikkinis, A.; Van Rhoon, G.; Crezee, H.; Aebersold, D.M.; Bodis, S.; Beck, M.; Nadobny, J.; Budach, V.; Wust, P.; et al. Hyperthermia-related clinical trials on cancer treatment within the Clinical Trials.gov registry. *Int. J. Hyperth.* **2015**, *31*, 609–614. [CrossRef]

11. Van der Zee, J.; Van Rhoon, G.C. Hyperthermia with radiotherapy and with system therapies. In *Brest Cancer*; Veronesi, U., Ed.; Springer: Berlin/Heidelberg, Germany, 2017; pp. 855–862.

12. Lassche, G.; Crezee, J.; Van Herpen, C.M.L. Whole-body hyperthermia in combination with systemic therapy in advanced solid malignancies. *Crit. Rev. Oncol. Hemat.* **2019**, *139*, 67–74. [CrossRef] [PubMed]

13. Lee, H.; Park, H.J.; Park, C.-S.; Oh, E.-T.; Choi, B.-H.; Brent, W.; Le, C.K.; Song, C.W. Response of Breast Cancer Cells and Cancer Stem Cells to Metformin and Hyperthermia Alone or Combined. *PLoS ONE* **2014**, *2*, e87979. [CrossRef]

14. Song, C.W.; Park, H.J.; Lee, C.K.; Griffin, R. Implications of increased tumor blood flow and oxygenation caused by mild temperature hyperthermia in tumor treatment. *Int. J. Hyperth.* **2005**, *21*, 761–767. [CrossRef]

15. Goéré, D.; Glehen, O.; Quenet, F.; Guilloit, J.-M.; Bereder, J.-M.; Lorimier, G.; Thibaudeau, E.; Ghouti, L.; Pinto, A.; Tuech, J.-J.; et al. Second-look surgery plus hyperthermic intraperitoneal chemotherapy versus surveillance in patients at high risk of developing colorectal peritoneal metastases (Prophylochip–Prodige 15): A randomised, phase 3 study. *Lancet Oncol.* **2020**, *21*, 1147–1154. [CrossRef]

16. Liang, Z.; Yang, D.; Cheng, W.; Cui, G. Clinical study on microwave deep hyperthermia combined with hepatic artery embolisation and portal vein perfusion in the treatment of advanced liver cancer. *Acta Med. Mediterr.* **2020**, *36*, 465–469.

17. Shabunin, A.V.; Tavobilov, M.M.; Grekov, D.N.; Drozdov, P.A. Combined modality treatment for patients with inoperable colorecral liver metastases. *Sib. J. Oncol.* **2018**, *17*, 34–40. [CrossRef]

18. Oei, A.L.; Kok, H.P.; Oei, S.B.; Horsman, M.R.; Stalpers, L.J.A.; Franken, N.A.P.; Crezee, J. Molecular and biological rationale of hyperthermia as radio- and chemosensitizer. *Adv. Drug Deliv. Rev.* **2020**, *163*, 84–97. [CrossRef] [PubMed]

19. Dobrodeev, A.Y.; Tuzikov, S.A.; Zavyalov, A.A.; Startseva, Z.A. Impact of preoperative thermochemoradiotherapy on surgical outcomes in patients with non-small cell lung cancer. *Vopr. Onkol.* **2020**, *66*, 143–147.

20. Bakker, A.; van der Zee, J.; van Tienhoven, G.; Kok, H.P.; Rasch, C.R.N.; Crezee, H. Temperature and thermal dose during radiotherapy and hyperthermia for recurrent breast cancer are related to clinical outcome and thermaltoxicity: A systematic review. *Int. J. Hyperth.* **2019**, *36*, 1023–1038. [CrossRef]

21. Westermann, A.; Mella, O.; van der Zee, J. Long term survival data of triple modality treatment of stage IIB-III-IVA cervical cancer with combination of radiotherapy, chemotherapy and hyperthermia—An update. *Int. J. Hyperth.* **2012**, *28*, 549–553. [CrossRef] [PubMed]

22. Dobšíček Trefná, H.; Schmidt, M.; van Rhoon, G.C.; Kok, H.P.; Gordeyev, S.S.; Lamprecht, U.; Marder, D.; Nadobny, J.; Ghadjar, P.; Abdel-Rahman, S.; et al. Quality assurance guidelines for interstitial hyperthermia. *Int. J. Hyperth.* **2019**, *36*, 277–294. [CrossRef]

23. Hildebrandt, B.; Wust, P.; Ahlers, O.; Dieing, A.; Sreenivasa, G.; Kerner, T.; Felix, R.; Riess, H. The cellular and molecular basis of hyperthermia. *Crit. Rev. Oncol. Hemat.* **2002**, *43*, 33–56. [CrossRef]

24. Roti Roti, J.L. Cellular responses to hyperthermia (40–46 C): Cell killing and molecular events. *Int. J. Hyperth.* **2008**, *24*, 3–15. [CrossRef] [PubMed]

25. Van den Tempel, N.; Horsman, M.R.; Kanaar, R. Improving efficacy of hyperthermia in oncology by exploiting biological mechanisms. *Int. J. Hyperth.* **2016**, *32*, 446–454. [CrossRef]

26. Takahashi, A. Molecular damage: Hyperthermia alone. In *Hyperthermic Oncology from Bench to Bedside*, 1st ed.; Kokura, S., Toshikazu, Y., Takeo, O., Eds.; Springer: Singapore, 2016; pp. 19–32.

27. Elming, P.B.; Sørensen, B.S.; Oei, A.L.; Frnken, N.A.P.; Crezee, J.; Overgaard, J.; Horsman, M.R. Hyperthermia: The optimal treatment to overcome radiation resistant hypoxia. *Cancers* **2019**, *11*, 60. [CrossRef] [PubMed]

28. Dobšíček, T.H.; Crezee, J.; Schmidt, M.; Marder, D.; Lamprecht, U.; Ehmann, M.; Nadobny, J.; Hartmann, J.; Lomax, N. Quality assurance guidelines for superficial hyperthermia clinical trials. *Strahlenther. Onkol.* **2017**, *193*, 351–366. [CrossRef] [PubMed]

29. Gilchrist, R.K.; Medal, R.; Shorey, W.D.; Hanselman, R.C.; Parrott, J.C.; Taylor, C.B. Selective inductive heating of lymph nodes. *Ann. Surg.* **1957**, *146*, 596–606. [CrossRef] [PubMed]

30. Johannsen, M.; Gneveckow, U.; Eckelt, L.; Feussner, A.; Waldofner, N.; Scholz, R.; Deger, S.; Wust, P.; Loening, S.A.; Jordan, A. Clinical hyperthermia of prostate cancer using magnetic nanoparticles: Presentation of a new interstitial technique. *Int. J. Hyperth.* **2005**, *21*, 637–647. [CrossRef]

31. Southern, P.; Pankhurst, Q. Commentary on the clinical and preclinical dosage limits of interstitially administered magnetic fluids for therapeutic hyperthermia based on current practice and efficacy models. *Int. J. Hyperth.* **2017**, *34*, 671–686. [CrossRef]

32. Maier-Hauff, K.; Ulrich, F.; Nestler, D.; Niehoff, H.; Wust, P.; Thiesen, B.; Orawa, H.; Budach, V.; Jordan, A. Efficacy and safety of intratumoral thermotherapy using magnetic iron-oxide nanoparticles combined with external beam radiotherapy on patients with recurrent glioblastoma multiforme. *J. Neuro-Oncol.* **2011**, *103*, 317–324. [CrossRef]

33. Kobayashi, T.; Kakimi, K.; Nakayama, E.; Jimbow, K. Antitumor immunity by magnetic nanoparticle-mediated hyperthermia. *Nanomedicine* **2014**, *9*, 1715–1726. [CrossRef]

34. Petryk, A.A.; Giustini, A.J.; Gottesman, R.E.; Kaufman, P.A.; Hoopes, P.J. Magnetic nanoparticle hyperthermia enhancement of cisplatin chemotherapy cancer treatment. *Int. J. Hyperth.* **2013**, *29*, 845–851. [CrossRef]

35. Attaluri, A.; Kandala, S.K.; Wabler, M.; Zhou, H.; Cornejo, C.; Armour, M.; Hedayati, M.; Zhang, Y.; DeWeese, T.L.; Herman, C.; et al. Magnetic nanoparticle hyperthermia enhances radiation therapy: A study in mouse models of human prostate cancer. *Int. J. Hyperth.* **2015**, *31*, 359–374. [CrossRef] [PubMed]

36. Spirou, S.V.; Costa Lima, S.A.; Bouziotis, P.; Vranješ-Djuric, S.; Efthimiadou, E.K.; Laurenzana, A.; Barbosa, A.I.; Garcia-Alonso, I.; Jones, C.; Jankovic, D.; et al. Recommendations for in vitro and in vivo testing of magnetic nanoparticle hyperthermia combined with radiation therapy. *Nanomaterials* **2018**, *8*, 306. [CrossRef]

37. Mahmoudi, K.; Bouras, A.; Bozec, D.; Ivkov, R.; Hadjipanayis, C. Magnetic hyperthermia therapy for the treatment of glioblastoma: A review of the therapy's history, efficacy and application in humans. *Int. J. Hyperth.* **2018**, *34*, 1316–1328. [CrossRef] [PubMed]

38. Rodrigues, H.F.; Capistrano, G.; Bakuzis, A.F. In vivo magnetic nanoparticle hyperthermia: A review on preclinical studies, low-field nano-heaters, noninvasive thermometry and computer simulations for treatment planning. *Int. J. Hyperth.* **2020**, *37*, 76–99. [CrossRef]

39. Brezovich, I.A. Low frequency hyperthermia: Capacitive and ferromagnetic thermoseed methods. In *Biological, Physical, and Clinical Aspects of Hyperthermia. Medical Physics Monograph 16*; Paliwal, B., Hetzel, F.W., Dewhirst, M.W., Eds.; American Institute of Physics: College Park, MI, USA, 1988; pp. 82–111.

40. Hergt, R.; Dutz, S. Magnetic particle hyperthermia—Biophysical limitations of a visionary tumour therapy. *J. Magn. Magn. Mater.* **2007**, *311*, 187–192. [CrossRef]

41. Dutz, S.; Herg, R. Magnetic nanoparticle heating and heat transfer on a microscale: Basic principles, realities and physical limitations of hyperthermia for tumor therapy. *Int. J. Hyperth.* **2013**, *29*, 790–800. [CrossRef]

42. Kozissnik, B.; Bohorquez, A.C.; Dobson, J.; Rinaldi, C. Magnetic fluid hyperthermia: Advances, challenges, and opportunity. *Int. J. Hyperth.* **2013**, *29*, 706–714. [CrossRef] [PubMed]

43. Dutz, S.; Clement, J.H.; Eberbeck, D.; Gelbrich, T.; Hergt, R.; Muller, R.; Wotschadlo, J.; Zeisberger, M. Ferrofluids of magnetic multicore nanoparticles for biomedical applications. *J. Magn. Magn. Mater.* **2009**, *321*, 1501–1504. [CrossRef]

44. Dutz, S.; Herg, R. Magnetic particle hyperthermia—A promising tumour therapy? *Nanotechnology* **2014**, *25*, 452001. [CrossRef] [PubMed]

45. Andreu, I.; Natividad, E. Accuracy of available methods for quantifying the heat power generation of nanoparticles for magnetic hyperthermia. *Int. J. Hyperth.* **2013**, *29*, 739–775. [CrossRef]

46. Garaio, E.; Collantes, J.M.; Plazaola, F.; Garsia, J.A.; Castellanos-Rubio, I.A. Multifrequency eletromagnetic applicator with an integrated AC magnetometer for magnetic hyperthermia experiments. *Meas. Sci. Technol.* **2014**, *25*, 115702. [CrossRef]

47. Coffey, W.T.; Kalmykov, Y.P. Thermal fluctuations of magnetic nanoparticles: Fifty years after Brown. *J. Appl. Phys.* **2010**, *112*, 121301. [CrossRef]

48. Endelmann, U.M.; Shasha, C.; Teeman, E.; Slabu, I.; Krishnan, K.M. Predicting size-dependent heating efficiency of magnetic nanoparticles from experi-ment and stochastic Néel-Brown Langevin simulation. *J. Magn. Magn. Mater.* **2019**, *471*, 450–456. [CrossRef]

49. Bordelon, D.E.; Cornejo, C.; Gruttner, C.; Westphal, F.; DeWeese, T.L.; Ivkov, R. Magnetic nanoparticle heating efficiency reveals magneto-structural differences when characterized with wide ranging and high amplitude magnetic fields. *J. Appl. Phys.* **2011**, *109*, 124904. [CrossRef]

50. Raouf, I.; Khalid, S.; Khan, A.; Lee, J.; Kim, H.S.; Kim, M.-H. A review on numerical modeling for magnetic nanoparticle hyperthermia: Progress and challenges. *J. Therm. Biol.* **2020**, *91*, 102644. [CrossRef]

51. Suleman, M.; Riaz, S.; Jalil, R. A mathematical modeling approach toward magnetic fluid hyperthermia of cancer and unfolding heating mechanism. *J. Therm. Anal. Calorim.* **2021**, *146*, 1193–1219. [CrossRef]

52. Wells, J.; Ortega, D.; Steinhoff, U.; Dutz, S.; Garaio, E.; Sandre, O.; Natividad, E.; Cruz, M.M.; Brero, F.; Southern, P.; et al. Challenges and recommendations for magnetic hyperthermia characterization measurments. *Int. J. Hyperth.* **2021**, *38*, 447–460. [CrossRef] [PubMed]

53. Kallumadil, M.; Tada, M.; Nakagawa, T.; Abe, M.; Southern, P.; Pankhurst, Q.A. Suitability of commercial colloids for magnetic hyperthermia. *J. Magn. Magn. Mater.* **2009**, *321*, 1509–1513. [CrossRef]

54. Hedayatnasab, Z.; Abnisa, F.; Daud, W. Review on magnetic nanoparticles for magnetic nanofluid hyperthermia application. *Mater. Des.* **2017**, *123*, 174–196. [CrossRef]

55. Barry, J.W.; Bookstein, J.J.; Alksne, J.F. Ferromagnetic embolization. *Radiology* **1981**, *138*, 341–349. [CrossRef]

56. Akuta, K.; Abe, M.; Kondo, M.; Yoshikawa, T.; Tanaka, Y.; Yoshida, M.; Miura, T.; Nakao, N.; Onoyama, Y.; Yamada, T.; et al. Combined effects of hepatic arterial embolization using degradable stach microspheres (DSM) in hyperthermia for liver cancer. *Int. J. Hyperth.* **1991**, *7*, 231–242. [CrossRef] [PubMed]

57. Mauer, C.A.; Renzulli, P.; Baer, H.U.; Mettler, D.; Uhlschmid, G.; Neuenschwander, P.; Suter, U.W.; Triller, J.; Zimmermann, A. Hepatic artery embolization with a novel radiopaque polymer causes extended liver necrosis in pigs due to occlusion of the concomitant portal vein. *J. Hepatol.* **2000**, *32*, 261–268. [CrossRef]

58. Moroz, P.; Jones, S.K.; Gray, B.N. Status of hyperthermia in the treatment of advanced liver cancer. *J. Surg. Oncol.* **2001**, *77*, 259–269. [CrossRef]

59. Wilhelm, S.; Tavares, A.J.; Dai, Q.; Ohta, S.; Audet, J.; Dvorak, H.F.; Chan, W.C.W. Analysis of nanoparticle delivery to tumours. *Nat. Rev. Mater.* **2016**, *1*, 16014. [CrossRef]

60. Moroz, P.; Jones, S.K.; Gray, B.N. Tumor response to arterial embolization hyperthermia and direct injection hyperthermia in a rabbit liver tumor model. *J. Surg. Oncol.* **2002**, *80*, 149–156. [CrossRef] [PubMed]

61. Takamatsu, S.; Matsui, O.; Gabata, T.; Kobayashi, S.; Okuda, M.; Ougi, T.; Ikehata, Y.; Nagano, I.; Nagaeet, H. Selective induction hyperthermia following transcatheter arterial embolization with a mixture of nano-sized magnetic particles (ferucarbotran) and embolic materials: Feasibility study in rabbits. *Radiat. Med.* **2008**, *26*, 179–187. [CrossRef]

62. Attaluri, A.; Seshadri, M.; Mirpour, S.; Wabler, M.; Marinho, T.; Furqan, M.; Zhou, H.; De Paoli, S.; Gruettner, C.; Gilson, W.; et al. Image-guided thermal therapy with dual-contrast magnetic nanoparticle formulation: A feasibility study. *Int. J. Hyperth.* **2016**, *32*, 543–557. [CrossRef] [PubMed]

63. Gunvén, P. Liver embolizations in oncology: A review. Part I. Arterial (chemo) embolizations. *Med. Oncol.* **2008**, *25*, 287–296. [CrossRef]

64. Renard, P.E.L.; Buchegger, F.; Petri-Fink, A.; Bosman, F.; Rufenacht, D.; Hofmann, H.; Doelker, E.; Jordan, O. Local moderate magnetically induced hyperthermia using an implant formed in situ in a mouse tumor model. *Int. J. Hyperth.* **2009**, *25*, 229–239. [CrossRef]

65. Granov, A.M.; Karelin, M.I.; Granov, D.A.; Tarazov, P.G.; Makovetskaya, K.N. Method for Treatment of Tumors of Parenchymatous Organs Tumors. RU Patent 2065734, 27 August 1996.

66. Karelin, M.I. Substantiation of X-ray Vascular Ferromagnetic Embolization and Local Hyperthermia in Stage IV Renal Cell Carcinoma. Ph.D. Thesis, The Russian Scientific Centre of Radiology and Surgical Technologies, Moscow, Russia, 1998. (In Russian).

67. Granov, A.M.; Davidov, M.I. *Interventional Radiology in Oncology—Ways of Development and Perspectives*; Tumors of Kidney: Ferromagnetic Embolization; LLC Publisher: Saint Petersburg, Russia, 2007; Chapter 8; pp. 289–297. (In Russian)

68. Smolkova, I.S.; Kazantseva, N.E.; Makoveckaya, K.N.; Smolka, P.; Saha, P.; Granov, A.M. Maghemite based silicone composite for arterial embolization hyperthermia. *Mat. Sci. Eng. C-Mater.* **2015**, *48*, 632–641. [CrossRef] [PubMed]

69. Makoveckaya, K.N.; Nikolaev, G.A.; Granov, A.M.; Tarazov, P.G.; Kazantseva, N.E.; Smolkova, I.S.; Saha, P.; Treshalina, E.M.; Yakynina, M.N.; Choroshavina, Y.A.; et al. Composition for Embolization and Hyperthermia of Vascular Tumors. RU Patent 26704464 (C1), 23 October 2018.

70. Soetaert, F.; Korangath, P.; Serantes, D.; Fiering, S.; Ivkov, R. Cancer therapy with iron oxide nanoparticles: Agents of thermal and immune therapies. *Adv. Drug Deliv. Rev.* **2020**, *163–164*, 65–83. [CrossRef] [PubMed]

71. Maier-Hauff, K.; Rothe, R.; Scholz, R.; Gneveckow, U.; Wust, P.; Thiesen, B.; Feussner, A.; Von Deimling, A.; Waldoefner, N.; Felix, R.; et al. Intracranial thermotherapy using magnetic nanoparticles combined with external beam radiotherapy: Results of a feasibility study on patients with glioblastoma multiforme. *J. Neuro-Oncol.* **2007**, *81*, 53–60. [CrossRef]

72. Babo, D.; Robinson, J.K.; Islam, J.; Thurecht, K.J.; Corrie, S.R. Nanoparticle-based medicine: A review of FDA-approved materials and clinical treals to date. *Pharm. Res.* **2016**, *33*, 2373–2387. [CrossRef] [PubMed]

73. Granov, A.M.; Muratov, O.V.; Frolov, V.F. Problems in the local hyperthermia of inductively heated embolized tissues. *Theor. Found. Chem. Eng.* **2002**, *36*, 63–66. [CrossRef]

74. Zhao, L.Y.; Liu, J.Y.; Ouang, W.W.; Li-Dan-Ye, L.L.; Tang, J.T. Magnetic mediated hyperthermia for cancer treatment: Research progress and clinical trials. *Chin. Phys. B* **2013**, *22*, 108104. [CrossRef]

75. Li, D.; Wang, K.; Wang, X.; Li, L.; Zhao, L.; Tang, J. Magnetic Arterial Embolization Hyperthermia Mediated by Carbonyl Iron Powder for Liver Carcinoma. In Proceedings of the World Congress on Medical Physics and Biomedical Engineering, Beijing, China, 26–31 May 2012.

76. Chikazumi, S.; Graham, C.D. Part VI Domainstructures. In *Physics of Ferromagnetism*, 2nd ed.; Oxford University Press: Oxford, UK, 1999; pp. 387–464.

77. Goodenough, J.B. Summary of losses in magnetic materials. *IEEE Trans. Magn.* **2002**, *38*, 3398–3408. [CrossRef]

78. Spaldin, N.A. *Magnetic Materials: Fundamentals and Applications*; Cambrige University Press: Cambridge, UK, 2011; pp. 3–270.

79. Dunlop, D.J. The rock magnetism of fine particles. *Phys. Earth Planet. Inter.* **1981**, *26*, 1–26. [CrossRef]

80. Moskowitz, B.M. Hithhiker's Guide to Magnetism. In *Environmental Magnetism Workshop (IRM)*; University of Minnesota: Minneapolis, MN, USA, 1991; Volume 279, pp. 1–48.

81. Roberts, A.P.; Almeida, T.P.; Church, N.S.; Harrison, R.J.; Heslop, D.; Li, Y.; Li, J.; Muxworthy, A.R.; Williams, W.; Zhao, X. Resolving the origin of pseudo-single domain magnetic behavior. *J. Geophys. Res.-Sol. Earth* **2017**, *122*, 9534–9558. [CrossRef]

82. Gatel, C.H.; Bonilla, F.J.; Meffre, A.; Snoeck, E.; Warot-Fonrose, B.; Chaudret, B.; Lacroix, L.M.; Blon, T. Size-Specific Spin Configurations in Single Iron Nanomagnet: From Flower to Exotic Vortices. *Nano Lett.* **2015**, *15*, 6952–6957. [CrossRef]

83. Coey, J.M.D. Ferromagnetism and exchange. In *Magnetism and Magnetic Materials*, 1st ed.; Cambridge University Press: Cambridge, UK, 2010; pp. 128–174.

84. Krishnan, K.M. Biomedical Nanomagnetics: A Spin through Possibilities in Imaging, Diagnostics, and Therapy. *IEEE Trans. Magn.* **2010**, *46*, 2523–2558. [CrossRef]
85. Dennis, C.L.; Ivkov, R. Physics of heat generation using magnetic nanoparticles for hyperthermia. *Int. J. Hyperth.* **2013**, *29*, 715–729. [CrossRef]
86. Mohapatra, J.; Xing, M.; Beatty, J.; Elkins, J.; Seda, T.; Mishra, S.R.; Liu, J.P. Enhancing the magnetic and inductive heating properties of Fe_3O_4 nanoparticles via morphology control. *Nanotechnology* **2020**, *31*, 275706. [CrossRef]
87. Nemati, Z.; Alonso, J.; Rodrigo, I.; Das, R.; Garaio, E.; García, J.A.; Orue, I.; Phan, M.H.; Srikanth, H. Improving the heating efficiency of iron oxide nanoparticles by turning their shape and size. *Phys. Chem. C* **2018**, *122*, 2367–2381. [CrossRef]
88. Das, R.; Alonso, J.; Porshokouh, Z.N.; Kalappattil, V.; Torres, D.; Phan, M.-H.; Garaio, E.; García, J.A.; Sanchez Llamazares, J.L.; Srikanth, H. Tunable High Aspect Ratio Iron Oxide Nanorods for Enhanced Hyperthermia. *Phys. Chem. C* **2016**, *120*, 10086–10093. [CrossRef]
89. Niraula, G.; Coaquira, J.A.H.; Zoppellaro, G.; Goya, G.F.; Sharma, S.K. Engineering shape anisotropy of Fe_3O_4-Fe_2O_3 hollow nanoparticles for magnetic hyperthermia. *ACS Appl. Nano Mat.* **2021**, *4*, 3148–3158. [CrossRef]
90. Gonzales-Weimuller, M.; Zeisberger, M.; Krishnan, K.M. Size-dependent heating rates of iron oxide nanoparticles for magnetic fluid hyperthermia. *J. Magn. Magn. Mater.* **2009**, *321*, 1947–1950. [CrossRef]
91. Castellanos-Rubio, I.; Rodrigo, I.; Munshi, R.; Oihane Arriortua, J.S.; Garitaonandia, A.M.-A.; Plazaola, F.; Iñaki, O.; Pralle, A.; Insausti, M. Outstanding heat loss via nano-octahedra above 20 nm in size: From wustite-rich nanoparticles to magnetite single-cristals. *Nanoscale* **2019**, *11*, 16635–16649. [CrossRef]
92. Khandhar, A.P.; Ferguson, R.M.; Krishnan, K.M. Monodispersed magnetite nanoparticles optimized for magnetic fluid hyperthermia: Implications in biological systems. *J. Appl. Phys.* **2011**, *109*, 7B310. [CrossRef]
93. Smolkova, I.S.; Kazantseva, N.E.; Babayan, V.; Vilcakova, J.; Pizurova, N.; Saha, P. The Role of Diffusion-Controlled Growth in the Formation of Uniform Iron Oxide Nanoparticles with a Link to Magnetic Hyperthermia. *Cryst. Growth Des.* **2017**, *17*, 2323–2332. [CrossRef]
94. Munoz-Menendez, C.; Conde-Leboran, I.; Baldomir, D.; Chubykalo-Fesenko, O.; Serantes, D. Role of size polydispersity in magnetic fluid hyperthermia: Average vs. local infra/over-heating effects. *Phys. Chem. Chem. Phys.* **2015**, *17*, 27812–27820. [CrossRef]
95. Ota, S.; Takemura, Y. Characterization of Néel and Brownian Relaxations Isolated from Complex Dynamics Influenced by Dipole Interactions in Magnetic Nanoparticles. *J. Phys. Chem. C* **2019**, *123*, 28859–28866. [CrossRef]
96. Kötitz, R.; Weitschies, W.; Trahms, L.; Brewer, W.; Semmler, W. Determination of the binding reaction between avidin and biotin by relaxation measurements of magnetic nanoparticles. *J. Magn. Magn. Mater.* **1999**, *194*, 62–68. [CrossRef]
97. Dieckhoff, J.; Eberbeck, D.; Schilling, M.; Ludwig, F. Magnetic-field dependence of Brownian and Néel relaxation times. *J. Appl. Phys.* **2016**, *119*, 043903. [CrossRef]
98. Hergt, R.; Hiergeist, R.; Hilger, I.; Kaiser, W.A.; Lapatnikov, Y.; Margel, S.; Richter, U. Maghemite nanoparticles with very high AC-losses for application in RF-magnetic hyperthermia. *J. Magn. Magn. Mat.* **2004**, *270*, 345–357. [CrossRef]
99. Bishop, K.J.M.; Wilmer, C.E.; Soh, S.; Grzybowski, B.A. Nanoscale forces and their uses in self-assembly. *Small* **2009**, *5*, 1600–1630. [CrossRef]
100. Lalatonne, Y.; Richardi, J.; Pileni, M.P. Van der Waals versus dipolar forces controlling mesoscopic organizations of magnetic nanocrystals. *Nat. Mater.* **2004**, *3*, 121–125. [CrossRef] [PubMed]
101. Serantes, D.; Baldomir, D. Nanoparticle Size Threshold for Magnetic Agglomeration and Associated Hyperthermia Performance. *Nanomaterials* **2021**, *11*, 2786. [CrossRef]
102. Held, G.A.; Grinstein, G.; Doyle, H.; Sun, S.H.; Murray, C.B. Competing interactions in dispersions of superparamagnetic nanoparticles. *Phys. Rev. B* **2001**, *64*, 124081–124084. [CrossRef]
103. Scholten, P.C.; Tjaden, D.L.A. Mutual attraction of superparamagnetic particles. *J. Colloid. Interf. Sci* **1980**, *73*, 254–255. [CrossRef]
104. Barman, A.; Mondal, S.; Sahoo, S.; De, A. Magnetization dynamics of nanoscale magnetic materials: A perspective. *J. Appl. Phys.* **2020**, *128*, 170901. [CrossRef]
105. Mørup, S.; Hansen, M.F.; Frandsen, C. Magnetic Nanoparticles. In *Comprehensive Nanoscience and Technology*, 2nd ed.; David, L., Andrews, G.D., Scholes, G.P., Eds.; Academic Press: Cambridge, MA, USA, 2011; pp. 437–491.
106. Schaller, V.; Wahnström, G.; Sanz-Velasco, A.; Enoksson, P.; Johansson, C. Monte Carlo simulation of magnetic multi-core nanoparticles. *J. Magn. Magn. Mater.* **2009**, *321*, 1400–1403. [CrossRef]
107. Smolková, I.S.; Kazantseva, N.E.; Babayan, V.; Smolka, P.; Parmar, H.; Vilčáková, J.; Schneeweiss, O.; Pizurová, N. Alternating magnetic field energy absorption in the dispersion of iron oxide nanoparticles in a viscous medium. *J. Magn. Magn. Mater.* **2015**, *374*, 508–515. [CrossRef]
108. Smolková, I.S.; Kazantseva, N.E.; Vitková, L.; Babayan, V.; Vilčáková, J.; Smolka, P. Size dependent heating efficiency of multicore iron oxide particles in low-power alternating magnetic fields. *Acta Phys. Pol. A* **2017**, *131*, 663–665. [CrossRef]
109. Bender, F.; Fock, J.; Frandsen, C.; Hansen, F.M.; Balceris, C.; Ludwig, F.; Posth, O.; Wetterskog, E.; Bogart, L.K.; Southern, P.; et al. Relating magnetic properties and high hyperthermia performance of iron oxide nanoflowers. *J. Phys. Chem. C* **2018**, *122*, 3068–3077. [CrossRef]
110. Dutz, S. Are magnetic nanoparticles promising candidates for biomedical applications? *IEEE Trans. Magn.* **2016**, *52*, 0200103. [CrossRef]

111. Ovejero, J.G.; Cabrera, D.; Carrey, J.; Valdivielso, T.; Salas, G.; Teran, F.J. Effects of inter- and intra-aggregate magnetic dipolar interactions on the magnetic heating efficiency of iron oxide nanoparticles. *Phys. Chem. Chem. Phys.* **2016**, *18*, 10954–10963. [CrossRef] [PubMed]
112. Landi, G.T. Role of dipolar interaction in magnetic hyperthermia. *Phys. Rev. B* **2014**, *89*, 011403. [CrossRef]
113. Coral, D.F.; Zelis, P.M.; Marciello, M.; Morales, M.D.; Craievich, A.A.; Sanchez, F.H.; Van Raap, M.B.F. Effect of nanoclustering and dipolar interactions in heat generation for magnetic hyperthermia. *Langmuir* **2016**, *32*, 1201–1213. [CrossRef]
114. Ivanov, A.O.; Kantorovich, S.S.; Elfimova, E.A.; Zverev, V.S.; Sindt, J.O.; Camp, P.J. The influence of interparticle correlations and self-assembly on the dynamic initial magnetic susceptibility spectra of ferrofluids. *J. Magn. Magn. Mater.* **2017**, *431*, 141–144. [CrossRef]
115. Branquinho, L.C.; Carriao, M.S.; Costa, A.S.; Zufelato, N.; Sousa, M.H.; Miotto, R.; Ivkov, R.; Bakuzis, A.F. Effect of magnetic dipolar interactions on nanoparticle heating efficiency: Implications for cancer hyperthermia. *Sci. Rep.* **2013**, *3*, 2887. [CrossRef]
116. Usov, N.A.; Serebryakova, O.N.; Tarasov, V.P. Interaction effects in assembly of magnetic nanoparticles. *Nanoscale Res. Lett.* **2017**, *12*, 489–497. [CrossRef]
117. Usov, N.A.; Nesmeyanov, M.S.; Tarasov, V.P. Magnetic vortices as efficient nano heaters in magnetic nanoparticle hyperthermia. *Sci. Rep.* **2018**, *8*, 1224–1233. [CrossRef]
118. Pourmiri, S.; Tzitzios, V.; Hadjipanayis, G.C.; Meneses Brassea, B.P.; El-Gendyet, A.A. Magnetic properties and hyperthermia behavior of iron oxide nanoparticle clusters. *AIP Adv.* **2019**, *9*, 125033–125038. [CrossRef]
119. Jonasson, C.H.; Schaller, V.; Zeng, L.; Olsson, E.; Frandsen, C.; Castro, A.; Nilsson, L.; Bogart, L.K.; Southern, P.; Pankhurst, Q.A.; et al. Modelling the effect of different core sizes and magnetic interactions inside magnetic nanoparticles on hyperthermia performance. *J. Magn. Magn. Mater.* **2019**, *477*, 198–202. [CrossRef]
120. Lee, J.H.; Jang, J.T.; Choi, J.S.; Moon, S.H.; Noh, S.H.; Kim, J.W.; Kim, J.G.; Kim, I.S.; Park, K.I.; Cheon, J. Exchange-coupled magnetic nanoparticles for efficient heat induction. *Nat. Nanotechnol.* **2011**, *6*, 418–422. [CrossRef] [PubMed]
121. López-Ortega, A.; Estrader, M.; Salazar-Alvarez, G.; Roca, A.G.; Nogués, J. Applications of exchange coupled bi-magnetic hard/soft and soft/hard magnetic core/shell nanoparticles. *Phys. Rep.* **2015**, *553*, 1–32. [CrossRef]
122. Phan, M.-H.; Alonso, A.; Khurshid, H.; Lampen-Kelley, P.; Chandra, S.; Repa, K.S.; Nemati, Z.; Das, R.; Iglesias, Ó.; Srikanthet, H. Exchange bias effects in iron oxide-based nanoparticle systems. *Nanomaterials* **2016**, *6*, 221. [CrossRef]
123. Balaev, D.A.; Semenova, S.V.; Dubrovskiia, A.A.; Krasikova, A.A.; Popkova, S.I.; Yakushkinb, S.S.; Kirillovb, V.L.; Mart'yanov, O.N. Synthesis and magnetic properties of the core–shell $Fe_3O_4/CoFe_2O_4$ nanoparticles. *Phys. Solid State* **2020**, *62*, 285–290. [CrossRef]
124. Rösch, J.; Keller, F.S.; Kaufman, J.A. The birth, early years, and future of interventional radiology. *J. Vasc. Interv. Radiol.* **2003**, *14*, 841–853. [CrossRef]
125. Muller, A.; Rouvière, O. Renal artery embolization-indications, technical approaches and outcomes. *Nat. Rev. Nephrol.* **2015**, *11*, 288–301. [CrossRef]
126. Le Renard, P.E.; Buchegger, F.; Petri-Fink, A.; Hofmann, H.; Doelker, E.; Jordan, O. Formulations for local, magnetically mediated hyperthermia treatment of solid tumors, and Dendritic nanostructures grown in hierarchical branched pores. In *Advances in Nanotechnology*, 1st ed.; Bartul, Z., Trenor, J., Eds.; Nova Science Publishes Inc.: New York, NY, USA, 2014; Volume 12, pp. 1–93, 123–155.
127. Poursaid, A.; Jensen, M.M.; Huoc, E.; Ghandehari, H. Polymeric materials for embolic and chemoembolic applications. *J. Control. Release* **2016**, *240*, 414–433. [CrossRef]
128. Xu, R.; Yu, H.; Zhang, M.; Chen, Z.; Wang, W.; Teng, G.; Ma, J.; Sun, X.; Gu, N. Three-dimensional model for determining inhomogenepus thermal dosage in a liver tumor during arterial embolization hyperthermia incorporating magnetic nanoparticles. *IEEE Trans. Magn.* **2009**, *45*, 3085–3091.
129. Sun, H.; Xu, L.; Fan, T.; Zhan, H.; Wang, H.; Zhou, Y.; Yang, R.-J. Targeted hyperthermia after selective embolization with ferromagnetic nanoparticles in a VX2 rabbit liver tumor model. *Int. J. Nanomed.* **2013**, *8*, 3795–3804. [CrossRef]
130. Idee, J.M.; Guiu, B. Use of Lipiodol as a drug delivery system for transcatheter arterial chemoembolization of hepatocellular carcinoma: A review. *Crit. Rev. Oncol. Hemat.* **2013**, *88*, 530–549. [CrossRef]
131. Cruise, G.M.; Constant, M.J.; Keely, E.M.; Greene, R.; Harris, C. Polymeric Treatment Composition. WO 2014062696 A1, 24 April 2014.
132. Kwabena Kan-Dapaah, K.; Rahbar, N.; Soboyejo, W. Implantable magnetic nanocomposites for the localized treatment of breast cancer. *J. Appl. Phys.* **2014**, *116*, 233505. [CrossRef]
133. Moroz, P.; Pardoe, H.; Jones, S.K.; St Pierre, T.G.; Song, S.; Gray, N.B. Arterial embolization hyperthermia: Hepatic iron particle distribution and its potential determination by magnetic resonance imaging. *Phys. Med. Biol.* **2002**, *47*, 1591–1602. [CrossRef] [PubMed]
134. Margolis, J.M. Elastomeric Materials and Processes. In *Handbook of Plastics, Elastomers and Composites*, 3rd ed.; Harper, C.A., Ed.; Oxford University Press: New York, NY, USA, 1996; Chapter 3; p. 844.
135. Smolkova, I.S.; Kazantseva, N.E.; Parmar, H.; Babayan, V.; Smolka, P.; Saha, P. Correlation between coprecipitation reaction course and magneto-structural properties of iron oxide nanoparticles. *Mater. Chem. Phys.* **2015**, *155*, 178–190. [CrossRef]
136. Kavanagh, G.M.; Ross-Murphy, S.B. Rheological characterization of polymer gels. *Prog. Polym. Sci.* **1998**, *23*, 533–562. [CrossRef]

137. Shen, W.; Zhang, J. Modeling and numerical simulation of bioheat transfer and biomechanics in soft tissue. *Math. Comput. Model.* **2005**, *41*, 1251–1265. [CrossRef]
138. Treshalina, H.M.; Yakunina, M.N.; Makovetskay, K.N.; Stangevskiy, A.A. Dynamics of tumor growth under the action of the new nano-ferrimagnetic nanoembosil with transarterial introduction. *Vopr. Onkol.* **2020**, *66*, 578–582.
139. Treshalina, H.M.; Zhukova, O.S.; Gerasimova, G.K. Guidelines for preclinical study of the antitumour activity of drugs. In *Manual for Conducting Preclinical Studies of Drugs*; Mironov, A.N., Bunatyan, N.D., Eds.; Grif and K: Moscow, Russia, 2012; Chapter 39; pp. 640–654.
140. Thomas, C.; Polin, L.; Lo Russo, P.; Valeriote, F.; Panchapor, C.; Pugh, S. In vivo methods for screening and preclinical testing. In *Anticancer Drug Development Guide: Preclinical Screening, Clinical Trials, and Approval*; Teicher, B.A., Andrews, P.A., Eds.; Humana Press: Totowa, NJ, USA, 2004; Chapter 6; pp. 99–123.
141. Nair, A.B.; Jacob, S. A simple practice guide for dose conversion between animals and human. *J. Basic Clin. Pharm.* **2016**, *7*, 27–31. [CrossRef]
142. Xia, H.; Li, F.; Hu, X.; Park, W.; Wang, S.; Jang, Y.; Du, Y.; Baik, S.; Cho, S.; Kang, T.; et al. pH-sensitive Pt nanoclusters assembly overcomes cisplatin resistance and heterogeneous stemness of hepatocellular carcinoma. *ACS Cent. Sci.* **2016**, *2*, 802–811. [CrossRef]
143. Goldman, A. *Mordent Ferrite Technology*; Springer Science & Business Media: Berlin/Heidelberg, Germany, 2006.
144. Mirković, M.; Radović, M.; Stanković, D.; Milanović, Z.; Janković, D.; Matović, M.; Jeremić, M.; Antić, B.; Vranješ-Ðurić, S. 99mTc–Bisphosphonate–Coated magnetic nanoparticles as potential theranostic nanoagent. *Mat. Sci. Eng. C* **2019**, *102*, 123–133. [CrossRef]
145. Abu-Bakr, A.F.; Iskakova, L.Y.; Zubarev, A.Y. Heat exchange within the surrounding biological tissue during magnetic hyperthermia. *Math. Model. Eng. Probl.* **2020**, *7*, 196–200. [CrossRef]

nanomaterials

MDPI

Article

Magnetic Study of CuFe$_2$O$_4$-SiO$_2$ Aerogel and Xerogel Nanocomposites

Alizé V. Gaumet [1,†], **Francesco Caddeo** [1,†], **Danilo Loche** [1], **Anna Corrias** [1], **Maria F. Casula** [2], **Andrea Falqui** [3] **and Alberto Casu** [4,*]

1   School of Physical Sciences, Ingram Building, University of Kent, Canterbury CT2 7NH, UK; avg32@bath.ac.uk (A.V.G.); fcaddeo@physnet.uni-hamburg.de (F.C.); daniloche@tiscali.it (D.L.); a.corrias@kent.ac.uk (A.C.)
2   Department of Mechanical, Chemical and Materials Engineering and INSTM, University of Cagliari, Via Marengo 2, 09123 Cagliari, Italy; casulaf@unica.it
3   Department of Physics "Aldo Pontremoli", University of Milan, Via Celoria 16, 20133 Milan, Italy; andrea.falqui@unimi.it
4   Biological and Environmental Sciences and Engineering (BESE) Division, King Abdullah University of Science and Technology (KAUST), Thuwal 23955-6900, Saudi Arabia
*   Correspondence: alberto.casu@kaust.edu.sa
†   These authors contributed equally.

check for **updates**

Citation: Gaumet, A.V.; Caddeo, F.; Loche, D.; Corrias, A.; Casula, M.F.; Falqui, A.; Casu, A. Magnetic Study of CuFe$_2$O$_4$-SiO$_2$ Aerogel and Xerogel Nanocomposites. *Nanomaterials* **2021**, *11*, 2680. https://doi.org/10.3390/nano11102680

Academic Editors: Raghvendra Singh Yadav and Zhidong Zhang

Received: 3 August 2021
Accepted: 4 October 2021
Published: 12 October 2021

Publisher's Note: MDPI stays neutral with regard to jurisdictional claims in published maps and institutional affiliations.

Abstract: CuFe$_2$O$_4$ is an example of ferrites whose physico-chemical properties can vary greatly at the nanoscale. Here, sol-gel techniques are used to produce CuFe$_2$O$_4$-SiO$_2$ nanocomposites where copper ferrite nanocrystals are grown within a porous dielectric silica matrix. Nanocomposites in the form of both xerogels and aerogels with variable loadings of copper ferrite (5 wt%, 10 wt% and 15 wt%) were synthesized. Transmission electron microscopy and X-ray diffraction investigations showed the occurrence of CuFe$_2$O$_4$ nanoparticles with average crystal size ranging from a few nanometers up to around 9 nm, homogeneously distributed within the porous silica matrix, after thermal treatment of the samples at 900 °C. Evidence of some impurities of CuO and α-Fe$_2$O$_3$ was found in the aerogel samples with 10 wt% and 15 wt% loading. DC magnetometry was used to investigate the magnetic properties of these nanocomposites, as a function of the loading of copper ferrite and of the porosity characteristics. All the nanocomposites show a blocking temperature lower than RT and soft magnetic features at low temperature. The observed magnetic parameters are interpreted taking into account the occurrence of size and interaction effects in an ensemble of superparamagnetic nanoparticles distributed in a matrix. These results highlight how aerogel and xerogel matrices give rise to nanocomposites with different magnetic features and how the spatial distribution of the nanophase in the matrices modifies the final magnetic properties with respect to the case of conventional unsupported nanoparticles.

Keywords: copper ferrite; magnetic properties; sol-gel; aerogels; xerogels

1. Introduction

Spinel ferrite nanoparticles with general formula MFe$_2$O$_4$ (where M is a bi-valent transition metal ion such as Mn^{2+}, Ni^{2+}, Co^{2+}, Cu^{2+}, Zn^{2+}, etc.) have been the object of intense investigation due to their interesting optical, magnetic and catalytic properties for their potential application in storage devices [1], photo-catalysis [2], magnetic fluids [3], sensors [4] and biomedicine [5]. In particular, the introduction of copper in the general spinel ferrite crystal structure gives rise to a soft magnet, which has been deemed an interesting candidate for multiple applications in many diverse fields, ranging from sensors to catalysis [6–8]. This has led to a large number of studies in recent years, focusing on attaining an always increasing degree of control over the shape, size, distribution and phase of the copper ferrite nanoparticles. In fact, copper ferrite represents a special case among ferrites, since it displays two distinct crystalline phases, namely a high temperature disordered cubic *Fd-3m* phase (c-$CuFe_2O_4$) and an ordered tetragonal *I4I/amd* phase (t-$CuFe_2O_4$) that can be obtained during slow cooling in air [9]. The tetragonal phase arises from the collective Jahn–Teller distortion along one of the axes of the octahedral sites, which is typical for Cu^{2+} ions (d9) as a consequence of the removal of the e_g orbital degeneracy [10,11]. For this reason, many studies have been focused on synthesizing copper ferrite nanoparticles with a wide number of methods including thermal decomposition [12], hydrothermal [13], solvothermal [14], co-precipitation [15], electrospinning [16] and sol-gel [17], also taking into account that the desired crystalline phase might be obtained by selecting an appropriate thermal treatment. Moreover, in a recent report it has been shown that, independently of the final heat treatment, the cubic phase and tetragonal phase can be obtained by employing sol-gel and co-precipitation methods, respectively [18]. All the aforementioned synthetic methods yield nanoparticles with dimensions of several tens of nanometers and a poor degree of size homogeneity, mainly due to the aggregation and sintering of the nanoparticles during the thermal treatments needed to obtain the desired crystalline phase. In turn, this also affects the magnetic properties and makes it difficult to achieve a precise control over the different parameters at play in giving rise to the final magnetic behavior [19].

A fruitful strategy to produce crystalline magnetic nanoparticles with a controlled dimension at the nanometer range and to control their spatial distribution and aggregation relies on embedding them into a suitable support or matrix capable of keeping them durably dispersed, thus limiting interparticle magnetic interactions. [20–26] As a consequence, the magnetic response of the resulting magnetic nanocomposites is due to the degree of control over the monodispersity (in size, shape, and phase) of the nanoparticles population, and to the loading and dispersion of the magnetic phase within the matrix. In particular, magnetic interactions and hardening (or softening) of the magnetic behavior of the nanocomposite is directly influenced by the loading and dispersion of the magnetic phase within the matrix. The capability of obtaining small-sized nanoparticles is of interest because it guarantees that the whole magnetic phase will comprise single domain nanoparticles, i.e., nanoparticles that act as one single magnetic domain under an applied magnetic field, not dividing their atomic moments along different orientations and thus maximizing their overall magnetization.

In this framework, our group has successfully applied a sol-gel route to produce Mn, Zn, Co, Ni, and Cu ferrite nanoparticles which have been crystallized within a porous silica matrix [20–26]. The sol-gel technique offers the possibility to obtain different forms of porous materials, such as xerogels and aerogels, by adopting specific procedures to obtain the dry gel. It was found that the crystallization of the ferrite nanoparticles within a porous silica matrix represents a valid method to obtain nanocomposites with a homogeneous distribution of the nanoparticles within the matrix, and to limit size growth. In particular, xerogel and aerogel nanocomposites with 10 wt% loading of copper ferrite enabled us to gain insights on the structural properties of $CuFe_2O_4$ nanocrystals, such as the cation distribution within the crystal structure, using X-ray absorption spectroscopy [26]. Interestingly, this study showed that the Jahn–Teller distortion is present, which is consistent with the formation of the tetragonal phase. However, the distortion only interests the copper environment while the iron environment is the same of the disordered cubic phase. Here, we study the magnetic properties of $CuFe_2O_4$-SiO_2 nanocomposites obtained by sol-gel synthesis in the form of aerogels and xerogels by either supercritical or conventional drying, respectively. The nanocomposites were prepared with ferrite loading ranging between 5 wt% and 15 wt% and submitted to thermal treatments at increasing temperature up to 900 °C. This choice has allowed us to investigate two aspects that come into play in giving rise to the magnetic response of the nanocomposites, namely the presence of small magnetic nanoparticles below the single domain size threshold, and their distribution according to different loadings within matrices with two kinds of porosity.

To perform this study, $CuFe_2O_4$-SiO_2 aerogel and xerogel nanocomposites with ferrite loading of 5 wt%, 10 wt% and 15 wt% were synthetized and investigated from a structural and magnetometric point of view, to assess the effect of different matrices on the $CuFe_2O_4$ nanoparticles and on the net magnetic features of the nanocomposites.

2. Materials and Methods

2.1. Synthesis of the Aerogel and Xerogel Nanocomposites

The synthesis of the $CuFe_2O_4$–SiO_2 aerogel and xerogel nanocomposites follows a two-step catalyzed sol-gel procedure, whose general details are reported in our previously published work [26]. Concisely, a solution of 7.9 mL of tetraethyl orthosilicate (TEOS, Aldrich, UK, 98%) in 3 mL of absolute ethanol (Fischer Chemicals, UK) was pre-hydrolyzed under acidic catalysis through addition of 3.965 mL of a HNO_3 stock solution prepared by mixing 2 mL of HNO_3 (Fisher Chemicals, 70%), 80 mL of absolute ethanol, as well as 130 mL of distilled water, and heating at 50 °C for 30 min. After cooling at room temperature, 7.5 mL of an ethanolic solution containing appropriate amounts of $Cu(NO_3)_2 \cdot 2.5H_2O$ (Aldrich, 98%) and $Fe(NO_3)_3 \cdot 9H_2O$ (Aldrich, 98%) was added to the TEOS solution. The appropriate amounts of copper and iron nitrates had been calculated in order to obtain nanocomposites with a final $CuFe_2O_4/(CuFe_2O_4+SiO_2)$ composition of 5 wt%, 10 wt% and 15 wt%. For the purpose of promoting the condensation reactions and therefore induce gelation, a solution containing 3.513 g of urea (Aldrich, >99%) in 9 mL of absolute ethanol and 4.92 mL of distilled water was added to the TEOS solution, which was then refluxed at 85 °C for 128 min corresponding to the time needed to initiate the gelation. The sol was transferred in vials and left at 40 °C for 20 h to complete gelation.

The aerogels were obtained by ethanol supercritical drying of the gels. For this purpose, the gels were placed in a Parr 300 mL stainless-steel autoclave filled with 70 mL of ethanol and thoroughly flushed with pure N_2 gas. Once an inert atmosphere was ensured, the sealed autoclave was heated up to 330 °C with a corresponding autogenous pressure of over 70 atm before slowly venting while keeping the temperature constant.

The xerogels were obtained through evaporation of the solvent in an open container at 40 °C for 70 h.

Thermal treatments were performed with a heating ramp of 10 °C·min^{-1} to a variable final temperature followed by a variable dwell time. Aerogel and xerogel samples are indicated, respectively, with "A" or "X" as the first letter in their acronym. The samples have also been labelled with the indication of composition (wt% of $CuFe_2O_4$), temperature and duration of the thermal treatment. For instance, A5_450_1 is used for an aerogel sample with a 5 wt% loading of $CuFe_2O_4$ nanoparticles, thermally treated at 450 °C for 1 h.

2.2. Characterization

Powder X-ray diffraction (XRD) patterns were measured by using a Rigaku MiniFlex 600 with a D/teX Ultra high-speed one-dimensional detector (Rigaku, Tokyo, Japan), in the range of 10–90° (2θ), using Cu Kα radiation. The Scherrer equation was used to determine the average size of crystallite domains.

Transmission electron spectroscopy (TEM) images were obtained in bright field (BF) and dark field (DF) modes using a Hitachi H-7000 (Hitachi, Tokyo, Japan) and a Jeol JEM 1400 Plus (Jeol, Tokyo, Japan) equipped with a W thermoionic gun operating at 100 kV respectively and equipped with an AMT DVC (AMT, Danvers, MA, USA) and Ruby2 CCD Camera (Jeol, Tokyo, Japan).

Magnetic characterization was performed in a Quantum Design MPMS SQUID magnetometer (Quantum Design, San Diego, CA, USA), equipped with a superconducting magnet producing fields up to 50 kOe (5 T). Zero field-cooled (ZFC) and field-cooled (FC) magnetizations were collected in the range of temperatures 5–400 K. ZFC curves were measured by cooling samples in a zero magnetic field and subsequently increasing the temperature under an applied field of 100 Oe. FC curves were recorded by cooling the samples while maintaining the applied field at 100 Oe. Hysteresis loops were recorded up to ± 50 kOe (5 T) at 5 K. The samples were placed in gelatin capsules enclosed inside a pierced straw with a uniform diamagnetic background.

Blocking temperature (T_B) and irreversibility temperature (T_{IRR}) indicate the temperature corresponding to the maximum of the ZFC curve and the minimum temperature of superposition between the ZFC and FC curves, respectively. Hysteresis loops were analyzed according to their typical parameters, which are indicated as follows: maximum magnetization ($M_{5T} = (|M_{+5T}| + |M_{-5T}|)/2$, where M_{+5T} and M_{-5T} indicate magnetization recorded with and applied field of $+/-5$ Tesla, respectively), mean remanence ($M_R = (|M_{R+}| + |M_{R-}|)/2$, where M_{R+} and M_{R-} indicate upper and lower remanence magnetization), mean coercivity ($H_C = (|H_{C1}| + |H_{C2}|)/2$, where H_{C1} and H_{C2} are the negative and positive coercive fields, respectively) and exchange bias coercivity ($H_E = (|H_{C1}| - |H_{C2}|)/2$). Saturation magnetization (M_S) values were determined from the hysteresis loops through extrapolation of M values vs. $1/H$ for $1/H \rightarrow 0$. All the components of the SQUID samples were weighed for mass of magnetic phase normalization.

3. Results and Discussion

In Figure 1, the powder XRD patterns of the aerogel samples are reported, whereas the XRD patterns of the corresponding xerogel samples are reported in Figure 2. In Figure 1a–c, the XRD patterns compare the evolution of the copper ferrite crystalline phase within the aerogel matrix, as a function of the calcination temperature for any given loading, whereas in Figure 1d, the XRD patterns of the three different loadings are compared after the same thermal treatment at 900 °C. In the case of the 5 wt% nanocomposites (Figure 1a), the XRD pattern is dominated by the presence of the amorphous SiO_2 phase, highlighted by the typical halo centered at ~22° recurring in all the XRD patterns. Only in the case of the sample treated at 900 °C, some very broad reflections appear, the most intense one centered at ~35.5°, which can be assigned to the formation of either the tetragonal or the cubic crystalline phase of the $CuFe_2O_4$ [27,28].

In the case of the aerogels with a 10 wt% and 15 wt% of dispersed phase (Figure 1b,c), some peaks are also detectable in the samples treated at 450 °C and 750 °C, providing some insights on the formation of $CuFe_2O_4$ nanoparticles within the aerogels. In particular, peaks that can be ascribed to CuO [29] are visible in the samples treated at 450 °C, especially in the aerogel with 15 wt% of dispersed phase. As previously found in the sol-gel synthesis of other ferrites dispersed in silica aerogels [30], iron is very likely present in the form of ferrihydrite, whose peaks are hidden within the amorphous silica background because of the poor crystallinity of this phase. When the same samples with a 10 wt% and 15 wt% dispersed phase are submitted to 750 °C some additional peaks appear, which further evolve with increasing the temperature of the thermal treatment. Peaks due to either tetragonal or cubic $CuFe_2O_4$ increase progressively while the peaks due to CuO tend to progressively disappear. Moreover, a peak centered around 33° appears with thermal treatment at 750 °C and then tends to disappear with thermal treatment at 900 °C, indicating the formation of an intermediate phase. Based on the peak position and the composition of the sample, it seems very likely that this peak is due to some hematite forming as intermediate phase from ferrihydrite [31]. The XRD patterns of the samples treated at 750 °C for 1 h and 6 h seem very similar, which proves that thermal treatments longer than 1 h do not induce further crystallization and/or evolution of the initial and intermediate phases. The differences in terms of loading of the crystalline phase are highlighted in Figure 1d, where the intensity of the reflections corresponding to the $CuFe_2O_4$ increases with loading, as was expected.

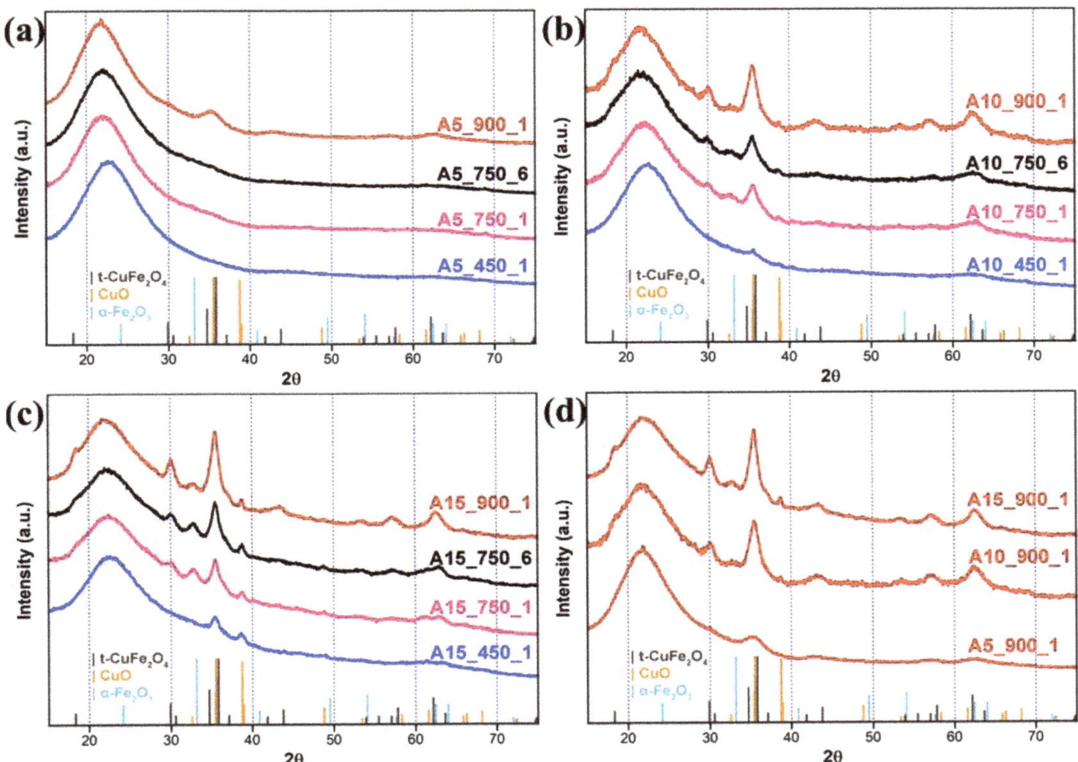

Figure 1. X-ray diffraction (XRD) patterns of the aerogel nanocomposites with (**a**) 5 wt%; (**b**) 10 wt%; (**c**) 15 wt% loading of the dispersed nanophase, reported as a function of the calcination temperature; (**d**) comparison of the XRD patterns of the aerogel samples after thermal treatment at 900 °C, at different $CuFe_2O_4$ loadings.

It should be noted that in the case of the aerogels with a 10 wt% and 15 wt% dispersed phase, two small peaks due to some unreacted α-Fe_2O_3 and CuO are still detectable together with the peaks of $CuFe_2O_4$, which appears to be the predominant phase.

The XRD patterns corresponding to the xerogel samples (Figure 2a–d) show a few differences in the evolution with the thermal treatment, with respect to the aerogel samples. Apart from the typical silica halo, no peaks are detectable in any of the samples treated at 450 °C, regardless of the composition, indicating that very poorly crystalline phases must be present at this stage. When the samples are treated at 750 °C, some peaks appear that become more evident as the loading increases. These peaks are ascribed to the formation of either tetragonal or cubic $CuFe_2O_4$ and further evolve with increasing the temperature of the thermal treatment. The main difference between the xerogel and aerogel samples treated at 900 °C is the width of the peaks, which is larger for the xerogel samples, indicating smaller crystallite sizes. Moreover, in the case of the xerogel samples, all the detectable peaks are due to $CuFe_2O_4$, and no sign of CuO and/or hematite is visible. As mentioned above, the XRD patterns of the nanocomposites do not allow to distinguish between the tetragonal and the cubic crystalline phase of $CuFe_2O_4$ due to the nanocrystalline nature of the samples that generates broad reflections and the XRD patterns of these phases sharing most of the peaks. However, an X-ray absorption investigation on related samples suggested that, under the adopted conditions, $CuFe_2O_4$ nanoparticles crystallize in the tetragonal phase [26]. The crystallite sizes for the samples thermally treated at 900 °C are:

3, 6 and 9 nm for the 5 wt%, 10 wt% and 15 wt% aerogel nanocomposites, respectively, and 3, 4 and 5 nm for the 5 wt%, 10 wt% and 15 wt% xerogel nanocomposites, respectively. Although the sizes increase as a function of the loading, the effect is slower in the xerogel samples versus the aerogel ones.

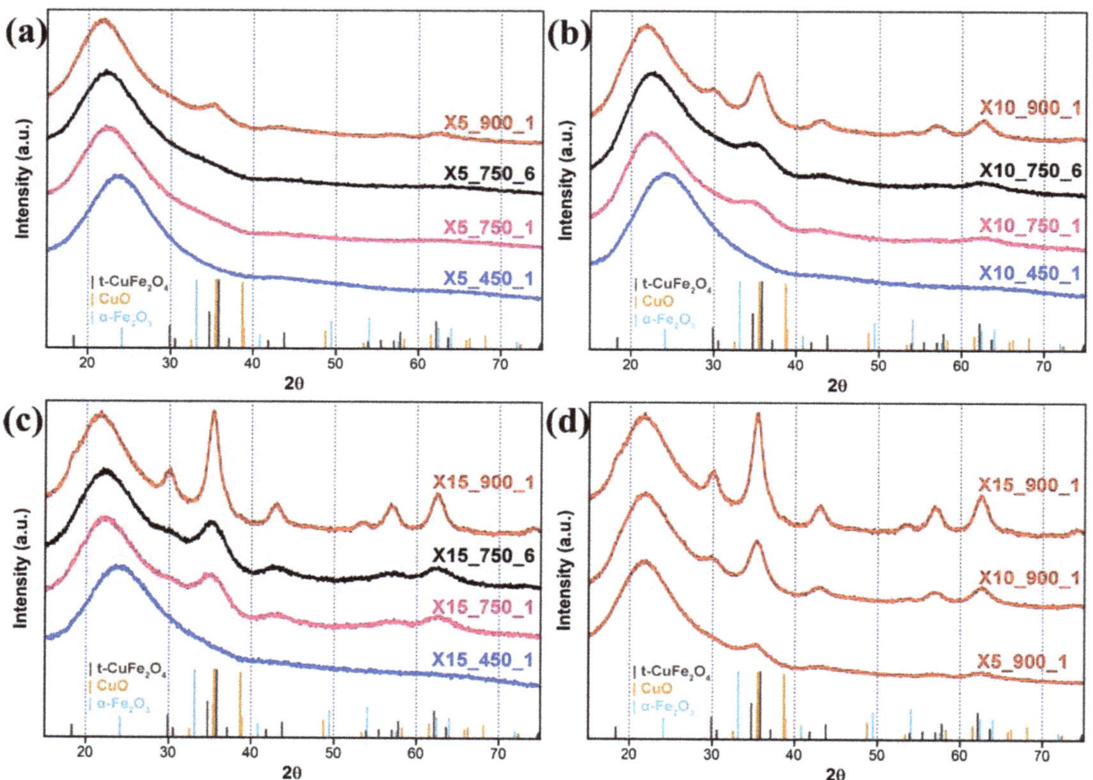

Figure 2. XRD patterns of the xerogel nanocomposites with (**a**) 5 wt%; (**b**) 10 wt%; (**c**) 15 wt% loading of the dispersed nanophase, reported as a function of the calcination temperature; (**d**) comparison of the XRD patterns of the xerogel samples after thermal treatment at 900 °C, at different $CuFe_2O_4$ loadings.

In Figures 3 and 4, bright-field (BF) and dark-field (DF) images for the aerogel and xerogel samples treated at 900 °C are shown. The images provide clear evidence that the nanoparticles are well dispersed in the SiO_2 matrix, which is hindering their agglomeration and growth. Their presence can be more clearly observed in the DF images, where the nanoparticles appear as bright dots over a darker background, except for the samples with 5 wt% loading, due to the low amount of ferrite, combined with the very small size of crystallites (3 nm for both samples according to Scherrer calculations).

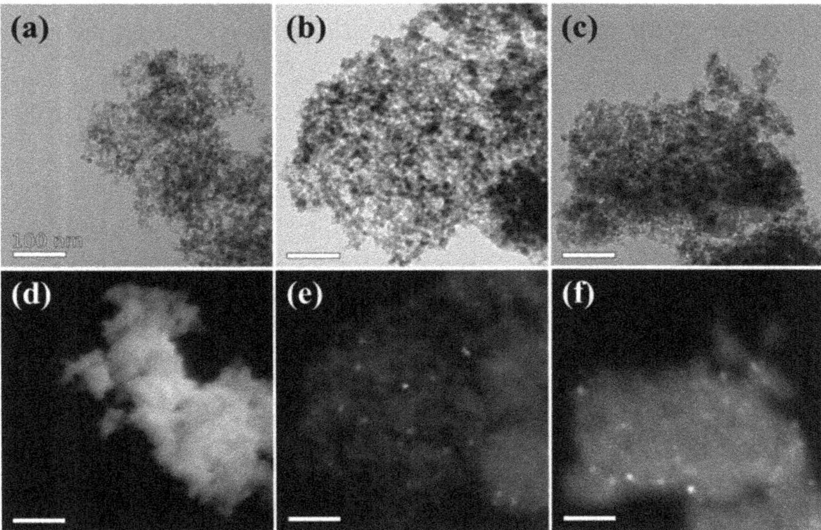

Figure 3. Representative BF (**a–c**) and DF (**d–f**) transmission electron microscopy (TEM) images of aerogel samples treated at 900 °C with increasing loading of 5 wt% (**a,d**), 10 wt% (**b,e**) and 15 wt% (**c,f**). All scalebars correspond to 100 nm.

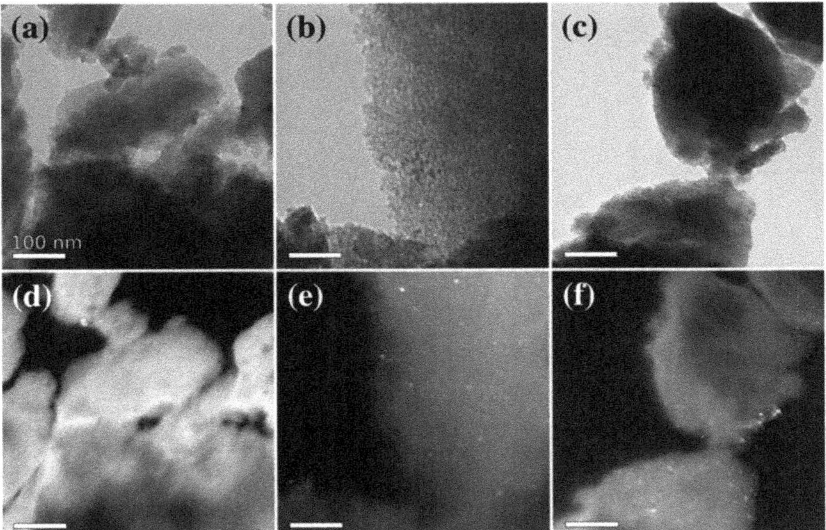

Figure 4. Representative bright field (BF) (**a–c**) and dark field (DF) (**d–f**) TEM images of xerogel samples treated at 900 °C with increasing loading of 5 wt% (**a,d**), 10 wt% (**b,e**) and 15 wt% (**c,f**). All scalebars correspond to 100 nm.

The images also show a finer and denser texture for the xerogels, as compared to the aerogels, where a more open texture is evident, as was expected [32]. In fact, aerogels obtained via specific gel drying techniques exhibit an open porous network presenting a significant mesopores contribution. On the other hand, xerogels production implies partial collapse of the pores and generates a relatively denser matrix with smaller pores. The size of the nanocrystals increases with ferrite loading and is slightly larger in the aerogel as compared to the xerogel, in agreement with the average nanocrystal size as determined by XRD, likely due to the different matrix characteristics.

DC magnetometry was used to investigate the effect of the ferrite loadings on the overall magnetic properties of the aerogel and xerogel nanocomposites treated at 900 °C, as XRD analysis provides the clearest evidence that nanocrystalline $CuFe_2O_4$ is the most prevalent phase for the samples treated at this temperature. All the relevant parameters are reported in Tables 1 and 2 for aerogels and xerogels, respectively.

Table 1. Magnetic parameters for the 5 wt%, 10 wt% and 15 wt% $CuFe_2O_4$-SiO_2 nanocomposite aerogel samples. Relevant parameters are indicated as T_B (blocking temperature), T_{IRR} (irreversibility temperature), H_C (coercivity), H_E (exchange bias), M_R (remanent magnetization, M_{5T} (maximum magnetization, M_S (saturation magnetization). Hysteresis loops were recorded at 5 K. Values were determined according to what was reported in the Materials and Methods section.

Sample	T_B (K)	T_{IRR} (K)	H_C (Oe)	H_E (Oe)	M_R (emu/g)	M_{5T} (emu/g)	M_S (emu/g)	M_R /M_S	M_{5T} /M_S
A5_900_1	7	13	32	1	4	47	51	0.08	0.93
A10_900_1	83	225	623	18	15	38	39	0.38	0.97
A15_900_1	127	294	825	12	15	34	34	0.41	0.94

Table 2. Magnetic parameters for the 5 wt%, 10 wt% and 15 wt% $CuFe_2O_4$ -SiO_2 nanocomposite xerogel samples. Relevant parameters are indicated as T_B (blocking temperature), T_{IRR} (irreversibility temperature), H_C (coercivity), H_E (exchange bias), M_R (remanent magnetization, M_{5T} (maximum magnetization, M_S (saturation magnetization). Hysteresis loops were recorded at 5 K. Values were determined according to what was reported in the Materials and Methods section.

Sample	T_B (K)	T_{IRR} (K)	H_C (Oe)	H_E (Oe)	M_R (emu/g)	M_{5T} (emu/g)	M_S (emu/g)	M_R /M_S	M_{5T} /M_S
X5_900_1	5	11	28	1	9	45	53	0.16	0.85
X10_900_1	9	32	80	0	9	54	58	0.15	0.93
X15_900_1	17	53	128	0	13	51	51	0.25	0.99

The three aerogel samples are all magnetically unblocked at room temperature, as shown by ZFC-FC curves (Figure 5a–c) with different T_B and T_{IRR} values that tend to increase together with the loading of magnetic phase and crystallite size. Similar trends, i.e., the variation of magnetic parameters with the loading of magnetic phase, can also be appreciated in the hysteresis loops recorded at 5 K (Figure 5d–f). Both the magnetic hardness and the maximum measured magnetization, indicated by coercivity H_C and M_{5T}, respectively, are affected by the quantity of magnetic phase present in the SiO_2 matrix. These trends point at a progressive hardening with the increase of Cu ferrite, while the lowered magnetization is consistent with the presence of different crystal phases, as already indicated by the traces of α-Fe_2O_3 and CuO observed in the XRD patterns, that are expected to contribute negatively to the net magnetization due to their antiferromagnetic nature. Also, even considering the slight variation in terms of crystallite size between the two samples with highest loadings, the combination of lower magnetization and higher coercivity in the 15 wt% sample with respect to the 10 wt% suggests that a lower number of magnetic moments are available for the reorientation under an applied magnetic field (hence the lower M_S) and that the said reorientation requires higher applied magnetic fields (hence the higher H_C).

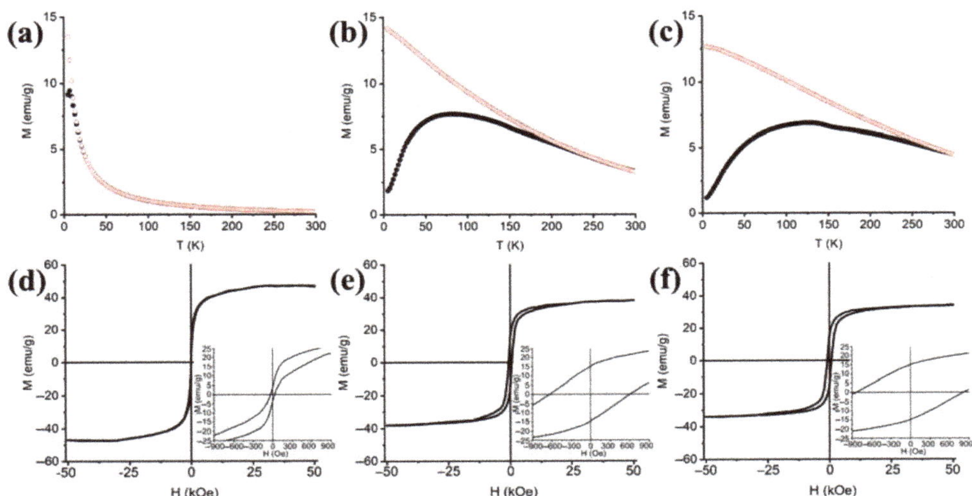

Figure 5. Zero field-cooled–field cooled (ZFC-FC) magnetization curves (**a–c**) and hysteresis cycles (**d–f**) of aerogel samples treated at 900 °C with increasing loading of 5 wt% (**a,d**), 10 wt% (**b,e**) and 15 wt% (**c,f**).

This finding is in good accordance with the presence of spontaneous exchange bias (H_E) in the hysteresis loops of the 10 wt% and 15 wt% aerogel samples, whose presence usually indicates structural disorder that depletes the number of available magnetic moments and affects negatively the net magnetization of the samples [33–36].

On the other hand, the low coercivity, higher magnetization and null exchange bias (H_E) observed for the 5 wt% sample can be explained by considering the 3 nm size of the crystallites, which puts them in the lower portion of the single domain regime. Such small-sized crystallites correspond to small single magnetic domains, which are easily aligned to applied magnetic fields and undergo superparamagnetic relaxation at very low temperatures, as shown by the low T_B and T_{IRR} values observed in the ZFC-FC curves in the present case. Also, no impurities were observed in the corresponding XRD pattern and in this range of sizes no clear distinctions between volume- and surface-based effects should be expected, so the absence of H_E and the high values of magnetization can be expected.

The magnetic features of the xerogel nanocomposites present some differences from their aerogel counterparts and should be compared in full, taking into account the differences in crystallite size occurring among xerogels as a function of the loading, and between aerogel and xerogel samples with the same loading, as well as the absence of the CuO and α-Fe$_2$O$_3$ impurities, whose presence was only evidenced in the aerogel samples (Figure 6). At any given loading, blocking and irreversibility temperatures T_B and T_{IRR} and coercivity H_C have smaller values in the xerogels than in the aerogels. As for the aerogel samples, the lowest percentage of magnetic phase (5 wt%) corresponds to the smallest crystallites, with the xerogel sample having the same average size of the aerogel sample and forming small single magnetic domains that can be easily reoriented along with the external magnetic field. Additionally, spontaneous exchange bias H_E is negligible in all the xerogel samples, suggesting a low degree of structural disorder. Since structural disorder can affect the symmetric shape of the hysteresis loops and modify coercivity H_C and maximum magnetization values M_{5T} and M_S, the minimum H_C and maximum M_{5T} and M_S values observed in xerogel samples are all consistent with the fact that no impurities were observed by XRD in these samples.

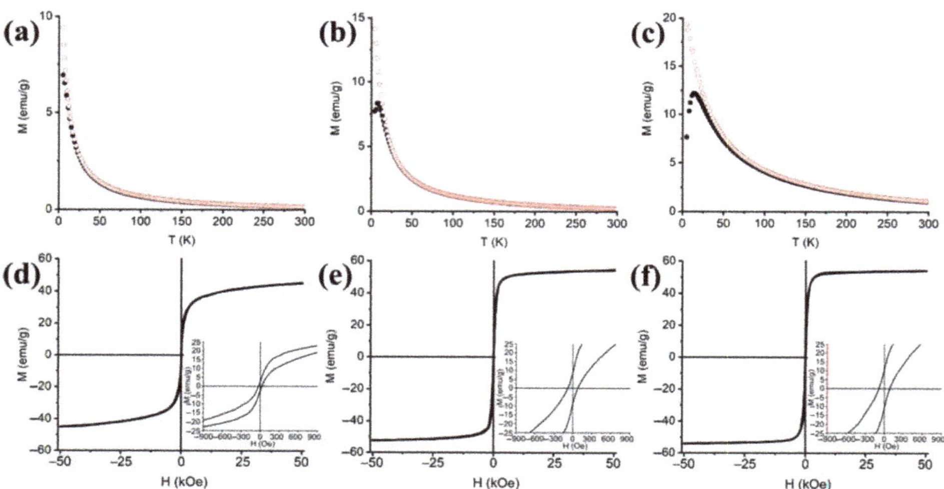

Figure 6. Zero-field cooled-field cooled magnetization curves (**a–c**) and hysteresis cycles (**d–f**) of xerogel samples treated at 900 °C with increasing loading of 5 wt% (**a,d**), 10 wt% (**b,e**) and 15 wt% (**c,f**).

The crystal size of the xerogels with 10 wt% and 15 wt% magnetic phase increases slightly with the loading, and it is safe to assume that the magnetic domains follow a similar trend. All the nanoparticles in xerogels also seem less affected by interparticle interactions with respect to the bigger-sized nanoparticles of the aerogels, as indicated by the lower blocking and irreversibility temperatures T_B and T_{IRR} observed in the ZFC curves and by their coercivity H_C. At any given loading, smaller values of all these parameters are observed in the xerogel samples compared to their aerogel counterparts, while their values grow with the loading. For non-interacting nanoparticles, variations in T_B and T_{IRR} values are expected, since bigger sizes require increasingly higher temperatures to overcome their blocked state and undergo superparamagnetic relaxation. However, the differences in size between the crystallites of aerogel and xerogel samples cannot solely account for the huge variation observed in the shape and parameters of the ZFC-FC curves. For these reasons, a large effect should be attributed to magnetic interactions affecting the aerogels since these generally work against the easy reorientation of the magnetic moments along an external magnetic field and manifest as a tendency towards magnetic hardening (i.e., higher coercivity) in the hysteresis loops and a broadening in the T_B peak of the ZFC curves. The former can be mostly attributed to intraparticle interaction-related structural disorder, while the latter is mostly driven by interparticle interactions [37].

The presence of magnetic interactions in the aerogels can be further proved by comparing the magnetic response of nanoparticles of similar size (6 nm and 5 nm, respectively), present in samples A_10_900_1 and X_15_900_1. Even in this case, despite the lower loading of the aerogel sample A_10_900_1 and the higher defectivity of its crystallites, indicated by the non-null H_E value, its T_B, T_{IRR} and H_C values are more than four times higher than those of X_15_900_1, indicating that interparticle magnetic interactions modify the magnetic response of the aerogel samples.

Thermal treatment is known to trigger a progressive reordering of the Cu^{2+} and Fe^{3+} ions in the octahedral and tetrahedral sites, which is proportional to the temperature and clearly modifies the magnetic characteristics of copper ferrite towards soft magnets already at 700 °C [38,39]. In fact, the 900 °C-treated samples are all characterized by higher values of saturation magnetization and lower coercivities than those observed in nanoparticles of similar size that did not undergo post-synthetic thermal treatments [40]. Overall, our DC magnetization studies confirm that the single-domain magnetic nanoparticles included in xerogel matrices are systematically smaller than in their aerogel counterparts and have a lower degree of superficial defectivity. Furthermore, nanoparticles in the xerogels better approximate the typical magnetic response of non-interacting nanoparticles, while magnetic interaction heavily affects the magnetic response of the nanoparticles included in aerogel matrices. In principle, although this difference might be the indication of a higher degree of dispersion of the magnetic nanoparticles in xerogel matrices, given the direct evidence provided by TEM images of the successful homogeneous distribution of the magnetic nanoparticles in both aerogel and xerogel matrices, it should be more likely ascribed to a matrix-based effect on the nanoparticles, which are also influenced by the presence of impurities in the aerogel samples that increase structural disorder in the nanoparticles.

When comparing xerogel and aerogel samples, in addition to the differences in the T_B and T_{IRR} values from the ZFC-FC curves, a striking dissimilarity can also be observed with regard to the magnetic hardness in the hysteresis loops recorded at 5 K, i.e., a low temperature where all the samples are in a magnetically blocked state. Here, even taking into account the fact that coercivity increases with the size of the magnetic domain, aerogel samples A10_900_1 and A15_900_1 display coercivity values that are more than five times higher than those of xerogel samples X10_900_1 and X15_900_1 with the same loadings, while also showing non-negligible spontaneous exchange bias H_E, which is absent for the xerogel samples. Exchange bias manifests in hysteresis loops as a horizontal shift and has been observed in nano-sized compounds with a high degree of superficial structural disorder. Here, the structurally disordered surface of the nanoparticles becomes heavily prone to phenomena such as spin-canting and behaves differently from the structurally ordered core. Then, the system as a whole reacts to applied magnetic fields as a makeshift magnetic core/shell, whose components undergo magnetic coupling during hysteresis loops, giving rise to horizontal shifts that manifest as a difference in the moduli of the negative and positive measured coercive fields, $|H_{C1}|$ and $|H_{C2}|$, respectively. When the horizontal shift is observed in zero field-cooled hysteresis loops (such as in the present case), it is usually indicated as spontaneous exchange bias (SEB) [33–35,41–44].

Considering the small size of the magnetic nanoparticles of all the aerogel and xerogel samples, a single domain behavior can be expected in all the nanocomposites, but taking in great care the indications provided by DC magnetometry and XRD, some additional assumptions can be made on both classes of samples. XRD indicates that the nanoparticles of A10_900_1 and A15_900_1 are bigger than those of X10_900_1 and X15_900_1, and also shows the presence of a few low intensity peaks that are attributed to impurities of CuO and α-Fe$_2$O$_3$. While the formation of small impurities is not expected to revolutionize the overall magnetic response of A10_900_1 and A15_900_1, their presence paired with increased magnetic hardness, spontaneous exchange bias and higher unblocking temperatures suggest that the magnetic nanoparticles included in the aerogel matrices tend to be more structurally disordered and that the presence of impurities might be located at the nanoparticles' surface, thus increasing the superficial disorder and consequently giving rise to a patched makeshift shell from the structural point of view. If one considers these "composed" nanoparticles, it is clear why they require higher temperatures to undergo superparamagnetic relaxation and can reach lower saturation magnetization: the structural disordered surface lowers the number of atomic moments that can be successfully reoriented under an external magnetic field, while opposing the thermal unblocking and the re-orientation of magnetic moments from the ordered "core" region. On the other hand, the lower unblocking temperatures and the narrow, symmetric hysteresis loops indicate that the xerogel samples contain smaller, well-formed crystal domains that do not show the disorder-related responses observed in aerogel samples and give rise to nanocomposites with a softer magnetic response and higher saturation magnetization.

Even considering the varied effect of the xerogel and aerogel matrices on the final magnetic features of the nanocomposites and the minor impurities observed in the aerogel samples, the controlled inclusion of copper ferrite nanoparticles in both xerogel and aerogel affects their magnetic features and distances them from the magnetic response of a typical population of similar sized unsupported nanoparticles [40,45]. In a comparison between evenly sized unsupported nanoparticles and nanoparticles included in nanocomposites, the former are magnetically harder, with higher exchange bias and tendentially lower saturation magnetization values. This means that the inclusion of copper ferrite nanoparticles inside aerogel and xerogel matrices helps in controlling their size and maintaining it well within the single domain regime, lowers the interparticle interaction and consequently boosts the saturation magnetization, while restoring the soft magnetic features typically observed in copper ferrite.

4. Conclusions

The sol-gel method used has shown to be effective in obtaining nanocomposites where copper ferrite nanoparticles are well dispersed in a silica matrix, both in the form of a highly porous aerogel or a denser xerogel. Crystallization of the copper ferrite phase within the matrix is achieved through thermal treatments of the xerogels and aerogels, and samples exhibit nanoparticles with average size within 10 nm after treatment at 900 °C. Even at this high temperature the nanocomposites maintain their porous structure as evidenced by the TEM images, with nanoparticles which are very well separated within the matrix. The average sizes of the ferrite nanoparticles depend on both the texture of the matrix and the loading of the ferrite, influencing the overall magnetic behavior of the samples. In particular, DC magnetometry indicates that the samples exhibit a superparamagnetic behavior and are unblocked at room temperature, with magnetic interactions partially coming into play in determining the final magnetic features of the nanocomposites. Moreover, the results obtained by DC magnetometry at low temperature, when the nanocomposites are in a magnetically blocked state, highlight how the spatial distribution of the magnetic phase operated by aerogel and xerogel matrices helps in boosting the saturation magnetization and lowering the coercivity in comparison with the magnetic response obtained by populations of unsupported nanoparticles of similar size.

Author Contributions: Formal analysis, A.V.G., F.C., D.L. and A.C. (Alberto Casu); writing—original draft preparation, A.V.G., F.C. and D.L.; writing—review and editing, A.C. (Alberto Casu), A.C. (Anna Corrias), M.F.C. and A.F.; supervision, A.C. (Anna Corrias); funding acquisition, A.C. (Anna Corrias). All authors have read and agreed to the published version of the manuscript.

Funding: This work was supported by the Engineering and Physical Sciences Research Council (EP/K50306X/1).

Data Availability Statement: The data presented in this study are available on request from the corresponding author.

Acknowledgments: We acknowledge the CeSAR (Centro Servizi Ricerca d'Ateneo) core facility of the University of Cagliari and A. Ardu for assistance with the generation of TEM data. We thank Paul Saines (University of Kent) for his help in the SQUID measurements.

Conflicts of Interest: The authors declare no conflict of interest.

References

1. Dai, Q.; Patel, K.; Donatelli, G.; Ren, S. Magnetic cobalt ferrite nanocrystals for an energy storage concentration cell. *Angew. Chem.* **2016**, *128*, 10595–10599. [CrossRef]
2. Kim, J.H.; Kim, H.E.; Kim, J.H.; Lee, J.S. Ferrites: Emerging light absorbers for solar water splitting. *J. Mater. Chem. A* **2020**, *8*, 9447–9482. [CrossRef]
3. Sharifi, I.; Shokrollahi, H.; Amiri, S. Ferrite-based magnetic nanofluids used in hyperthermia applications. *J. Magn. Magn.* **2012**, *324*, 903–915. [CrossRef]
4. Gumbi, S.W.; Mkwae, P.S.; Kortidis, I.; Kroon, R.E.; Swart, H.C.; Moyo, T.; Nkosi, S.S. Electronic and simple oscillatory conduction in ferrite gas sensors: Gas-sensing mechanisms, long-term gas monitoring, heat transfer, and other anomalies. *ACS Appl. Mater. Interfaces* **2020**, *12*, 43231–43249. [CrossRef] [PubMed]
5. Kandasamy, G. Recent advancements in manganite perovskites and spinel ferrite-based magnetic nanoparticles for biomedical theranostic applications. *Nanotechnology* **2019**, *30*, 502001. [CrossRef] [PubMed]

6. Sumangala, T.P.; Mahender, C.; Barnabé, A.; Venkataramani, N.; Prasad, S. Structural, magnetic and gas sensing properties of nanosized copper ferrite powder synthesized by sol gel combustion technique. *J. Magn. Magn. Mater.* **2016**, *418*, 48–53. [CrossRef]

7. Tan, J.; Xu, S.; Zhang, H.; Cao, H.; Zheng, G. Preparation of a porous bulk copper ferrite spinel with high performance in the electrolysis of water. *Electrochim. Acta* **2021**, *381*, 138199. [CrossRef]

8. Gohain, M.; Kumar, V.; van Tonder, J.H.; Swart, H.C.; Ntwaeaborwa, O.M.; Bezuidenhoudt, B.C. Nano $CuFe_2O_4$: An efficient, magnetically separable catalyst for transesterification of β-ketoesters. *RSC Adv.* **2015**, *5*, 18972–18976. [CrossRef]

9. Xing, Z.; Ju, Z.; Yang, J.; Xu, H.; Qian, Y. One-step solid state reaction to selectively fabricate cubic and tetragonal $CuFe_2O_4$ anode material for high power lithium ion batteries. *Electrochim. Acta* **2013**, *102*, 51–57. [CrossRef]

10. Kim, K.J.; Lee, J.H.; Lee, S.H. Magneto-optical investigation of spinel ferrite $CuFe_2O_4$: Observation of Jahn–Teller effect in Cu^{2+} ion. *J. Magn. Magn. Mater.* **2004**, *279*, 173–177. [CrossRef]

11. Nedkov, I.; Vandenberghe, R.E.; Marinova, T.; Thailhades, P.; Merodiiska, T.; Avramova, I. Magnetic structure and collective Jahn–Teller distortions in nanostructured particles of $CuFe_2O_4$. *Appl. Surface Sci.* **2006**, *253*, 2589–2596. [CrossRef]

12. Kenfack, F.; Langbein, H. Synthesis and thermal decomposition of freeze-dried copper–iron formates. *Thermochim. Acta* **2005**, *426*, 61–72. [CrossRef]

13. Phuruangrat, A.; Kuntalue, B.; Thongtem, S.; Thongtem, T. Synthesis of cubic $CuFe_2O_4$ nanoparticles by microwave-hydrothermal method and their magnetic properties. *Mater. Lett.* **2016**, *167*, 65–68. [CrossRef]

14. Kurian, J.; Mathew, M.J. Structural, optical and magnetic studies of $CuFe_2O_4$, $MgFe_2O_4$ and $ZnFe_2O_4$ nanoparticles prepared by hydrothermal/solvothermal method. *J. Magn. Magn. Mater.* **2018**, *451*, 121–130. [CrossRef]

15. Harzali, H.; Saida, F.; Marzouki, A.; Megriche, A.; Baillon, F.; Espitalier, F.; Mgaidi, A. Structural and magnetic properties of nano-sized NiCuZn ferrites synthesized by co-precipitation method with ultrasound irradiation. *J. Magn. Magn. Mater.* **2016**, *419*, 50–56. [CrossRef]

16. Ponhan, W.; Maensiri, S. Fabrication and magnetic properties of electrospun copper ferrite ($CuFe_2O_4$) nanofibers. *Solid State Sci.* **2009**, *11*, 479–484. [CrossRef]

17. López-Ramón, M.V.; Álvarez, M.A.; Moreno-Castilla, C.; Fontecha-Cámara, M.A.; Yebra-Rodríguez, Á.; Bailón-García, E. Effect of calcination temperature of a copper ferrite synthesized by a sol-gel method on its structural characteristics and performance as Fenton catalyst to remove gallic acid from water. *J. Colloid Interface Sci.* **2018**, *511*, 193–202. [CrossRef]

18. Calvo-de la Rosa, J.; Segarra Rubí, M. Influence of the synthesis route in obtaining the cubic or tetragonal copper ferrite phases. *Inorg. Chem.* **2020**, *59*, 8775–8788. [CrossRef]

19. Köferstein, R.; Walther, T.; Hesse, D.; Ebbinghaus, S.G. Crystallite-growth, phase transition, magnetic properties, and sintering behaviour of nano-$CuFe_2O_4$ powders prepared by a combustion-like process. *J. Solid State Chem.* **2014**, *213*, 57–64. [CrossRef]

20. Loche, D.; Casula, M.F.; Falqui, A.; Marras, S.; Corrias, A. Preparation of Mn, Ni, Co ferrite highly porous silica nanocomposite aerogels by an urea-assisted sol–gel procedure. *J. Nanosci. Nanotech.* **2010**, *10*, 1008–1016. [CrossRef]

21. Carta, D.; Casula, M.F.; Falqui, A.; Loche, D.; Mountjoy, G.; Sangregorio, C.; Corrias, A. A structural and magnetic investigation of the inversion degree in ferrite nanocrystals MFe_2O_4 (M = Mn, Co, Ni). *J. Phys. Chem. C* **2009**, *113*, 8606–8615. [CrossRef]

22. Carta, D.; Loche, D.; Mountjoy, G.; Navarra, G.; Corrias, A. $NiFe_2O_4$ nanoparticles dispersed in an aerogel silica matrix: An X-ray absorption study. *J. Phys. Chem. C* **2008**, *112*, 15623–15630. [CrossRef]

23. Carta, D.; Mountjoy, G.; Navarra, G.; Casula, M.F.; Loche, D.; Marras, S.; Corrias, A. X-ray absorption investigation of the formation of cobalt ferrite nanoparticles in an aerogel silica matrix. *J. Phys. Chem. C* **2007**, *111*, 6308–6317. [CrossRef]

24. Carta, D.; Casula, M.F.; Mountjoy, G.; Corrias, A. Formation and cation distribution in supported manganese ferrite nanoparticles: An X-ray absorption study. *Phys. Chem. Chem. Phys.* **2008**, *10*, 3108–3117. [CrossRef]

25. Carta, D.; Marras, C.; Loche, D.; Mountjoy, G.; Ahmed, S.I.; Corrias, A. An X-ray absorption spectroscopy study of the inversion degree in zinc ferrite nanocrystals dispersed on a highly porous silica aerogel matrix. *J. Chem. Phys.* **2013**, *138*, 054702. [CrossRef] [PubMed]

26. Caddeo, F.; Loche, D.; Casula, M.F.; Corrias, A. Evidence of a cubic iron sub-lattice in t-$CuFe_2O_4$ demonstrated by X-ray Absorption Fine Structure. *Sci. Rep.* **2018**, *8*, 1–12. [CrossRef]

27. Prince, E.; Treuting, R.G. PDF card, reference code: 00-034-0425. *Acta Crystallogr.* **1956**, *9*, 1025–1028. [CrossRef]

28. Verwey, E.J.W.; Heilmann, E.L. PDF card, reference code: 01-077-0010. *J. Chem. Phys.* **1947**, *15*, 174–180. [CrossRef]

29. Asbrink, S.; Norrby, L. PDF card, reference code: 00-005-0661. *Acta Crystallogr. B.* **1970**, *26*, 8–15.

30. Carta, D.; Casula, M.F.; Corrias, A.; Falqui, A.; Loche, D.; Mountjoy, G.; Wang, P. Structural and magnetic characterization of Co and Ni silicate hydroxides in bulk and in nanostructures within silica aerogels. *Chem. Mater.* **2009**, *21*, 945–953.

31. Powder Diffraction File Database, ICDD, PDF-2 00-033-0664 (Natl. Bur. Stand. (U.S.). *Monograph* **1981**, *25*, 18, 37.

32. Cutrufello, M.G.; Rombi, E.; Ferino, I.; Loche, D.; Corrias, A.; Casula, M.F. Ni-based xero- and aerogels as catalysts for nitroxidation processes. *J. Sol-Gel Sci. Technol.* **2011**, *60*, 324–332. [CrossRef]

33. Nogués, J.; Sort, J.; Langlais, V.; Skumryev, V.; Suriñach, S.; Muñoz, J.S.; Baró, M.D. Exchange bias in nanostructures. *Phys. Rep.* **2005**, *422*, 65–117. [CrossRef]

34. Sahoo, R.C.; Paladhi, D.; Dasgupta, P.; Poddar, A.; Singh, R.; Das, A.; Nath, T.K. Antisite-disorder driven large exchange bias effect in phase separated $La_{1.5}Ca_{0.5}CoMnO_6$ double perovskite. *J. Magn. Magn. Mater.* **2017**, *428*, 86–91. [CrossRef]

35. Lentijo-Mozo, S.; Deiana, D.; Sogne, E.; Casu, A.; Falqui, A. Unexpected insights about cation-exchange on metal oxide nanoparticles and its effect on their magnetic behavior. *Chem. Mater.* **2018**, *30*, 8099–8112. [CrossRef]

36. Zuddas, E.; Lentijo-Mozo, S.; Casu, A.; Deiana, D.; Falqui, A. Building composite iron—manganese oxide flowerlike nanostructures: A detailed magnetic study. *J. Phys. Chem. C* **2017**, *121*, 17005–17015. [CrossRef]

37. Blanco-Gutierrez, V.; Saez-Puche, R.; Torralvo-Fernandez, M.J. Superparamagnetism and interparticle interactions in $ZnFe_2O_4$ nanocrystals. *J. Mater. Chem.* **2012**, *22*, 2992–3003. [CrossRef]

38. Rani, B.J.; Saravanakumar, B.; Ravi, G.; Ganesh, V.; Ravichandran, S.; Yuvakkumar, R. Structural, optical and magnetic properties of $CuFe_2O_4$ nanoparticles. *J. Mater. Sci.: Mater. Electron.* **2018**, *29*, 1975–1984. [CrossRef]

39. Naseri, M.G.; Saion, E.B.; Ahangar, H.A.; Shaari, A.H. Fabrication, characterization, and magnetic properties of copper ferrite nanoparticles prepared by a simple, thermal-treatment method. *Mater. Res. Bull.* **2013**, *48*, 1439–1446. [CrossRef]

40. Goya, G.F.; Rechenberg, H.R.; Jiang, J.Z. Structural and magnetic properties of ball milled copper ferrite. *J. Appl. Phys.* **1998**, *84*, 1101–1108. [CrossRef]

41. Kim, Y.I.; Kim, D.; Lee, C.S. Synthesis and characterization of $CoFe_2O_4$ magnetic nanoparticles prepared by temperature-controlled coprecipitation method. *Physica B Condens. Matter* **2003**, *337*, 42–51. [CrossRef]

42. Goya, G.F.; Berquó, T.S.; Fonseca, F.C.; Morales, M.P. Static and dynamic magnetic properties of spherical magnetite nanoparticles. *J. Appl. Phys.* **2003**, *94*, 3520–3528. [CrossRef]

43. Brabers, V.A.M. *Handbook of Magnetic Materials*; Buschow, K.H.J., Ed.; Elsevier: Amsterdam, The Netherlands, 1995; Volume 8, p. 212.

44. Punnoose, A.; Magnone, H.; Seehra, M.S.; Bonevich, J. Bulk to nanoscale magnetism and exchange bias in CuO nanoparticles. *Phys. Rev. B* **2001**, *64*, 174420–174428. [CrossRef]

45. Jiang, J.Z.; Goya, G.F.; Rechenberg, H.R. Magnetic properties of nanostructured $CuFe_2O_4$. *J. Phys. Condens. Matter* **1999**, *11*, 4063. [CrossRef]

*nanomaterials*

MDPI

Article

Remotely Self-Healable, Shapeable and pH-Sensitive Dual Cross-Linked Polysaccharide Hydrogels with Fast Response to Magnetic Field

Andrey V. Shibaev [1,*], Maria E. Smirnova [1], Darya E. Kessel [1], Sergey A. Bedin [2], Irina V. Razumovskaya [2] and Olga E. Philippova [1]

[1]   Physics Department, Lomonosov Moscow State University, 119991 Moscow, Russia; smirnova.me15@physics.msu.ru (M.E.S.); kes@polly.phys.msu.ru (D.E.K.); phil@polly.phys.msu.ru (O.E.P.)
[2]   Institute of Physics, Technology and Informational Systems, Moscow Pedagogical State University, 119435 Moscow, Russia; sa.bedin@mgpu.su (S.A.B.); irinarasum9@mail.ru (I.V.R.)
*   Correspondence: shibaev@polly.phys.msu.ru; Tel.: +7-495-939-1464

Abstract: The development of actuators with remote control is important for the construction of devices for soft robotics. The present paper describes a responsive hydrogel of nontoxic, biocompatible, and biodegradable polymer carboxymethyl hydroxypropyl guar with dynamic covalent cross-links and embedded cobalt ferrite nanoparticles. The nanoparticles significantly enhance the mechanical properties of the gel, acting as additional multifunctional non-covalent linkages between the polymer chains. High magnetization of the cobalt ferrite nanoparticles provides to the gel a strong responsiveness to the magnetic field, even at rather small content of nanoparticles. It is demonstrated that labile cross-links in the polymer matrix impart to the hydrogel the ability of self-healing and reshaping as well as a fast response to the magnetic field. In addition, the gel shows pronounced pH sensitivity due to pH-cleavable cross-links. The possibility to use the multiresponsive gel as a magnetic-field-triggered actuator is demonstrated.

Keywords: polymer hydrogel; self-healing; magnetic gel; actuator

Citation: Shibaev, A.V.; Smirnova, M.E.; Kessel, D.E.; Bedin, S.A.; Razumovskaya, I.V.; Philippova, O.E. Remotely Self-Healable, Shapeable and pH-Sensitive Dual Cross-Linked Polysaccharide Hydrogels with Fast Response to Magnetic Field. *Nanomaterials* **2021**, *11*, 1271. https://doi.org/10.3390/nano11051271

Academic Editor: Raghvendra Singh Yadav

Received: 21 April 2021
Accepted: 9 May 2021
Published: 12 May 2021

Publisher's Note: MDPI stays neutral with regard to jurisdictional claims in published maps and institutional affiliations.

1. Introduction

Polymer hydrogels demonstrate highly responsive properties to various stimuli including temperature, pH-, magnetic field, and so on [1–4]. Of particular interest are magnetic-field-sensitive hydrogels (so-called ferrogels) [5–10], since they demonstrate rapid response to the stimulus and remote-control ability [11,12]. To provide magnetic-field responsiveness, magnetic particles like magnetite Fe_3O_4 [13], maghemite γ-Fe_2O_3, or cobalt ferrite $CoFe_2O_4$ [11,14,15] should be incorporated into the gel matrix. The resulting material combines the magnetic properties inherent to magnetic filler and the elastic properties inherent to hydrogels. When the magnetic field is applied, it induces the interaction between the particles, which leads to the change of the shape and of the elastic properties of the entire gel in a fast and reversible manner. It makes the magnetic-field responsive gels especially important for the development of remotely manipulated actuators for soft robotics [16], controlled valves [6], or plugs [8], and so on.

Magnetic hydrogels based on natural polymers are preferable over the other gels because of their enhanced sustainability, biocompatibility, biodegradation, and nontoxicity [12]. Biocompatible magnetic hydrogels are very promising as catheters for remotely controlled manipulation systems [17], microgrippers for excising cells and intravascular surgery [18], vehicles for magnetically guided drug delivery [19–22], and so forth. Biopolymer magnetic gels not only provide the possibility of biomedical applications, but also (which is even more important) significantly reduce the environmental pollution. Usually, such hydrogels are produced from various polysaccharides including sodium alginate [23,24], cellulose [22,25] and its derivatives [26], κ-carrageenan [27], chitosan [28,29] and its derivatives [30], agarose [31], starch [21], and some others.

Much less attention has been paid to guar gum polysaccharide. At the same time, it is cheap and produced in very large amounts from the seeds of Cyamopsis tetragonolobus [32], that is, from renewable biomass resources. Guar gum is a high-molecular weight non-ionic polysaccharide, which consists of α-(1→4)-linked-D-mannopyranose backbone with α-D-galactopyranosyl side groups attached by (1→6) linkages [32]. It easily forms gels upon the addition of various cross-linkers such as glutaraldehyde, phosphate, urea-formaldehyde, and borate [33]. Among them, borate is especially interesting, because it provides dynamic covalent cross-links with 1,2-cis-diol groups of polymer [34]. The enthalpy of formation of such bonds is rather low and equals 5–20 kT [35,36], which is much lower than the energy of "standard" covalent bonds [37]. As a consequence, borate cross-links are "labile" and reversibly break and recombine even at room temperature. Such labile cross-links provide faster response to external triggers as well as reshaping and self-healing ability that are highly requested in soft robotics.

Currently, there are only few papers describing guar gum magnetic gels. Most of them consider the combination of guar gum with synthetic polymers like polyacrylamide [38,39]. To the best of our knowledge, the magnetic guar gum gels without other polymer components were studied only in one paper [33], but these gels were formed only at elevated temperatures. This is useful for injection applications but significantly restricts the usage of these gels in many other areas, for instance, in the production of actuators.

The aim of the present paper is to prepare at room temperature the magnetically actuated hydrogels with fast response to the magnetic field on the basis of a guar gum derivative—carboxymethyl hydroxypropyl guar (CMHPG). To provide fast response, we use dynamic covalent cross-links, which allow the polymer matrix to easily follow the displacement of magnetic particles under the action of a magnetic field. Additionally, a dynamic nature of the cross-links imparts to the gel reshaping and self-healing ability. In order to enhance the magnetic response, we use cobalt ferrite $CoFe_2O_4$ nanoparticles (NPs), which have higher magnetization than, for instance, magnetite NPs of the same size [40,41]; therefore, they are preferable for obtaining magnetic gels with strong response to magnetic field.

2. Materials and Methods

2.1. Materials

CMHPG POLYFLOS CH410P was provided by Lamberti (Gallarate, Italy). It is an anionic polyelectrolyte, derivative of a polysaccharide guar, bearing hydroxypropyl and carboxymethyl groups (Figure 1).

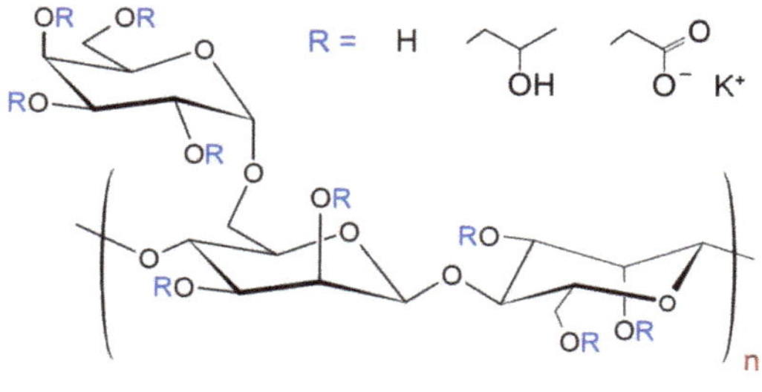

Figure 1. Chemical structure of carboxymethyl hydroxypropyl guar (CMHPG).

From NMR data, the degree of substitution of side α-D-galactopyranosyl units is equal to 0.8; the degrees of hydroxypropylation and carboxymethylation (numbers of hydroxypropyl and carboxymethyl groups per one monosaccharide residue) are 0.41 and 0.1, respectively [42]. According to viscometry, the molecular weight of CMHPG macromolecules is 1.6×10^6 g/mol [42]. Hydroxypropyl guar (HPG) Jaguar HP105 from Solvay

(Brussels, Belgium), boric acid (>99.9% purity) from ACROS (Geel, Belgium), potassium hydroxide (>98% purity) from Sigma Aldrich (St. Louis, MO, USA), and cobalt ferrite ($CoFe_2O_4$) NPs (99% trace metal basis) from Sigma Aldrich (St. Louis, MO, USA) were used as received. Solutions were prepared with deionized distilled water obtained on a Milli-Q device (Millipore, Burlington, MA, USA).

2.2. Polymer Purification

Aqueous CMHPG solution (0.5 wt%) was subjected to vacuum filtration through ceramic filters (ROBU GmbH, Hattert, Germany) with a pore diameter of 16–40 µm to remove water-insoluble impurities. Then, CMHPG was reprecipitated in ethanol (polymer solution:ethanol = 1:10 v/v). The precipitate was dissolved in water and lyophilized.

2.3. Preparation of the Samples

First, a desired number of NPs was suspended in KOH aqueous solution (pH 11). At this pH, $CoFe_2O_4$ NPs are negatively charged as a result of deprotonation of their surface -OH groups [43], and, therefore, they are electrostatically stabilized in the solution. In order to obtain a homogeneous dispersion, the solutions were ultrasonicated for 30 min with a Sonorex RK102H ultrasonic bath (Bandelin GmbH, Berlin, Germany) at 40 °C. Immediately after ultrasonication, dry polymer was added, and the solutions were mixed by a magnetic stirrer overnight. As a result, stable and homogeneous dispersions of NPs in polymer solutions (at polymer concentrations higher than 0.5 wt%) were obtained. Afterwards, a cross-linker (borax) solution (3.9 wt%, pH 11) was added, and the mixture was vigorously stirred with a spatula for 10–20 s and allowed to rest for a day, which resulted in the formation of cross-linked gels. Final pH of the samples was kept at 10.7 ± 0.1 by adding a small amount (1–3 µL) of 5M KOH solution when necessary.

2.4. Transmission Electron Microscopy (TEM)

Transmission electron microscopy (TEM) and high resolution (HR-TEM) images were obtained using JEM 2100 F/Cs (Jeol, Tokyo, Japan) operated at 200 kV and equipped with UHR pole tip as well as spherical aberration corrector (CEOS, Heidelberg, Germany) and EEL spectrometer (Gatan, Munich, Germany). To prepare the TEM specimens, a drop (10 µL) of a pre-sonicated NPs aqueous solution was deposited onto a copper grid and air-dried under ambient conditions. Typical preparation of the TEM specimens is described elsewhere [44,45]. According to the TEM data, NPs are almost spherical and rather monodisperse, and their mean size equals to 20 nm (Figure 2A). HR-TEM images show that they are monocrystalline.

2.5. Magnetization Measurements

Magnetization curve of $CoFe_2O_4$ NPs was measured by a vibrating-sample magnetometer (VSM) 7407 (Lake Shore Cryotronics, Westerville, OH, USA)at room temperature in the range of magnetic field strengths ± 16 kOe. Dry NPs powder (5.4 mg) was sealed in a polyethylene container (4 × 4 × 0.5 mm) and laminated before measurements. NPs show hard ferrimagnetic behavior (Figure 2B) [46] and are characterized by a saturation magnetization M_s equal to 56.0 ± 0.5 emu/g, a rather large hysteresis loop with a high remanent magnetization M_r of 30 ± 0.5 emu/g, and a coercive force H_C of 2.2 ± 0.1 kOe. These magnetic properties are typical for single magnetic domain particles and are consistent with the data reported previously for $CoFe_2O_4$ NPs of similar size at room temperature [40,47].

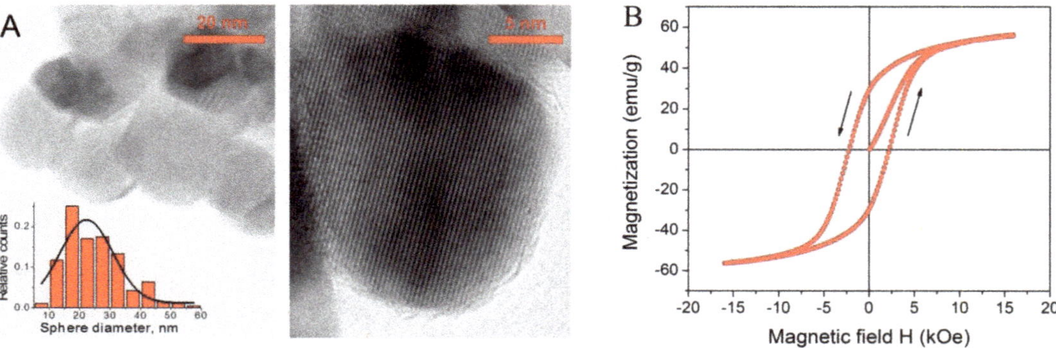

Figure 2. (**A**) TEM and HR-TEM images and histogram of the size distribution of CoFe$_2$O$_4$ NPs; (**B**) magnetic hysteresis curve of CoFe$_2$O$_4$ NPs at 297 K.

2.6. Rheometry

Mechanical properties were investigated on a Physica MCR 301 rotational rheometer (Anton Paar, Graz, Austria). The experimental procedures are described in detail in [48,49]. Plate–plate geometry of the measuring cell was used. For the measurements in the absence of magnetic field, an upper plate had a diameter of 25 mm; and cylindrical gel samples with diameter of 25 mm and height of 3–4 mm were synthesized. The temperature, which was controlled using Peltier elements, was set to 20.00 ± 0.05 °C. To avoid evaporation of the solvent from the sample during measurements, a casing with Peltier elements was used. For the measurements in the presence of the magnetic field, a special plate–plate cell (MRD70/1T, Anton Paar, Graz, Austria) made from nonmagnetic material was utilized. The diameter of the upper measuring plate was 20 mm, and the gap was set at 0.7 mm. The magnetic field in the range of 0–1 T was generated in the measuring cell in the direction perpendicular to shear. The generated magnetic field was almost homogeneous at the sample length scale.

Experiments were carried out in the oscillation mode, in which the frequency dependences of the storage modulus G' and the loss modulus G'' were measured in the range of external impact frequency $\omega = 0.04$–50 s^{-1}. All measurements were performed in the linear viscoelasticity mode (at a strain amplitude of $\gamma = 3$–5%), in which the storage and loss moduli were amplitude independent. Linear viscoelastic range was determined by amplitude sweep measurements for CMHPG solution without borax and NPs, as well as for gels with borax, and with both borax and NPs (Figure 3).

2.7. Actuation Force Measurements

Force generated by the gels in the magnetic field was measured by a custom-made experimental setup equipped with a digital force meter 53002 (Megeon, Moscow, Russia) with measurement limits 0–2 H and accuracy 1 mH. A magnetic field (0–0.26 Tl) was generated by a permanent NdFeB magnet. During measurements, the sample and the magnet were moved away for at least 15 cm from the force meter, and the sample was connected to it by a non-magnetic spacer in order to avoid the influence of magnetic field on the force measurements.

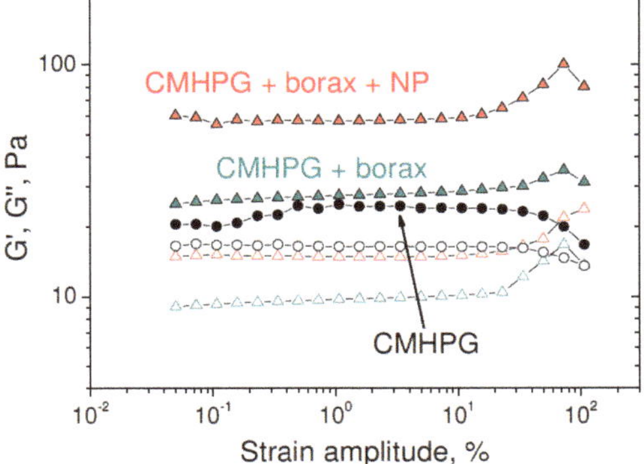

Figure 3. Amplitude dependences of storage G′ (filled symbols) and loss G″ (open symbols) moduli for aqueous solutions containing CMHPG (black symbols), CMHPG and borax (dark cyan symbols), CMHPG, borax, and $CoFe_2O_4$ NPs (red symbols). Concentrations: CMHPG—1 wt%; potassium borate—0.058 wt% (molar ratio (borate)/(monomer units) = 0.5); NPs—10 wt%. Temperature: 20 °C.

3. Results and Discussion

In this article, we investigate hydrogels of CMHPG cross-linked by borate ions (borax) with added 20-nm magnetic $CoFe_2O_4$ nanoparticles. The concentration of CMHPG is fixed at 1 wt%, which corresponds to the semi-dilute regime [42]. In this regime, borate can cross-link different CMHPG macromolecules into a network due to the formation of dynamic covalent bonds with hydroxypropyl groups of the polymer [50]. Such cross-links are formed only at high pH > 9.2, e.g., above the pKa of boric acid, when it is transformed to borate [51]. Taking this into account, in the present work, the gels were prepared at pH 10.7. The concentration of NPs was varied in the range of 0–10 wt% (0–1.9 vol%).

3.1. Effect of NP Content

First, we investigate the effect of NPs concentration on the properties of CMHPG/borate gels. In the absence of NPs, the gels show pronounced viscoelastic behavior (Figure 4A), which confirms the formation of a cross-linked polymer network: a cross-over point between G′(ω) and G″(ω) is observed due to the dynamic nature of borate cross-links, as a result of which the gels can flow. Such a behavior is typical for hydroxypropyl guar (HPG) gels cross-linked by borate [52].

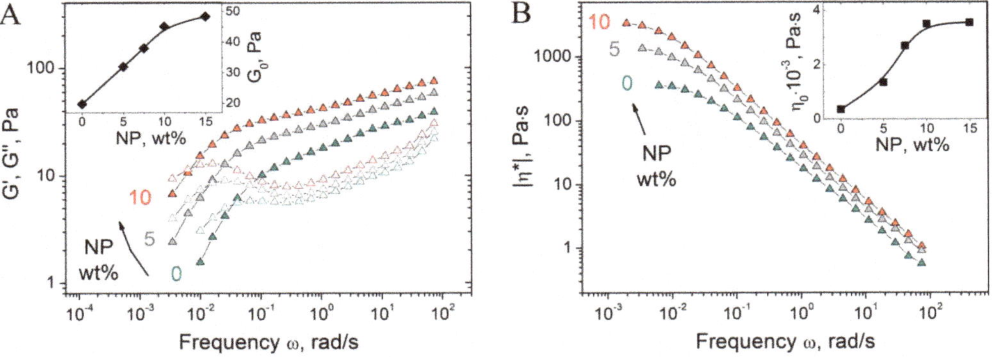

Figure 4. Frequency dependences of storage G′ (filled symbols) and loss G″ (open symbols) moduli (**A**) and modulus of complex viscosity |η*| (**B**) for hydrogels containing 1 wt% CMHPG, 0.058 wt% borax (molar ratio (borate)/(monomer units) = 0.5) and various concentrations of $CoFe_2O_4$ NPs indicated in the figure. Temperature: 20 °C. Inlets show the dependences of G_0 (storage modulus at ω = 1.7 rad/s) and zero-shear viscosity $η_0$ on NPs concentration.

Added NPs induce the increase of the storage modulus G′ (Figure 4A), suggesting that NPs form cross-links between CMHPG chains in addition to borate cross-links. At the same time, the gels retain viscoelastic behavior, meaning that the cross-links formed by NPs are reversible, similar to the borate cross-links. Addition of NPs also results in the shift of the cross-over point between G′(ω) and G″(ω) to the left, e.g., to the increase of the relaxation time (Figure 4A). This is explained by the fact that the movement of polymer chains is slowed down by their interaction with NPs, leading to the retardation of stress relaxation. Both these processes, formation of additional cross-links and slowing down of stress relaxation, contribute to the increase of viscosity (Figure 4B). As seen from the inlet in Figure 4B, zero-shear viscosity η_0 (calculated from the low-frequency plateau value of the modulus of complex viscosity $|\eta^*|$, which coincides with η_0 for borate-crosslinked fluids [53,54]) increases with the rise of NP content. As mentioned above, this is related to the diffusion of polymer chains, which is slowed down by their interaction with NPs.

Therefore, in the presence of NPs, dual cross-linked gels are formed, in which borate ions and NPs serve as two types of dynamic cross-links (Figure 5). Cross-links by NPs are formed due to non-covalent interactions of macromolecules with their surface. CMHPG chains contain two types of functional groups—carboxylic and hydroxylic—which may be involved in the interaction with NPs. In order to elucidate which groups are responsible for cross-linking of polymer by NPs, a similar polymer—hydroxypropyl guar (HPG)—was taken, which bears only hydroxyl groups. As seen from Figure 6, NPs induce similar rheology enhancement of HPG solutions as for CMHPG. Therefore, it can be assumed that hydroxylic groups mainly participate in the polymer–NP interactions. The fact that carboxylic groups do not play a role in the binding of polymer to NPs may be explained by the fact that, at high pH 10.7, all carboxylic groups are deprotonated [42]. At the same time, cobalt ferrite has an isoelectric point at pH ~ 6.5; therefore, at pH 10.7, the NPs are also strongly negatively charged [55]. Thus, electrostatic repulsion exists between –COO⁻ polymer groups and NPs. Consequently, one can assume that the formation of cross-links by NPs is due to the interaction of hydroxylic groups of polymer with their surface. Literature data suggest that this interaction may be due to the formation of hydrogen bonds. Previously, guar absorption was observed on alumina, titania, and some other mineral-bearing surface –OH groups [56]. The formation of hydrogen bonds was observed in the case of HPG interacting with TiO_2 NPs even at high pH ~ 10, when the NPs are negatively charged [57]. Though individual bonds between polymer and NP are weak, the resulting cross-links are rather strong and contribute to the storage modulus G′, since many groups can interact with one NP with the formation of a "multipoint" cross-link.

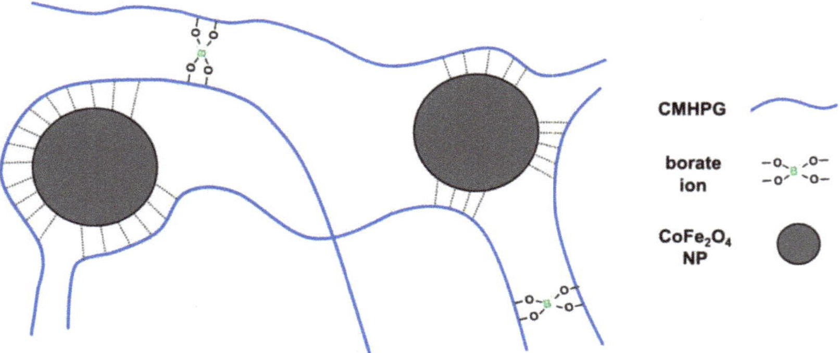

Figure 5. Scheme of formation of a CMHPG hydrogel dual-crosslinked by borate ions and $CoFe_2O_4$ magnetic NPs.

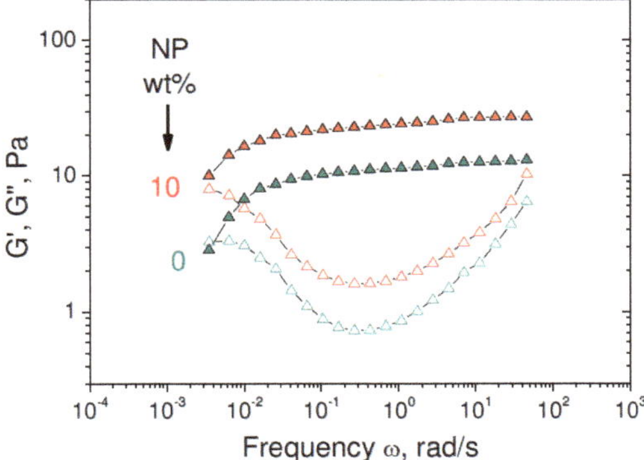

Figure 6. Frequency dependences of storage G′ (filled symbols) and loss G″ (open symbols) moduli for hydrogels containing 1 wt% HPG, 0.058 wt% borax (molar ratio (borate)/(monomer units) = 0.5) and various concentrations of $CoFe_2O_4$ NPs indicated in the figure. Temperature: 20 °C.

The formation of cross-links by NPs is possible because of their rather small size (20 nm). Indeed, the degree of polymerization of CMHPG is ca. 2000 (assuming that one monomer unit is constituted by two monosaccharide residues in the main chain, with some side galactose units attached). Taking into account that the projection of the repeat saccharide unit on the main chain is ~0.52 nm [58], the contour length of a CMHPG macromolecule equals to ca. 2 μm. Water is a good solvent for CMHPG [59]; therefore, macromolecules are in expanded coil conformation. Their end-to-end distance R can be estimated by using the equation [60]

$$R = l_k \times n_k^{3/5} = 316 \, \text{nm} \tag{1}$$

where l_k = 20 nm is the Kuhn segment and $n_k = L/l_k$ = 100 is the number of Kuhn segments in the chain. For the Kuhn segment, an intrinsic value of 20 nm for guar was taken [61]. If polyelectrolyte effect, resulting from the presence of carboxymethyl groups, is taken into account, this would increase the length of Kuhn segment, and, correspondingly, the value of R. At 10 wt% of NPs, the mean distance between their surfaces is ca. 40 nm. Therefore, the size of the polymer coils is much larger than the distance between NPs. Consequently, one polymer chain can interact with several NPsand "connect" them together, which results in the increase of mechanical properties. Previously, in the literature, it was mentioned that small size of non-magnetic TiO_2 particles is crucial for thickening of guar solutions [57].

Moreover, NPs are effective for the enhancement of viscoelastic properties of the gels, since one NP can possibly connect multiple macromolecules and serve as a "multifunctional" cross-link, contrary to one borate ion, which can link no more than two macromolecules. Indeed, the hydrodynamic radius of the polymer coil is [58]

$$R_h = (3\pi/128)^{1/2} \times R \approx 85 \, \text{nm} \tag{2}$$

The volume around one NP with a radius r = 10 nm, which can be occupied by macromolecules having contact with the NP surface, can be estimated as [62]

$$V = 4/3\pi \left[(2R_h + r)^3 - r^3\right] \tag{3}$$

The number of macromolecules attached to one NP may be approximately calculated (assuming that polymer coils near the surface of NP do not overlap) as $n_a = V/V_1 \approx 9$, where $V_1 = 4/3\pi R_h^3$ is the hydrodynamic volume of a single polymer coil.

Therefore, the use of rather small NPs as compared to the size of polymer coils allows obtaining dual physically cross-linked gels of CMHPG, where "single" cross-links are formed by borate ions, and "multiple" cross-links are represented by NPs. Addition of more NPs results in the enhancement of the mechanical properties of the gels.

3.2. Effect of Borate Content

In order to find the optimal conditions for the formation of CMHPG/borate/NP gels, the effect of borate concentration on the mechanical properties was studied. As seen from Figure 7A, the increase of borate concentration induces a significant enhancement of viscoelastic properties. Indeed, in the absence of borax, only very weak viscoelastic properties are seen, which are due to cross-links formed by NPs and some entanglements between polymer chains. Upon addition of borate, a high-frequency plateau at $G'(\omega)$ dependence appears and widens, whereas the cross-over point between $G'(\omega)$ and $G''(\omega)$ moves to the left, meaning the increase of the relaxation time. This shows that, in addition to NP cross-links, borate cross-links are formed, and their amount rises with increasing borate concentration, contributing to the formation of a more tightly cross-linked network.

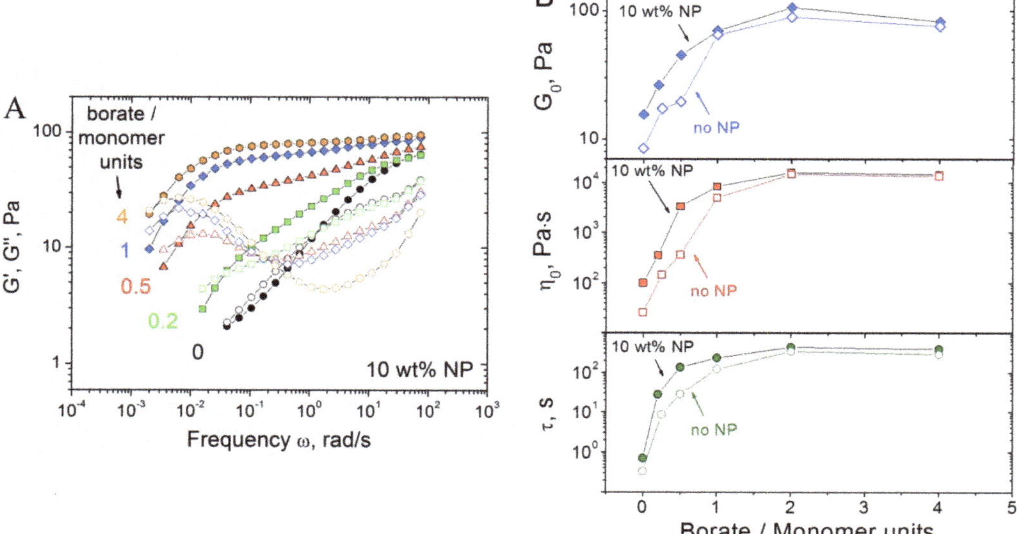

Figure 7. (**A**) Frequency dependences of storage G' (filled symbols) and loss G'' (open symbols) moduli for hydrogels containing 1 wt% CMHPG, various concentrations of borate (indicated in the figure) and 10 wt% $CoFe_2O_4$ NPs. (**B**) Dependences of G_0 (storage modulus at $\omega = 1.7$ rad/s), η_0 (zero-shear viscosity), and relaxation time (τ) on the molar ratio of borate to CMHPG monomer units in the absence (open symbols) and in the presence of 10 wt% $CoFe_2O_4$ NPs (filled symbols). Temperature: 20 °C.

From Figure 7B, it is seen that in a wide range of borate concentrations (molar ratios of borate to monomer units from 0 to ca. 2), the rheological characteristics (high-frequency storage modulus G_0, zero-shear viscosity η_0, and relaxation time τ) of the gels grow with increasing borate content. Additionally, these characteristics are larger in the presence of NPs than without them. It indicates that NPs form cross-links between polymer chains.

At very high borate concentrations (molar ratios of borate to monomer units higher than 2), the rheological properties level off, which suggests that all accessible 1,2-cis-diol groups are involved in the interaction with borax (with the formation of both di-diol cross-links between one borate ion and two polymer chains, as well as mono-diol complexes between one borate ion and one macromolecule). Note that at these conditions, the rheological properties with and without NPs almost coincide. This may indicate that borate ions compete with NPs for 1,2-cis-diol groups of polymer; therefore, at the excess of borate, NPs cannot form cross-links. These results count in favor of our suggestion that hydroxyl groups of CMHPG are involved in the interaction with NPs.

Thus, the gels dually cross-linked by NPs and borax are formed in a wide range of borate concentrations, where the total number of accessible hydroxyl groups is rather large.

3.3. Effect of pH

Since borate cross-links are pH-cleavable [34], it is of interest to study the effect of pH on the rheological properties of CMHPG/borate/NP gels. Figure 8 shows the frequency dependencies of storage and loss moduli of the gels at different pH of the solutions: 8 and 10.7. It is seen that at pH 10.7 the frequency range corresponding to predominantly elastic response ($G' > G''$) is very broad and there is a wide plateau of the $G'(\omega)$ curve. At pH 8, the plateau disappears, and the frequency range corresponding to elastic response diminishes. At this pH, the $G'(\omega)$ and $G''(\omega)$ curves approach those for the same system in the absence of borate, suggesting the disruption of most of the borate cross-links. This is consistent with the literature data for galactomannan/borax gels, indicating that the cross-links persist only at pH higher than 9.0 [51]. At the same time, even in the absence of borate cross-links, the system keeps viscoelastic behavior, which is due to some entanglements between macromolecules and the remaining cross-links by NPs. Therefore, one can suggest that lowering pH from 10.7 to 8 disrupts most borate cross-links, whereas cross-links by NPs still persist in the system. Such behavior can be quite helpful for the application of these gels as actuators, since it helps to reversibly tune the mechanical properties of the gel, making its matrix softer and more responsive to external triggers like magnetic field.

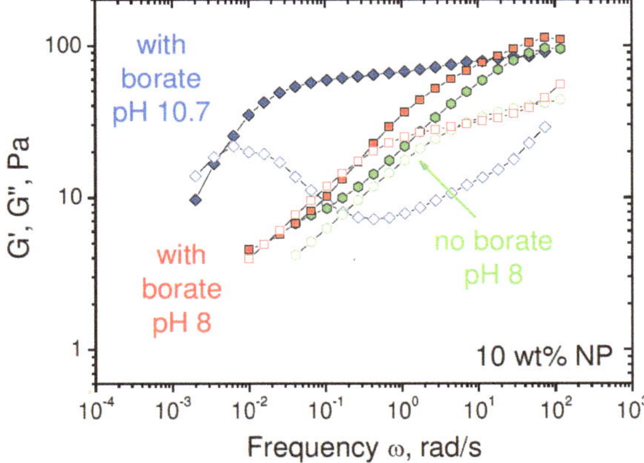

Figure 8. Frequency dependences of storage G' (filled symbols) and loss G'' (open symbols) moduli for hydrogels containing 1 wt% CMHPG, 10 wt% CoFe$_2$O$_4$ NPs in the presence of 0.116 wt% borax (molar ratio borate/monomer units = 1) at pH 10.7 (diamonds) or 8 (squares) and in the absence of borax at pH 8 (hexagons). Temperature: 20 °C.

3.4. Effect of Magnetic Field on Mechanical Properties

For the application of CMHPG/borate/NP gels as magnetic-field-triggered actuators, it is important to elucidate the effect of magnetic field on their mechanical properties. In order to investigate this effect, the gels were put into a uniform magnetic field, and their mechanical properties were studied at shear deformations perpendicular to the field. It is seen (Figure 9) that the magnetic field has a strong effect on the mechanical properties: the high-frequency storage modulus G_0 significantly increases with the rise of the field intensity. These results can be explained as follows. When a uniform magnetic field is applied, NPs acquire magnetization and interact by magnetic forces, moving towards each other and forming chain-like structures, which are oriented along the field force lines (Figure 9). The chains formed by magnetic NPs are oriented perpendicular to shear; therefore, they inhibit the deformation, which is manifested in the increase of the elastic modulus. This effect was previously observed for various magnetic materials, including magnetic elastomers [63], polymer [64] and surfactant hydrogels [65]; however, to the best of our knowledge, it has never been reported for borate cross-linked magnetic gels. Borate cross-links play an important role in high sensitivity of the gels to magnetic field: indeed, borate cross-links, as well as the bonds between polymer chains and NPs, are dynamic and reversible; therefore, NPs are able to move inside the borate-cross-linked polymer matrix to follow the magnetic field.

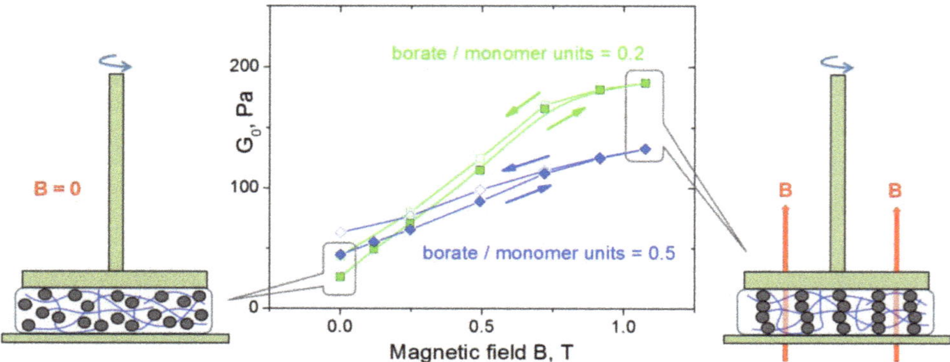

Figure 9. Dependences of G_0 (storage modulus at $\omega = 1.7$ rad/s) on magnetic field strength for hydrogels containing 1 wt% CMHPG, 10 wt% $CoFe_2O_4$ NPs in the presence of different borax concentrations: 0.023 wt% (corresponding to borate/monomer units = 0.2 − squares) and 0.058 wt% (corresponding to borate/monomer units = 0.5 − diamonds). Temperature: 20 °C.

In the case of a loosely cross-linked polymer matrix (borate/monomer units = 0.2), the effect of magnetic field on the elastic modulus is much stronger than for a tightly cross-linked CMPGH (borate/monomer units = 0.5). This is explained by the fact that at low elasticity of the polymer matrix, NPs can easily move and form well-assembled chains, while for a strongly cross-linked matrix, elastic forces acting on NPs from the polymer network resist their movement and formation of NP chains. In the latter case, the mesh size of the network ξ was estimated from the value of the elastic modulus G_0 by using the equation [60]

$$G_0 = kT / \xi^3 \tag{4}$$

where k is the Boltzmann constant, and T is the absolute temperature. ξ equals to ca. 45 nm, which is comparable to the size of NPs (20 nm). Due to this, the particles are not tightly fixed and are able to move inside the network, but polymer coils serve as obstacles and hamper free movement.

The dependences of G_0 on the magnetic field show hysteresis: when the magnetic field is reduced, the elastic modulus is higher than for the case of increasing field (Figure 9). This is presumably due to the ferrimagnetic properties of cobalt ferrite NPs: they possess high remanent magnetization (Figure 2B), and when the field is removed, they still interact due to remanent magnetic moments, and chain-like structures are not completely broken. Their partial conservation in the absence of the field may be assisted by the elasticity of the polymer matrix preventing disorganization of NPs.

Therefore, the mechanical properties of CMHPG/borate/NP magnetic gels can be enhanced by magnetic field, which is due to the formation of NP chains made possible due to the dynamic nature of cross-links in the gels.

3.5. Reshaping and Self-Healing

A possibility to modify the shape in a controlled manner can play an essential role in the development of functional actuators. Reshaping of CMHPG/borate/NP gels was investigated at room temperature on an example of hydrogel containing 1 wt% CMHPG, 0.116 wt% borax (molar ratio borate/monomer units = 1), and 10 wt% $CoFe_2O_4$ NPs. To modify the shape, the sample was slowly reformed by a spatula. It was observed that the hydrogels can be easily reshaped many times, as shown in Figure 10. The shape deformation occurs instantaneously. It is obviously due to the labile character of cross-links both by borax and by NPs, which under the external force can quickly break and reform in order to keep the new shape.

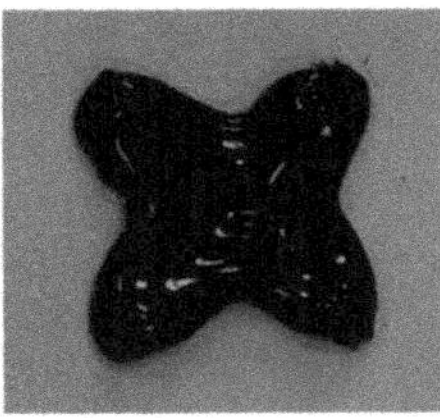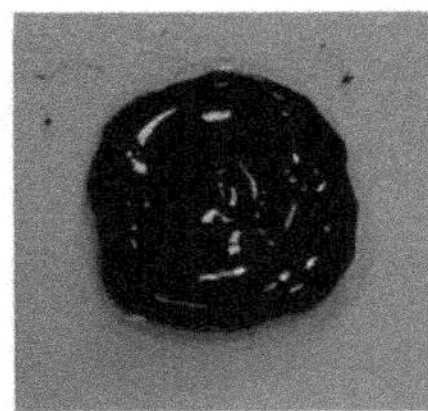

Figure 10. Reshaping of hydrogels containing 1 wt% CMHPG, 10 wt% $CoFe_2O_4$ NPs in the presence of 0.116 wt% borax (molar ratio borate/monomer units = 1) at pH 10.7. Temperature: 20 °C.

Self-healing of CMHPG/borate/NP gels was studied at room temperature by cutting a gel sample in half and placing the two parts in contact for 10 min. During this time, the gel was completely restored. Being stretched, the healed gel did not break at the cut (Figure 11A and Video S1 of Supplementary materials). After 10 min, the gel exhibited 100% healing efficiency with regard to the elongation ratios. Remote self-healing by the magnetic field was also achieved (Figure 11B and Video S2 of Supplementary materials). In this case, two parts of the cut gel were placed in the field of a magnet (0.26 T), and they were brought together remotely by magnetic forces acting on the NPs. Since NPs are bound to the polymer matrix, this resulted in the movement of the gel as a whole. The gel regained conformity in less than 1 min. After 15 min, the self-healed gel was stretched by a factor of 7 without breaking. The observed self-healing properties are due to fast restoration of both types of labile cross-links between polymer chains.

Thus, the CMHPG/borate/NP gels demonstrate the ability to reshape and self-heal, which mimics the healing of muscles and will help to prolong the lifetime of the actuators.

3.6. Actuation

In order to study the actuation capability of CMHPG/borate/NP gels, a strip of the gel (80 mm length, 5×5 mm wide) containing 1 wt% CMHPG, 0.116 wt% borate (molar ratio borate/monomer units = 1), and 10 wt% $CoFe_2O_4$ NPs was prepared. This strip was put in the non-uniform magnetic field of a permanent magnet (Figure 12A), so that one end of the strip was located close to the magnet pole. When the other end of the strip, located far from the pole, was released, the gel contracted very rapidly—in 0.5 s, it decreased its length by a factor of 5, meaning that the gel shows a very fast actuation capability. This is due to the movement of magnetic NPs in the non-uniform magnetic field, which tend to displace along the field gradient, e.g., attract to the magnet pole. Since NPs are bound to the polymer matrix, this causes the movement of the gel as a whole. Large and fast deformations of the gel are possible due to the dynamic nature of borate and NPs cross-links, which reversibly break and adapt to the movement of the material.

(A) SELF-HEALING BY MECHANICAL CONTACT

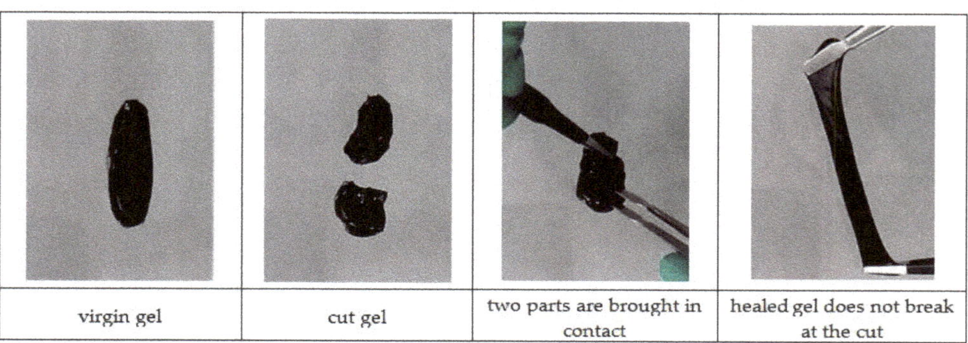

virgin gel	cut gel	two parts are brought in contact	healed gel does not break at the cut

(B) REMOTE SELF-HEALING BY MAGNETIC FIELD

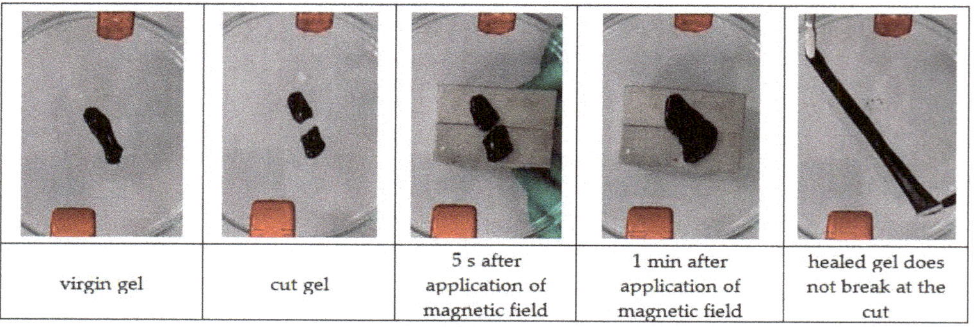

virgin gel	cut gel	5 s after application of magnetic field	1 min after application of magnetic field	healed gel does not break at the cut

Figure 11. Self-healing of hydrogels containing 1 wt% CMHPG, 10 wt% $CoFe_2O_4$ NPs and 0.116 wt% borax (molar ratio borate/monomer units = 1) at pH 10.7 by bringing two parts in contact (**A**) and remotely in the presence of magnetic field (**B**). Temperature: 20 °C.

To quantify the actuation capability of the gels, the force generated by them in the magnetic field was measured. Both ends of the gel stripe were fixed, and the magnetic field acting on the gel was varied by changing the distance from the magnet (Figure 12B). Another end of the gel stripe was connected to a force meter. From Figure 12C, it is seen that the force generated by the gel rises with the increase of the magnetic field gradient, which is evidently due to stronger magnetic forces acting on NPs. It should be noted that the actuation force appears in rather low magnetic fields (up to 0.26 T) and low field gradients (15–20 T/m). Such fields may be created by small magnets, which makes CMHPG/borax/NP gels promising for use as compact actuators with controllable actuation force. In order to generate the force, direct contact of the gel with the magnet is not necessary, which means that the remote actuation (through some surfaces or non-magnetic materials) may be achieved.

A combination of high deformability and strong and fast response is possible due to the utilization of very small cobalt ferrite NPs, which have high magnetization and at the same time increase the mechanical strength of the gels, and the use of dynamic borate cross-links between polymer molecules.

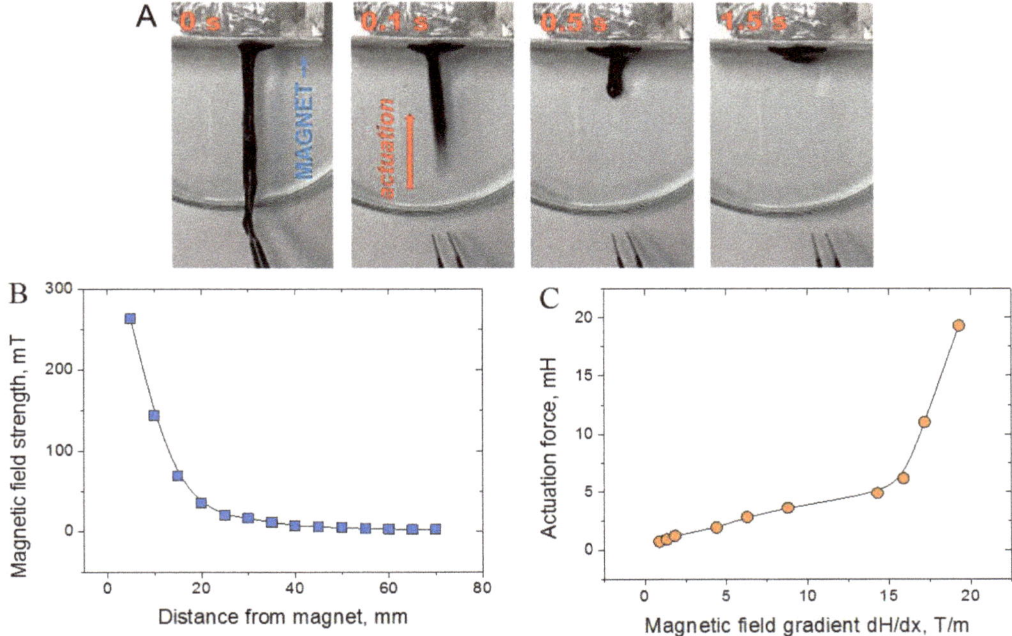

Figure 12. (**A**) Movement of the hydrogel (1 wt% CMHPG, 0.116 wt% borax (molar ratio borate/monomer units = 1), and 10 wt% $CoFe_2O_4$ NPs) in a non-uniform magnetic field of a permanent magnet; (**B**) dependence of the magnetic field strength on the distance from the magnet; (**C**) dependence of the force generated by the gel in the non-uniform magnetic field on the magnetic field gradient (calculated in the center of the gel).

4. Conclusions

Hydrogels with strong and fast response to the magnetic field were prepared from CMHPG polysaccharide dually cross-linked by cobalt ferrite NPs and borate ions, both types of cross-links being labile and easily reformable. It is shown that multifunctional NP cross-links enhance the mechanical properties of the gels. At the same time, the lability of cross-links imparts such useful functional properties as self-healing and reshaping as well as fast response to external stimuli. It is discovered that the mechanical properties of the gels are magnetically tunable: their elastic modulus increases in the external magnetic field, and the strongest increase is observed for the case of the polymer matrix loosely cross-linked by borate. This is explained by the fact that NPs can move inside the polymer matrix cross-linked by labile borate bonds and adapt to the change of the external field. In addition to magnetoresponsive properties provided by NPs, the gels exhibit pH-sensitivity provided by pH-cleavable borate cross-links. Multiresponsive properties and fast reaction to the triggers make the gel prospective for various actuator and sensing applications. Future efforts in this area may be directed at the cross-linking of polymer carboxylic functional groups (for instance, by multivalent metal ions) and obtaining mechanically tough and multiresponsive gels triply cross-linked by metal ions, borate, and NPs.

Supplementary Materials: The following are available online at https://www.mdpi.com/article/10.3390/nano11051271/s1, Video S1: self-healing of gels by mechanical contact, Video S2: self-healing of gels by magnetic field.

Author Contributions: conceptualization, A.V.S. and O.E.P.; methodology, A.V.S.; data acquisition, M.E.S. and D.E.K.; formal analysis, A.V.S., M.E.S., and O.E.P.; investigation, A.V.S., M.E.S., D.E.K., and S.A.B.; resources, S.A.B. and I.V.R.; visualization, A.V.S. and M.E.S.; writing—original draft preparation, A.V.S. and O.E.P.; writing—review and editing, A.V.S. and O.E.P.; supervision, A.V.S. and O.E.P.; project administration, A.V.S.; funding acquisition, A.V.S. All authors have read and agreed to the published version of the manuscript.

Funding: The work was financially supported by Russian Science Foundation (project № 18-73-10162).

Data Availability Statement: The data presented in this study are available on request from the corresponding author.

Acknowledgments: The authors thank Yulia A. Alekhina for her help with the measurement of NPs magnetization curve. We are grateful for Solvay for providing HPG sample (Jaguar HP105) as a generous gift.

Conflicts of Interest: The authors declare no conflict of interest.

Abbreviations

CMHPG	carboxymethyl hydroxypropyl guar
HPG	hydroxypropyl guar
NP	nanoparticle
TEM	transmission electron microscopy
HR-TEM	high resolution transmission electron microscopy

References

1. Dušek, K. *Responsive Gels: Volume Transitions 1*; Springer: Berlin/Heidelberg, Germany, 1993. [CrossRef]
2. Philippova, O.E.; Khokhlov, A.R. Polymer gels. In *Polymer Science: A Comprehensive Reference*; Matyjaszewski, K., Möller, M., Eds.; Elsevier: Amsterdam, The Netherlands, 2012; Volume 1, pp. 339–366.
3. Kuckling, D.; Doering, A.; Krahl, F.; Arndt, K.-F. Stimuli-responsive polymer systems. In *Polymer Science: A Comprehensive Reference*; Matyjaszewski, K., Möller, M., Eds.; Elsevier: Amsterdam, The Netherlands, 2012; Volume 1, pp. 377–413.
4. Gong, J.P. Materials both tough and soft. *Science* **2014**, *344*, 161–162. [CrossRef]
5. Zrinyi, M.; Barsi, L.; Szabo, D.; Kilian, H.G. Direct observation of abrupt shape transition in ferrogels induced by nonuniform magnetic field. *J. Chem. Phys.* **1997**, *106*, 5685–5692. [CrossRef]
6. Filipcsei, G.; Csetneki, I.; Szilgyi, A.; Zrinyi, M. Magnetic field-responsive smart polymer composites. *Adv. Polym. Sci.* **2007**, *206*, 137–189. [CrossRef]
7. Bhattacharya, S.; Eckert, F.; Boyko, V.; Pich, A. Temperature-, pH-, and magnetic-field-sensitive hybrid microgels. *Small* **2007**, *3*, 650–657. [CrossRef] [PubMed]
8. Philippova, O.E.; Barabanova, A.I.; Molchanov, V.S.; Khokhlov, A.R. Magnetic polymer beads: Recent trends and developments in synthetic design and applications. *Eur. Polym. J.* **2011**, *47*, 542–559. [CrossRef]
9. Blyakhman, F.A.; Safronov, A.P.; Makarova, E.B.; Fadeyev, F.A.; Shklyar, T.F.; Shabadrov, P.A.; Armas, S.F.; Kurlyandskaya, G.V. Magnetic properties of iron oxide nanoparticles do not essentially contribute to ferrogel biocompatibility. *Nanomaterials* **2021**, *11*, 1041. [CrossRef]
10. Petrenko, V.I.; Aksenov, V.L.; Avdeev, M.V.; Bulavin, L.A.; Rosta, L.; Vekas, L.; Garamus, V.M.; Willumeit, R. Analysis of the structure of aqueous ferrofluids by the small-angle neutron scattering method. *Phys. Solid State* **2010**, *52*, 974–978. [CrossRef]
11. Messing, R.; Frickel, N.; Belkoura, L.; Strey, R.; Rahn, H.; Odenbach, S.; Schmidt, A. Cobalt ferrite nanoparticles as multifunctional cross-linkers in PAAm ferrohydrogels. *Macromolecules* **2011**, *44*, 2990–2999. [CrossRef]
12. Liao, J.; Huang, H. Review on magnetic natural polymer constructed hydrogels as vehicles for drug delivery. *Biomacromolecules* **2020**, *21*, 2574–2594. [CrossRef] [PubMed]
13. Nagornyi, A.V.; Shlapa, Y.Y.; Avdeev, M.V.; Solopan, S.O.; Belous, A.G.; Shulenina, A.V.; Ivankov, O.I.; Bulavin, L.A. Structural characterization of aqueous magnetic fluids with nanomagnetite of different origin stabilized by sodium oleate. *J. Mol. Liq.* **2020**, *312*, 113430. [CrossRef]
14. Torres-Diaz, I.; Rinaldi, C. Recent progress in ferrofluids research: Novel applications of magnetically controllable and tunable fluids. *Soft Matter* **2014**, *10*, 8584–8602. [CrossRef] [PubMed]
15. Iatridi, Z.; Georgiadou, V.; Menelaou, M.; Dendrinou-Samara, C.; Bokias, G. Application of hydrophobically modified water-soluble polymers for the dispersion of hydrophobic magnetic nanoparticles in aqueous media. *Dalton Trans.* **2014**, *43*, 8633–8643. [CrossRef]
16. Chung, H.-J.; Parsons, A.M.; Zheng, L. Magnetically controlled soft robotics utilizing elastomers and gels in actuation: A review. *Adv. Intell. Syst.* **2021**, *3*, 2000186. [CrossRef]
17. Haider, H.; Yang, C.H.; Zheng, W.J.; Yang, J.H.; Wang, M.X.; Yang, S.; Zrinyi, M.; Osada, Y.; Suo, Z.; Zhang, Q.; et al. Exceptionally tough and notch-insensitive magnetic hydrogels. *Soft Matter* **2015**, *11*, 8253–8261. [CrossRef] [PubMed]

18. Breger, J.C.; Yoon, C.; Xiao, R.; Kwag, H.R.; Wang, M.O.; Fisher, J.P.; Nguyen, T.D.; Gracias, D.H. Self-folding thermo-magnetically responsive soft microgrippers. *ACS Appl. Mater. Interfaces* **2015**, *7*, 3398–3405. [CrossRef]
19. Liu, T.Y.; Hu, S.H.; Liu, K.H.; Liu, D.M.; Chen, S.Y. Preparation and characterization of smart magnetic hydrogels and its use for drug release. *J. Magn. Magn. Mater.* **2006**, *304*, e397–e399. [CrossRef]
20. Qin, J.; Asempah, I.; Laurent, S.; Fornara, A.; Muller, R.N.; Muhammed, M. Injectable superparamagnetic ferrogels for controlled release of hydrophobic drugs. *Adv. Mater.* **2009**, *21*, 1354–1357. [CrossRef]
21. Campbell, S.; Maitland, D.; Hoare, T. Enhanced pulsatile drug release from injectable magnetic hydrogels with embedded thermosensitive microgels. *ACS Macro Lett.* **2015**, *4*, 312–316. [CrossRef]
22. Lin, F.; Zheng, J.; Guo, W.; Zhu, Z.; Wang, Z.; Dong, B.; Lin, C.; Huang, B.; Lu, B. Smart cellulose-derived magnetic hydrogel with rapid swelling and deswelling properties for remotely controlled drug release. *Cellulose* **2019**, *26*, 6861–6877. [CrossRef]
23. Teleki, A.; Haufe, F.L.; Hirt, A.M.; Pratsinisa, S.E.; Sotiriou, G.A. Highly scalable production of uniformly-coated superparamagnetic nanoparticles for triggered drug release from alginate hydrogels. *RSC Adv.* **2016**, *6*, 21503–21510. [CrossRef]
24. Gila-Vilchez, C.; Duran, J.D.G.; Gonzalez-Caballero, F.; Zubarev, A.; Lopez-Lopez, M.T. Magnetorheology of alginate ferrogels. *Smart Mater. Struct.* **2019**, *28*, 035018. [CrossRef]
25. Zhang, H.; Luan, Q.; Tang, H.; Huang, F.; Zheng, M.; Deng, Q.; Xiang, X.; Yang, C.; Shi, J.; Zheng, C.; et al. Removal of methyl orange from aqueous solutions by adsorption on cellulose hydrogel assisted with Fe_2O_3 nanoparticles. *Cellulose* **2017**, *24*, 903–914. [CrossRef]
26. Kurdtabar, M.; Nezam, H.; Rezanejade Bardajee, G.; Dezfulian, M.; Salimi, H. Biocompatible magnetic hydrogel nanocomposite based on carboxymethylcellulose: Synthesis, cell culture property and drug delivery. *Polym. Sci.* **2018**, *60*, 231–242. [CrossRef]
27. Mahdavinia, G.R.; Mosallanezhad, A.; Soleymani, M.; Sabzi, M. Magnetic- and pH-responsive κ-carrageenan/chitosan complexes for controlled release of methotrexate anticancer drug. *Int. J. Biol. Macromol.* **2017**, *97*, 209–217. [CrossRef] [PubMed]
28. Wang, Y.; Li, B.; Zhou, Y.; Jia, D. In situ mineralization of magnetite nanoparticles in chitosan hydrogel. *Nanoscale Res. Lett.* **2009**, *4*, 1041–1046. [CrossRef]
29. Zhang, D.; Sun, P.; Li, P.; Xue, A.; Zhang, X.; Zhang, H.; Jin, X. A magnetic chitosan hydrogel for sustained and prolonged delivery of Bacillus Calmette-Guérin in the treatment of bladder cancer. *Biomaterials* **2013**, *34*, 10258–10266. [CrossRef]
30. Chen, X.; Fan, M.; Tan, H.; Ren, B.; Yuan, G.; Jia, Y.; Li, J.; Xiong, D.; Xing, X.; Niu, X.; et al. Magnetic and self-healing chitosan-alginate hydrogel encapsulated gelatin microspheres via covalent cross-linking for drug delivery. *Mater. Sci. Eng. C* **2019**, *101*, 619–629. [CrossRef]
31. Liu, K.; Chen, L.; Huang, L.; Ni, Y.; Xu, Z.; Lin, S.; Wang, H. A facile preparation strategy for conductive and magnetic agarose hydrogels with reversible restorability composed of nanofibrillated cellulose, polypyrrole, and Fe_3O_4. *Cellulose* **2018**, *25*, 4565–4575. [CrossRef]
32. George, M.; Abraham, T.E. pH sensitive alginate–guar gum hydrogel for the controlled delivery of protein drugs. *Int. J. Pharm.* **2007**, *335*, 123–129. [CrossRef] [PubMed]
33. Murali, R.; Vidhya, P.; Thanikaivelan, P. Thermoresponsive magnetic nanoparticle-aminated guar gum hydrogel system for sustained release of doxorubicin hydrochloride. *Carbohydr. Polym.* **2014**, *110*, 440–445. [CrossRef] [PubMed]
34. Guan, Y.; Zhang, Y. Boronic acid-containing hydrogels: Synthesis and their applications. *Chem. Soc. Rev.* **2013**, *42*, 8106–8121. [CrossRef]
35. Schultz, R.K.; Myers, R.R. The chemorheology of poly(vinyl alcohol)-borate gels. *Macromolecules* **1969**, *2*, 281–285. [CrossRef]
36. Bishop, M.; Shahid, N.; Yang, J.; Barron, A.R. Determination of the mode and efficacy of the cross-linking of guar by borate using MAS 11B NMR of borate cross-linked guar in combination with solution 11B NMR of model systems. *Dalton Trans.* **2004**, 2621–2634. [CrossRef]
37. Leal, J.P.; Marques, N.; Takats, J.J. Bond dissociation enthalpies of U(IV) complexes. An integrated view. *Organomet. Chem.* **2001**, *632*, 209–214. [CrossRef]
38. Soppirnath, K.S.; Aminabhavi, T.M. Water transport and drug release study from cross-linked polyacrylamide grafted guar gum hydrogel microspheres for the controlled release application. *Eur. J. Pharm. Biopharm.* **2002**, *53*, 87–98. [CrossRef]
39. Narayanamma, A.; Raju, K.M. Development of guar gum based magnetic nanocomposite hydrogels for the removal of toxic Pb^{2+} ions from the polluted water. *Int. J. Sci. Res.* **2016**, *5*, 1588–1595.
40. Maaz, K.; Mumtaz, A.; Hasanain, S.K.; Ceylan, A. Synthesis and magnetic properties of cobalt ferrite ($CoFe_2O_4$) nanoparticles prepared by wet chemical route. *J. Magn. Magn. Mater.* **2007**, *308*, 289–295. [CrossRef]
41. Zhang, W.; Jia, S.; Wu, Q.; Ran, J.; Wu, S.; Liu, Y. Convenient synthesis of anisotropic Fe_3O_4 nanorods by reverse co-precipitation method with magnetic field-assisted. *Mater. Lett.* **2011**, *65*, 1973–1975. [CrossRef]
42. Shibaev, A.V.; Doroganov, A.P.; Larin, D.E.; Smirnova, M.E.; Cherkaev, G.V.; Kabaeva, N.M.; Kitaeva, D.K.; Buyanovskaya, A.G.; Philippova, O.E. Hydrogels of polysaccharide carboxymethyl hydroxypropyl guar crosslinked by multivalent metal ions. *Polym. Sci. Ser. A* **2021**, *63*, 24–33. [CrossRef]
43. Sanpo, N.; Tharajak, J.; Li, Y.; Berndt, C.C.; Wen, C.; Wang, J. Biocompatibility of transition metal-substituted cobalt ferrite nanoparticles. *J. Nanopart. Res.* **2014**, *16*, 2510. [CrossRef]
44. Gervits, L.L.; Shibaev, A.V.; Gulyaev, M.V.; Molchanov, V.S.; Anisimov, N.V.; Pirogov, Y.A.; Khokhlov, A.R.; Philippova, O.E. A facile method of preparation of polymer-stabilized perfluorocarbon nanoparticles with enhanced contrast for molecular magnetic resonance imaging. *BioNanoScience* **2017**, *7*, 456–463. [CrossRef]

45. Shibaev, A.V.; Shvets, P.V.; Kessel, D.E.; Kamyshinsky, R.A.; Orekhov, A.S.; Abramchuk, S.S.; Khokhlov, A.R.; Philippova, O.E. Magnetic-field-assisted synthesis of anisotropic iron oxide particles: Effect of pH. *Beilstein J. Nanotechnol.* **2020**, *11*, 1230–1241. [CrossRef] [PubMed]

46. Rani, B.J.; Ravina, M.; Saravanakumar, B.; Ravi, G.; Ganesh, V.; Ravichandran, S.; Yuvakkumar, R. Ferrimagnetism in cobalt ferrite ($CoFe_2O_4$) nanoparticles. *Nano-Struct. Nano-Objects* **2018**, *14*, 84–91. [CrossRef]

47. Franco, A.; e Silva, F.C. High temperature magnetic properties of cobalt ferrite nanoparticles. *Appl. Phys. Lett.* **2010**, *96*, 172505. [CrossRef]

48. Shibaev, A.V.; Aleshina, A.L.; Arkharova, N.A.; Orekhov, A.S.; Kuklin, A.I.; Philippova, O.E. Disruption of cationic/anionic viscoelastic surfactant micellar networks by hydrocarbon as a basis of enhanced fracturing fluids clean-up. *Nanomaterials* **2020**, *10*, 2353. [CrossRef]

49. Shibaev, A.V.; Philippova, O.E. Viscoelastic properties of new mixed wormlike micelles formed by a fatty acid salt and alkylpyridinium surfactant. *Nanosyst. Phys. Chem. Math.* **2017**, *8*, 732–739. [CrossRef]

50. Berlangieri, C.; Poggi, G.; Murgia, S.; Monduzzi, M.; Dei, L.; Carretti, E. Structural, rheological and dynamics insights of hydroxypropyl guar gel-like systems. *Colloids Surf. B Biointerfaces* **2018**, *168*, 178–186. [CrossRef]

51. Maier, H.; Anderson, M.; Karl, C.; Magnuson, K.; Whistler, R.L. Guar, locust bean, tara, and fenugreek gums. In *Industrial Gums. Polysaccharides and Their Derivatives*, 3rd ed.; Academic Press: New York, NY, USA, 1993; pp. 181–226. [CrossRef]

52. Kesavan, S.; Prud'homme, R.K. Rheology of guar and HPG cross-linked by borate. *Macromolecules* **1992**, *25*, 2026–2032. [CrossRef]

53. Inoue, T.; Osaki, K. Rheological properties of poly(vinyl alcohol)/sodium borate aqueous solutions. *Rheol. Acta* **1993**, *32*, 550–555. [CrossRef]

54. Huang, G.; Zhang, H.; Liu, Y.; Chang, H.; Zhang, H.; Song, H.; Xu, D.; Shi, T. Strain hardening behavior of poly(vinyl alcohol)/borate hydrogels. *Macromolecules* **2017**, *50*, 2124–2135. [CrossRef]

55. Drašler, B.; Drobne, D.; Novak, S.; Valant, J.; Boljte, S.; Otrin, L.; Rappolt, M.; Sartori, B.; Iglič, A.; Kralj-Iglič, V.; et al. Effects of magnetic cobalt ferrite nanoparticles on biological and artificial lipid membranes. *Int. J. Nanomed.* **2014**, *9*, 1559–1581. [CrossRef] [PubMed]

56. Ma, X.; Pawlik, M. Role of background ions in guar gum adsorption on oxide minerals and kaolinite. *J. Colloid Interface Sci.* **2007**, *313*, 440–448. [CrossRef]

57. Hurnaus, T.; Plank, J. Behavior of titania nanoparticles in cross-linking hydroxypropyl guar used in hydraulic fracturing fluids for oil recovery. *Energy Fuels* **2015**, *29*, 3601–3608. [CrossRef]

58. Korchagina, E.V.; Philippova, O.E. Multichain aggregates in dilute solutions of associating polyelectrolyte keeping a constant size at the increase in the chain length of individual macromolecules. *Biomacromolecules* **2010**, *11*, 3457–3466. [CrossRef]

59. Szopinski, D.; Kulicke, W.-M.; Luinstra, G.A. Structure–property relationships of carboxymethyl hydroxypropyl guar gum in water and a hyperentanglement parameter. *Carbohydr. Polym.* **2015**, *119*, 159–166. [CrossRef] [PubMed]

60. Khokhlov, A.R.; Grosberg, A.Y.; Pande, V.S. *Statistical Physics of Macromolecules*; AIP-Press: New York, NY, USA, 1994; p. 378.

61. Morris, G.A.; Patel, T.R.; Picout, D.R.; Ross-Murphy, S.B.; Ortega, A.; de la Torre, J.G.; Harding, S.E. Global hydrodynamic analysis of the molecular flexibility of galactomannans. *Carbohydr. Polym.* **2008**, *72*, 356–360. [CrossRef]

62. Barabanova, A.I.; Shibaev, A.V.; Khokhlov, A.R.; Philippova, O.E. Nanocomposite hydrogels with multifunctional cross-links. *Dokl. Phys. Chem.* **2021**, *497*, 41–47. [CrossRef]

63. Sorokin, V.V.; Stepanov, G.V.; Shamonin, M.; Monkman, G.J.; Khokhlov, A.R.; Kramarenko, E.Y. Hysteresis of the viscoelastic properties and the normal force in magnetically and mechanically soft magnetoactive elastomers: Effects of filler composition, strain amplitude and magnetic field. *Polymer* **2015**, *76*, 191–202. [CrossRef]

64. Nardecchia, S.; Jiménez, A.; Morillas, J.R.; de Vicente, J. Synthesis and rheological properties of 3D structured self-healing magnetic hydrogels. *Polymer* **2021**, *218*, 123489. [CrossRef]

65. Molchanov, V.S.; Pletneva, V.A.; Klepikov, I.A.; Razumovskaya, I.V.; Philippova, O.E. Soft magnetic nanocomposites based on adaptive matrix of wormlike surfactant micelles. *RSC Adv.* **2018**, *8*, 11589. [CrossRef]

nanomaterials

Article

Magnetic Properties of Iron Oxide Nanoparticles Do Not Essentially Contribute to Ferrogel Biocompatibility

Felix A. Blyakhman [1,2], Alexander P. Safronov [2,3], Emilia B. Makarova [1,4], Fedor A. Fadeyev [1,5], Tatyana F. Shklyar [1,2], Pavel A. Shabadrov [1,2], Sergio Fernandez Armas [6] and Galina V. Kurlyandskaya [2,7,*]

1   Institute of Natural Sciences and Mathematics, Ural State Medical University, 620028 Ekaterinburg, Russia; feliks.blyakhman@urfu.ru (F.A.B.); emilia1907@yandex.ru (E.B.M.); fdf79@mail.ru (F.A.F.); t.f.shkliar@urfu.ru (T.F.S.); pavel.shabadrov@urfu.ru (P.A.S.)
2   Institute of Natural Sciences and Mathematics, Ural Federal University, 620002 Ekaterinburg, Russia; alexander.safronov@urfu.ru
3   Institute of Electrophysics UB RAS, 620016 Ekaterinburg, Russia
4   Ural Scientific Institute of Traumatology and Orthopaedics, 620014 Ekaterinburg, Russia
5   Institute for Medical Cell Technologies, 620026 Ekaterinburg, Russia
6   SGIKER, Universidad del País Vasco UPV/EHU, 48080 Bilbao, Spain; sergio.fernandez@ehu.eus
7   Departamento de Electricidad y Electrónica, Universidad del País Vasco UPV/EHU, 48940 Leioa, Spain
*   Correspondence: kurlyandskaya.gv@ehu.eus; Tel.: +34-9460-13237; Fax: +34-9460-13071

Citation: Blyakhman, F.A.; Safronov, A.P.; Makarova, E.B.; Fadeyev, F.A.; Shklyar, T.F.; Shabadrov, P.A.; Armas, S.F.; Kurlyandskaya, G.V. Magnetic Properties of Iron Oxide Nanoparticles Do Not Essentially Contribute to Ferrogel Biocompatibility. *Nanomaterials* **2021**, *11*, 1041. https://doi.org/10.3390/nano1104 1041

Academic Editor: Raghvendra Singh Yadav

Received: 29 March 2021
Accepted: 17 April 2021
Published: 19 April 2021

Abstract: Two series of composite polyacrylamide (PAAm) gels with embedded superparamagnetic Fe_2O_3 or diamagnetic Al_2O_3 nanoparticles were synthesized, aiming to study the direct contribution of the magnetic interactions to the ferrogel biocompatibility. The proliferative activity was estimated for the case of human dermal fibroblast culture grown onto the surfaces of these types of substrates. Spherical non-agglomerated nanoparticles (NPs) of 20–40 nm in diameter were prepared by laser target evaporation (LTE) electrophysical technique. The concentration of the NPs in gel was fixed at 0.0, 0.3, 0.6, or 1.2 wt.%. Mechanical, electrical, and magnetic properties of composite gels were characterized by the dependence of Young's modulus, electrical potential, magnetization measurements on the content of embedded NPs. The fibroblast monolayer density grown onto the surface of composite substrates was considered as an indicator of the material biocompatibility after 96 h of incubation. Regardless of the superparamagnetic or diamagnetic nature of nanoparticles, the increase in their concentration in the PAAm composite provided a parallel increase in the cell culture proliferation when grown onto the surface of composite substrates. The effects of cell interaction with the nanostructured surface of composites are discussed in order to explain the results.

Keywords: hydrogel; Fe_2O_3 and Al_2O_3 nanoparticles; gel-based composites; magnetic properties; cells; biocompatibility

1. Introduction

In the past several decades, a number of biologically inert materials were introduced as useful composites for cell culturing for the needs of tissue engineering [1,2]. Among others, magnetic soft materials such as ferrogels (FG) have demonstrated promising applications in biomedicine due to the ability to change their physical properties in response to an external magnetic field [3,4] or to be used as the components of the addressed delivery by the gradient external magnetic field both the encapsulated drugs and soft implants [5]. Furthermore, a magnetic field *per se* may stimulate the biological activity of certain types of cells [6,7] by the enhancement of cell adhesion, proliferation, differentiation as far as modifying properties of the fluids used for cell cultivation [8–10].

In our earlier studies, we tested the biocompatibility of ferrogel based on the polyacrylamide (PAAm) polymer network with embedded maghemite (γ-Fe_2O_3) magnetic nanoparticles (MNPs) fabricated by the electrophysical technique of the laser target evaporation (LTE) [11]. Such parameters as adhesion and proliferation of human peripheral leucocytes or human dermal fibroblasts when grown onto the surface of the FG samples were studied. It became clear that the addition of magnetic nanoparticles to the PAAm gel network always resulted in an increase in cell adhesion and proliferation when grown onto the surface of the gel-based composites [12,13]. In particular, the gradual increase in magnetic nanoparticle concentration in PAAm gel from 0 to 2 wt.% was accompanied by the increase in cell monolayer density by a factor of five for the culture grown onto the FG surface [14].

Meanwhile, the nature of the positive impact of MNPs on the biocompatibility of ferrogels is still not clear. Many direct and/or indirect factors can contribute to this phenomenon. In particular, the addition of magnetic nanoparticles to the network of PAAm gel significantly changes the electrical and mechanical properties of FG that strongly control the cell adhesion and proliferation on the ferrogel substrate [13–15]. The arrangement of MNPs in ferrogel structure is also a potential source of stray magnetic fields [5,16]. This feature, could hypothetically influence biological cell activity as well. At a time, the existence of stray magnetic fields of the gel-based composites can be used for the definition of implant position or degradation state using magnetic field sensors [5,16].

The aim of the present study was to check whether the magnetic MNPs per se affect the ferrogel biocompatibility. In particular, we intended to compare the biocompatibility of series of composite gels with embedded superparamagnetic and diamagnetic nanoparticles, which were almost the same in terms of their characteristic dimensions, morphology, and the properties of their suspensions in water. We analyzed the mechanical properties of these series of composite gels, their electrical potential, and the proliferative activity of human dermal fibroblasts on the surface of these two series of gel-based substrates.

Here, we show that regardless of the superparamagnetic or diamagnetic nature of nanoparticles, the increase in NPs concentration in the nanostructured PAAm gel was accompanied by a similar increase in the cell proliferation on the surface of the gel-based substrates.

2. Materials and Methods

2.1. Synthesis and Characterization of Nanoparticles

Maghemite superparamagnetic (γ-Fe$_2$O$_3$) and alumina (Al$_2$O$_3$) diamagnetic nanoparticles, denoted as MNPs (maghemite) and ANPs (alumina), were accordingly synthesized by means of laser target evaporation (LTE) method. The main details of the fabrication technology and the apparatus of LTE were described in detail in our previous reports [5,11,13]. As a source of laser irradiation for iron oxide rotating target evaporation, we used a Ytterbium (Yb) fiber laser at a 1.07 μm wavelength.

As-synthesized air-dry MNPs and ANPs were studied using transmission electron microscopy (JEOL JEM2100, JEOL Corporation, Tokyo, Japan) operated at 200 kV. For TEM studies MNPs and ANPs were spread onto a Cu grid. Characterization of phase composition of MNPs and ANPs was done using X-ray diffraction technique (Bruker D8 Discover, Bruker Corporation, Billerica, MA, USA) with a graphite monochromator (Cu K$_\alpha$ radiation, wavelength λ = 1.5418 A) and a scintillation detector. The diffractograms were processed by Rietveld full-profile refinement using built-in Bruker software TOPAS-3. The specific surface area (S_{sp}) of nanoparticles was measured using Micromeritics TriStar3000 analyzer (Micromeritics, Norcross, GA, USA).

Magnetic characterization of MNPs and ANPs was done using a superconducting quantum device, SQUID (Quantum Design MPMS-7, Quantum Design Inc., San Diego, CA, USA). All magnetic measurements were made at room temperature. Apart from the measurements of the magnetic hysteresis loops M(H) for MNPs and ANPs, M(H) loops were also measured for all kinds of gels (blank gel, ferrogel field with MNPs, and gels filled with ANPs). For magnetic measurements of MNPs and ANPs, samples of about 5 mg for MNPs and 10 mg for AMPs were used. For gel-based composites, samples of about 60 mg were employed.

The mean hydrodynamic diameter of particles/aggregates in suspensions was measured by the dynamic light scattering (DLS) using Brookhaven ZetaPlus analyzer (Brookhaven

Instruments, Holtsville, NY, USA). The same instrument was used for the measurement of the zeta-potential in suspensions by the electrophoretic light scattering (ELS).

In addition, scanning electron microscopy (SEM) of dried MNPs and ANPs filled gel composites were performed with 20 kV accelerating voltage (JEOL JSM-640, JEOL Corporation, Tokyo, Japan). In order to avoid the surface charging of the polymer composite, a carbon film using a sputtering technique was deposited onto the composite surfaces with a thickness of about 20 nm.

2.2. Synthesis of Composite Gels

Synthesis of composite gels with embedded MNPs and ANPs was performed by free-radical polymerization of acrylamide monomers (AAm, AppliChem, Darmstadt, Germany) dissolved in water suspensions of MNPs and ANPs. The suspensions were prepared by the dispersion of nanoparticles in 5 mM water solution of sodium citrate, which was an electrostatic stabilizer of the suspension. Dispersions were de-aggregated by ultrasound treatment with permanent cooling for 30 min using Cole-Parmer CPX-750 (Cole-Parmer, Vernon Hills, IL, USA) processor operated at 250 W. Remained large aggregates were precipitated by centrifuging for 5 min at 8000 rpm using Hermle Z383 centrifuge (Hermle AG, Gosheim, Germany). De-aggregation in suspensions was controlled using DLS. The concentration of nanoparticles in these stock suspensions was determined using the weight of a dry residue after drying at 90 °C to the constant weight (with the correction for the dissolved sodium citrate). The content of MNPs in the stock suspension was as high as 4.8% by weight, and the content of ANPs was as high as 4.0%.

Stock suspensions were diluted by 5 mM water solution of sodium citrate to provide variation of nanoparticle content in the resulted composite gels. Monomer AAm was dissolved in MNPs and ANPs suspensions in 1.6 M concentration. Cross-linking agent N,N′-methylene bisacrylamide (MBAA, Merck Schuchardt, Hohenbrunn, Germany) was added in 1:100 molar ratio to AAm. An initiator, ammonium persulfate (APS), was used at a 3 mM concentration. Polymerization was performed at room temperature employing N,N,N′,N′-tetra-ethylene-methylenediamine (TEMED, Merck Schuchardt, Hohenbrunn, Germany) in 5 mM concentration as a catalyst.

For the implication of substrates for cell cultivation, the composite gels were synthesized in the shape of thin sheets. Therefore, polymerization was done between two polished glass plates, separated by 0.8 mm spacers and using the mold sealing by a silicon resin. The reaction mixture was poured between the plates using a syringe. It took approximately 5 min for the gelation of the reaction mixture in the mold. The mold was kept for an extra 60 min to complete the polymerization, and then it was disassembled. The resulted sheets of composite gels with embedded MNPs and ANPs were extensively washed in distilled water with daily water renewal in order to remove salts and unreacted monomer until the equilibrium water uptake was achieved. The final contents of nanoparticles in the composite gels swollen to equilibrium were as high as 0.33%, 0.63%, and 1.19% in the gel series with embedded MNPs, and 0.34%, 0.61%, and 1.23% in the gel series with embedded ANPs.

Afterward, the gel sheets were equilibrated for 2 days in Hanks Balanced Salt Solution (HBSS) pH = 6.8–7.2 (PanEco Ltd. Moscow, RF, Russia) with gentamicin (100 mg/L) with a daily renewal of the solution. Then for 2 days, they were kept in 199 solution pH = 7.0–7.4, osmolality 300 ± 20 mmol/kg, and a buffering capacity of ≤ 1.5 mL (PanEco Ltd. Moscow, RF, Russia) with gentamicin (100 mg/L) with daily renewal. Then the substrates in the shape of disks (13 mm in diameter) were cut from the gel sheets to fit the wells of the standard 24-well polystyrene plate for cell culturing. Prior to their use in the cell culture, the gel-based substrates were sterilized in an autoclave at 121 °C for 20 min.

2.3. Mechanical Properties and Electrical Potential of Nanostructured Gels

The elastic properties of gels and FGs were determined using a laboratory setup for mechanical tests [12,15]. Cylindrical samples were placed between two plates. One was connected rigidly to the actuator of a linear electromagnetic motor, and the other was connected to a precision strain-gage sensor. The motor induced compression strain with a magnitude of up to 20% in steps of 2% of the initial gel length. Stress–strain dependences were plotted as a result of these tests, and their linear sections were used to determine the Young modulus for the investigated materials.

The electrical potential of gels was determined using the standard technique, which is routinely applied to living cells. Specifically, two identical silver-chloride electrodes in glass micropipettes TW150F-6 (World Precision Instruments, Sarasota, FL, USA) with a tip diameter of ~1 μm filled with a 3 M KCl solution were used. One electrode was placed in the solution surrounding the gel, and the other one was introduced into the studied sample. The potential difference was measured with an INA 129 instrumentation amplifier (Burr-Brown, Dallas, TX, USA).

2.4. Human Dermal Fibroblasts Culture

The lines of human dermal fibroblasts were obtained from donor skin, as described previously [10,13,14]. Briefly, tissue biopsies were collected from patients (who have given the informed contest) during surgery. The study was approved by the Ethics Committee of the Institute of Medical Cell Technologies, Ekaterinburg. Skin biopsies were cut into small pieces, and cells were extracted by tissue dissociation method. Extracted fibroblasts were grown in culture flasks (Nunc, Roskilde, Denmark) at 37 °C and 5% CO_2. Cells were passaged when they covered 80% of the flask surface area. Then sub-culturing cells were treated with 0.25% trypsin-EDTA solution (Gibco, Thermo Fisher Scientific, Inc., Waltham, MA, USA). Cell number was estimated by cell counter TC-20 device (Bio-rad, Hercules, CA, USA). The cell viability was measured by staining with trypan blue. Fibroblasts were stored in liquid nitrogen.

2.5. Cells Proliferation Assays on Nanostructured Gel-Based Composites

Thawed cells were passaged 2 times before they were used in experiments. Fibroblasts (fifth passage) were detached from the flask surface by trypsin. After detachment of cells, trypsin was neutralized by fetal bovine serum. Gel discs with MNPs, ANPs, and blank gel discs were placed into the wells of 24-well tissue-treated culture plates (Nunc, Roskilde, Denmark). Some wells were left empty to be used as controls of cell growth on tissue-treated culture plastic. Fibroblasts were re-suspended in a growth medium. The suspension was dispensed in wells with a seeding density of 3000 viable cells/cm^2 (viability $\geq$ 95%) for all types of substrates. Plates were incubated at 37 °C, in the 5% CO_2 atmosphere for 96 h. All cell experiments were performed without the application of the external magnetic field. However, we did not shield possible laboratory magnetic fields, which usually do not exceed one Oersted strength. After incubation, cells in the monolayer were fixed with 2.5% glutaric aldehyde. Fibroblasts on the substrate surfaces were visualized by staining cell nuclei with 4′,6-diamidino-2-phenylindole (DAPI, Sigma-Aldrich, St. Louis, MO, USA) and cytoplasm with 0.3% pyrazolone yellow solution. Cells were counted using a fluorescent Axio Lab A1 FL (Carl Zeiss, Oberkochen, Germany) microscope. The analysis for nine fields of view for each sample at "$\times 100$" magnification was done. The number of cells in images was estimated using the ImageJ software (Wayne Rasband, NIH, Bethesda, MD, USA).

The experiment was performed in 6 replicates. We used the non-parametric Mann–Whitney U-test in order to compare the statistical significance of the difference between two independent groups with a level of significance set at 0.05. Statistical data processing was performed employing the application software package "STATISTICA 6.0" (Statsoft, Dell, Round Rock, TX, USA).

3. Results

3.1. Properties of Nanoparticles and Precursor Suspensions

Both MNPs and ANPs embedded into gel-based substrates were synthesized by the LTE method, which provided their basic similarity despite different chemical compositions. Figure 1 shows TEM images of MNPs and ANPs with histograms of particle size distributions (PSD) given in the insets, where Pn (d) is the number average lognormal distribution function.

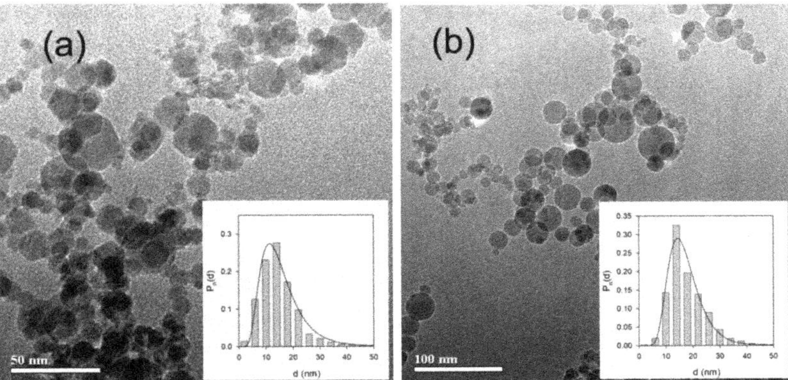

Figure 1. TEM images of LTE MNPs (**a**) and ANPs (**b**). Inserts: histograms of PSDs; lines give fitting of PSD with lognormal distribution function.

In both cases, the shape of nanoparticles was very close to being spherical, which was the result of their condensation from the vapor phase. The histograms were plotted as a result of the graphical analysis of 2572 images in the case of MNPs and 2206 images in the case of ANPs. In both cases, the diameter of nanoparticles fell within the 5–50 nm interval with a probability maximum of 14 nm. The histograms were nicely fitted by a lognormal distribution function:

$$P_n(d) = C\exp\left(-\frac{1}{2}\left[\frac{\ln(d/d_0)}{\sigma}\right]^2\right) \tag{1}$$

Parameters of the distributions are given in Table 1; they were very close to each other. Using the PSD parameters the values of number-average and weight-average diameter could be calculated as the first and the third moments of the distribution. These values are given in Table 1. Summarizing this set of data, we may conclude that from the viewpoint of their shape and characteristic dimensions, MNPs and ANPs were very similar. The similarity preserves concern their properties in water suspensions.

Table 1. Selected parameters of PSD for MNPs and ANPs in air-dry powder and water suspension.

Batch	Air-Dry Powder					Water Suspension		
	d_0 (nm)	σ	d_n (nm)	d_w (nm)	d_i (nm)	d_{hd} (nm)	ζ_0 (mV)	ζ (mV)
MNP	14.1 ± 0.4	0.46 ± 0.02	14.9	28.3	40.3	68 ± 3	35 ± 4	-66 ± 5
ANP	14.5 ± 0.4	0.35 ± 0.02	17.9	24.1	34.3	62.9 ± 0.4	40 ± 5	-59 ± 4

d_n: number-average diameter, calculated at the first moment of PSD; d_w: weight-average diameter, calculated at the third moment of PSD; d_i: intensity-average diameter, calculated at the fifth moment of PSD; d_{hd}: hydrodynamic diameter in suspension by DLS; ζ_0: zeta-potential in self-stabilized suspension by ELS; ζ: zeta-potential in suspension stabilized with sodium citrate by ELS.

The crystal structure of MNPs corresponded to single-phase maghemite with an inverse spinel cubic lattice (period a = 0.8357, 3 nm), space group Fd3m. The average size of the coherent diffraction domains was 12 nm estimated using the Scherrer approach [17]. The crystal structure of ANPs was γ-Al$_2$O$_3$ with a cubic lattice (period a = 0.7924, 7 nm). The average size of the coherent diffraction domains in ANPs was 17 nm.

The specific surface area (S_{sp}) of nanoparticles, determined using low-temperature adsorption of nitrogen (BET method), was 93 m^2/g for MNPs and 78 m^2/g for ANPs. These values could be used for the evaluation of the average diameter of particles according to the following Equation (2) [18], which relates the diameter of a sphere to its surface (S) and density (ρ):

$$d_{BET} = \frac{6}{\rho S_{sp}} \tag{2}$$

The crystallographic density of particles was 4.6 g/cm^3 in the case of MNPs, and it was 3.6 g/cm^3 in the case of ANPs. Given these values and the values of the specific surface area, the average diameter of MNPs was evaluated to be 14.0 nm, and the diameter of ANPs was 21.4 nm.

The specific feature of metal oxide nanoparticles synthesized by the method of gasphase physical dispersion is their self-stabilization in water suspensions [19]. It means that stable suspensions of these nanoparticles in water can be prepared without using special stabilizers. The basic reason for that is the formation of traces of nitrogen oxides during the evaporation of the precursor oxide target in the air. This process is initiated by high temperature (approximately 10^4 K) at the laser spot on the surface of the target during its evaporation. Nitrogen oxides then provide traces of metal nitrides on the surface of condensed nanoparticles. In water suspensions, nitrides dissociate, and the surface gets a net positive electrical charge due to the metal ions while nitride ions migrate to the water medium as counterions of the double electrical layer. It provides the colloidal stability of the suspension, which can be characterized by the value of the zeta-potential of the suspension. The values of zeta-potential for the self-stabilized suspensions of MNPs and ANPs are given in Table 1. They are positive both for MNPs and ANPs and close to each other. It indicates that the physical properties at the surface of these two batches of nanoparticles are likely common.

Meanwhile, self-stabilization of suspension of LTE nanoparticles is a limiting factor, and it can not provide the colloidal stability of these suspensions in the solutions with high ionic strength, which is typical in biological systems. Therefore, suspensions for biomedical applications were additionally stabilized by sodium citrate. Citrate ions adsorb on the surface and reverse their electrical charge from positive to negative; sodium cations become the counterions of the double electrical layer. The colloidal stability of the citrate-stabilized suspensions increases. Table 1 gives the values of zeta-potential in citrate-stabilized suspensions of MNPs and ANPs, which were used for the preparation of composite gel-based substrates for cell culture. In both cases, zeta-potential was highly negative, indicating that in both cases, the suspensions were very stable, and their colloidal properties were much alike.

Figure 2 shows the PSD plots obtained by DLS for the suspensions of MNPs and ANPs in water. It is noticeable that the plots were very close to each other. The median of the PSD, which is commonly referred to as the hydrodynamic diameter (d_{hd}) of a nanoparticle in a suspension, is given in Table 1 for the suspensions of MNPs and ANPs. The hydrodynamic diameter is an apparent value, which is the product of the Einstein equation for the diffusion coefficient. The diffusion coefficient is a value, which is measured by DLS. It is noticeable that the hydrodynamic diameters of MNPs and ANPs in their suspensions were close to each other.

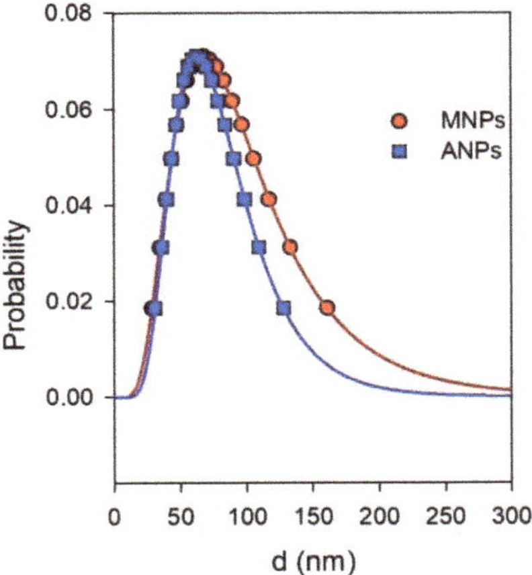

Figure 2. PSD of MNPs and ANPs in water suspensions measured by DLS. The probability function corresponds to the intensity of dynamic light scattering related to the fraction of nanoparticles with a certain diameter.

The average hydrodynamic diameter of the ensemble of nanoparticles is related to the fifth moment of PSD, which is given in Table 1 as the intensity-average diameter d_i. There was a certain difference between the values of d_{hd} and d_i. This difference partly stems from the solvation layers on the surface of nanoparticles in water medium. In addition, some small aggregates of nanoparticles still could remain in the suspension after centrifuging. The values of the fifth moment of distribution are highly sensitive even to the presence of a small number of aggregates. However, their total fraction is not dominant.

Thus, the analysis of the properties of MNPs and ANPs shows that these two batches of LTE nanoparticles are very similar concerning their shape, dimensions, surface properties in water suspension, and colloidal stability. The only difference between them is in their magnetic properties.

3.2. Magnetic Properties of Fe_2O_3 and Al_2O_3 Nanoparticles

Figure 3 shows magnetic hysteresis loops M(H) of dry as-prepared MNPs and ANPs. One can see a huge difference in their magnetic responses. γ-Fe_2O_3 magnetic nanoparticles are a well-known and well-studied biocompatible magnetic nanomaterial with high saturation magnetization. As expected for superparamagnetic nanoparticles of the size under consideration, the saturation magnetization (M_s) was significantly reduced in comparison with the bulk case [20]. The S-shape of the hysteresis loop and very low coercivity at room temperature are typical features of superparamagnetic MNPs. In fact, the saturation was not achieved in a magnetic field of 19 kOe, but a rough estimation of the magnetization value in the field of about 19 kOe confirmed that it was consistent with the values for M_s previously obtained for MNPs of this size.

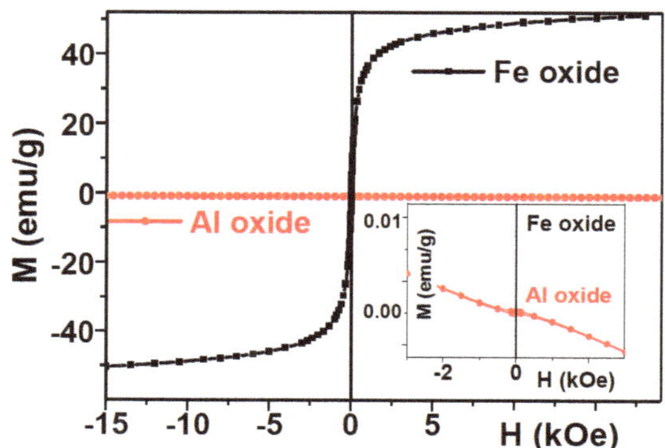

Figure 3. Magnetic hysteresis loops of Fe_2O_3 and Al_2O_3 LTE as-prepared nanoparticles. Inset shows the same responses in low magnetization scale.

ANPs had a clear diamagnetic response (linear magnetization dependence on the value of the applied magnetic field with a negative slope and very small value of the resulting magnetic moment), as expected for this kind of material. Again, the difference was huge for the scale appropriate for M(H) hysteresis loop of γ-Fe_2O_3 magnetic nanoparticles, and the magnetic field dependence of the magnetization of Al_2O_3 nanoparticles was not visible. However, it became noticeable at a low magnetization scale (Inset Figure 3).

3.3. Structure and Magnetic Properties of Composite Gels

We have focused on the properties of the suspensions of nanoparticles because these features strongly determine the arrangement of the embedded NPs in the composite gels. Currently, there is no reliable experimental method for the direct observation of the arrangement of the embedded nanoparticles in swollen composite gels. Any microscopic technique presumes certain dehydration of the gel, which inevitably would disturb its intact structure. In this respect, we have to rely somehow on the indirect features of the synthetic procedure and the macroscopic properties of the fabricated composite gels.

As shown above, the precursor suspensions of MNPs and ANPs were almost de-aggregated and contained mostly individual nanoparticles with a fraction of small aggregates. These suspensions were transparent, which meant that there were no moieties larger than the wavelength of the visual light—approximately 600 nm. Moreover, there was no visual opalescence in the suspensions, and consequently, the limit might be lowered down to a quarter of a wavelength, which was approximately 150 nm. This estimation of the upper limit in the characteristic dimension of moieties present in the suspensions correlated well with the PSD given in Figure 2.

During the synthesis, the visual appearance of the precursor suspensions did not change. They remain transparent at the stage of the mixing with reactants and at the stage of gelation as well. Visually, the gels look the same as the precursor suspensions. It made us assume that the distribution of the nanoparticles that were established in the precursor suspensions preserved, in general, in the resulted composite gels. Figure 4a shows the general view of all gel-based substrates.

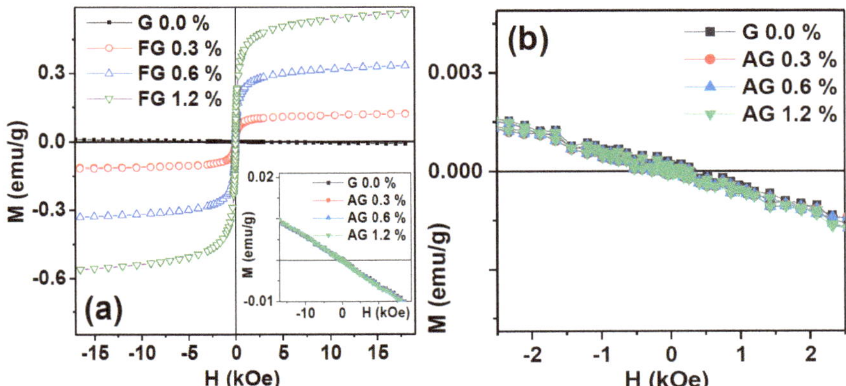

Figure 4. Magnetic hysteresis loops of gels and gel-based composites: MNP-based (**a**) and ANP-based (**b**) materials. G: blank gel; FG: Fe_2O_3-based ferrogels; AG: Al_2O_3-based gel composites.

The mesh size of the PAAm network in composite gels was estimated based on the value of the swelling ratio of gel. The swelling ratio (α) is given by Equation (3):

$$\alpha = \frac{m_g - m_{dr}}{m_{dr}} \tag{3}$$

where is m_g the mass of a gel sample and m_{dr} is the mass of the dry residue after the gel sample is dried to the constant weight.

However, in composite gels the value of m_{dr} includes both the mass of the dry polymeric network and the mass of the embedded solid nanoparticles. To obtain the swelling ratio of the polymeric network (α_p), the correction should be done according to Equation (4):

$$\alpha_p = \frac{\alpha}{1 - \omega} \tag{4}$$

where ω is the weight fraction of solid nanoparticles in the dry residue.

The swelling ratio of composite PAAm gels determined using Equations (3) and (4) was found to be 12.6 ± 0.4 both for gel series with MNPs and ANPs independently of the content of nanoparticles.

The swelling ratio of a hydrogel network is, in other words, the water uptake of the network, i.e., the amount of water that the network can hold within it. Its value, in principle, depends on the mesh size of the network and the energy of interaction between water and polymeric sub-chains. This dependence is given by the Flory–Rehner equation [21,22]. Using the equation, one can calculate the average number of monomer units in the sub-chain of gel network based on the value of the swelling ratio according to Equation (5):

$$N_c = \frac{V_1(0.5\alpha_0\alpha_p^{-1} - \alpha_0^{1/3}\alpha_p^{-1/3})}{V_2(\ln(1 - \alpha_p^{-1}) + \alpha_p^{-1} + \chi\alpha_p^{-2})} \tag{5}$$

where N_C is the number of monomer units among cross-links of the network; V_1 and V_2 are the molar volumes of water and of PAAm, respectively; χ is Flory–Huggins parameter for a binary interaction between water and PAAm; α_0 is the swelling degree of PAAm gel as provided by the composition of the reaction mixture in the synthesis. We used $V_1 = 18$ cm^3/mol (water), $V_2 = 56.2$ cm^3/mol (PAAm) and $\chi = 0.12$. The last two values were calculated by means of quantum mechanics molecular modeling software package CAChe7.5. As it was given above, the equilibrium swelling ratio was 12.6; the swelling ratio of as-synthesized gel was 10. Given these values, the average number of monomer units among cross-links of the network according to Equation (5) was found to be $N_C = 64$.

Considering that PAAm sub-chains of the network are random Gaussian coils with hindered rotation, one can evaluate the mean square distance between adjacent cross-links of the network according to Equation (6) [23]:

$$\left\langle R^2 \right\rangle = Na^2 \frac{(1 - \cos\,J)\,(1 - \cos\,f)}{(1 + \cos\,J)\,(1 + \cos\,f)} \tag{6}$$

where N is the number of bonds in the polymeric sub-chain, a is the bond length, θ is the bond angle, and φ is the angle of hindered rotation. In a PAAm polymeric chain, the bond length $a = 0.154$ nm for the ordinary C–C bond, the bond angle $\theta = 109.5°$, the angle of hindered rotation $\varphi = 120°$, and $N = 2N_C$. It is two-fold larger than N_C as it includes the bonds in monomer units and bonds between them. Given these values, the average distance among the cross-links according to Equation (6) was found to be 4.3 nm. This value was the effective mesh size of the PAAm network in the composite gels. It was evident that the mesh size of the network is much smaller than the characteristic dimensions of MNPs and ANPs (see Table 1). Thus, we may conclude that in both series, nanoparticles were entrapped in the gel network and were not able to leave it or to migrate along it.

The substantial difference between these two series of gel-based composites was in their magnetic properties. The main methodological advantage of the present study is the employment of the same synthesis technique for the fabrication of two different types of nanoparticles, which we denote as ferromagnetic (iron oxide) and non-ferromagnetic (aluminum oxide). Here we keep in mind the most general division of magnetics is in ferro-, antiferro-, para-, and dia-magnetics [20]. The bulk magnetite is a ferromagnetic material with rather high saturation magnetization [20]. However, in the case of nanosized particles, the saturation magnetization of iron oxide MNPs becomes strongly dependent on their size [11]. Alumina is a dia-magnetic material. The properties of as-prepared MNPs and ANPs were discussed above. Now let us analyze the magnetic properties of gel-based composites measured at room temperature (see Figure 4).

Figure 4 gives the comparison of magnetic properties of gel-based composites with MNPs and ANPs. An appropriate scale helps to visualize the difference in the best way. One can see that, as expected for the mainly water containing sample, hydrogel (G 0.0%) had quite a weak diamagnetic signal, which was much lower than the signals of MNPs in the concentrations under consideration.

Despite small concentrations of MNPs the magnetic properties of ferrogels were very different for the range of selected concentrations (Figure 4a); the higher is the concentration, the higher is the magnetic moment. The shape of the hysteresis loops of ferrogels was quite similar to the shape of the MNPs, typical for superparamagnetic particle responses with close to zero coercivity. S-shaped geometry but a lower saturation magnetization value was observed in comparison with MNPs.

Meanwhile, for the same concentrations of the ANPs in gel-based composites (Figure 4b), we could not see the same tendency; at the available accuracy of magnetic measurements, the magnetic responses of all composite gels with embedded ANPs showed very similar typical diamagnetic linear responses, which means that the magnetization declined with respect to the applied external magnetic field with close values of the negative slope.

3.4. Effect of NPs Concentration on the Young's Modulus and Electrical Potential in Composite Gels

Figure 5 shows the general view of the gel-based composites with different content of the embedded NPs (Figure 5a) and the effect of iron oxide or aluminum oxide NPs concentration on the value of Young's modulus in the PAAm gels (Figure 5b). One can see that the addition of particles at a minimum concentration (0.3%) to the polymer network resulted in a significant increase in the composite rigidity. Moreover, the contribution of MNPs or ANPs to the modulus at the lowest concentration in the gel network was approximately the same.

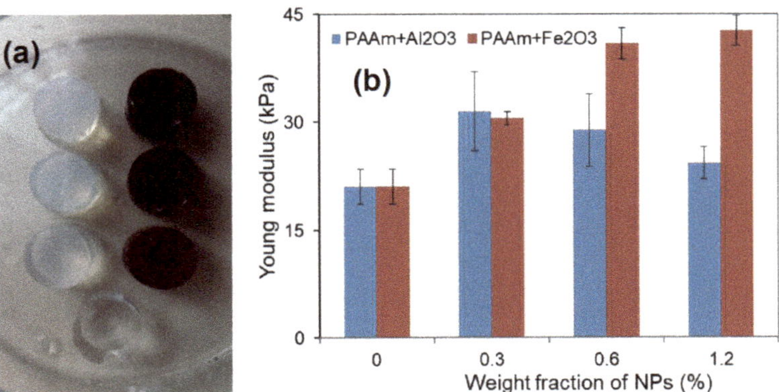

Figure 5. General view of gel-based samples: (**a**) ANPs-based composites, (**b**) MNPs-based composites, and the completely transparent sample is a blank gel; the diameter of the gel samples was equal to 13 mm. The values of Young's modulus for PAAm hydrogels filed with different concentrations of MNPs or ANPs. Data presented as mean value and standard deviation bar ($n = 5$).

At higher content of nanoparticles in composite gels, their concentration connection with Young's modulus was different. While in the case of MNPs the rigidity of composite gels increased further with the concentration increase, the gels filled with ANPs showed a tendency to decrease Young's modulus. The increase om the ANPs concentration from 0.3% to 1.2% was accompanied by the subsequent decay of rigidity of composite gels so that at ANPs concentration of 1.2%, the modulus was significantly smaller than that one at 0.3%. At the same range of MNPs concentrations, the rigidity of ferrogels was gradually increasing. Noteworthy, at nanoparticle concentrations of 0.6% or 1.2%, the Young's modulus of composite gels with embedded MNPs was significantly larger in comparison with the composite gels with embedded ANPs.

Figure 6 shows the general view of the gel and composite gels and the dependence of the electrical potential in gel-based composites on the concentration of MNPs or ANPs. One can see that the addition of nanoparticles to the PAAm gel at a minimum concentration (0.3%) resulted in an increase in the potential in absolute value so that both composite gels became more electronegative to approximately the same extent. A further increase in nanoparticle concentration led to a different change of the electrical potential for the gels with MNPs or ANPs. In gels filled with MNPs, the potential became progressively negative with the increase in MNPs content. At the same time, the opposite trend appeared in gels filled with ANPs. While the content of ANPs further increased, the negative values of the potential diminished. Furthermore, the differences in potential for the two series of composite gels were greater for the higher concentration of the particles.

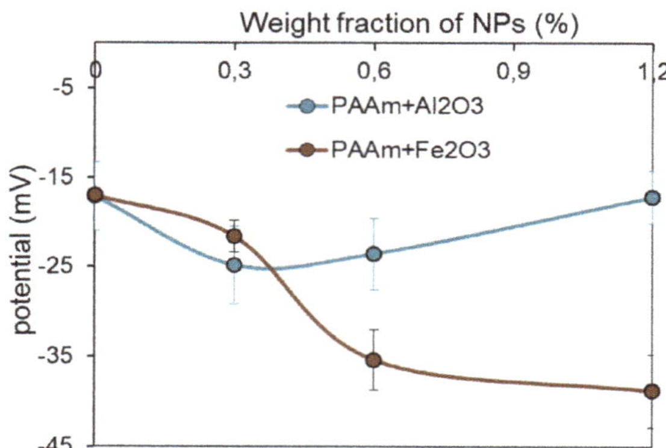

Figure 6. Dependence of electrical potential in gels on the concentration of Fe_2O_3 or Al_2O_3 nanoparticles. Data presented as mean value and SD bar ($n = 7$).

3.5. Effect of NPs Concentration on the Proliferation Activity of Cells

Figure 7 shows typical examples of the human dermal fibroblast cultures grown onto the surface of different PAAm gel-based substrates after appropriate staining. The first important observation is that the cell culture under consideration could be successfully grown onto all kinds of composite substrates described above. The second conclusion was that MNPs and AMPs made essential positive contributions to cellular proliferation in the described conditions.

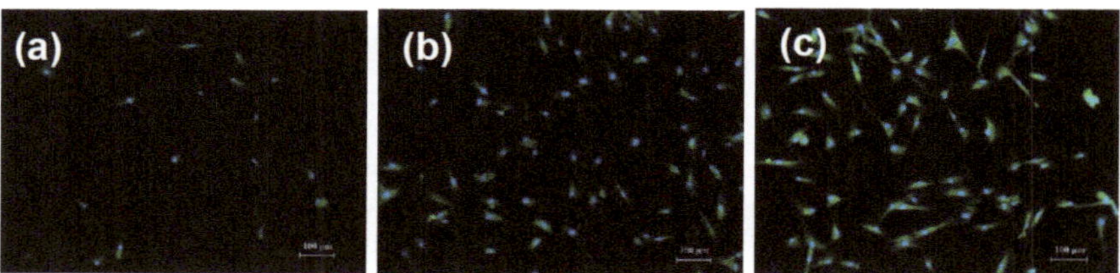

Figure 7. Fibroblasts on the surface of gel-based substrates. Cells were cultured for 96 h and fixed for fluorescent microscopy. Cell nuclei were stained with DAPI (4,6-diamidino-2-phenylindole), the cytoplasm was stained with pyrazolone yellow. (**a**) blank gel substrate; (**b**) gel-based substrate with 1.2% Al_2O_3 nanoparticles; (**c**) gel-based substrate with 1.2% Fe_2O_3 nanoparticles.

Table 2 displays the results of fibroblasts counting onto the surface of the tested gel-based substrates after four days of incubation. The data were obtained based on the analysis of nine fields of view for each biological sample, and they are presented as X and m, where m is the error of the mean (X, $n = 54$). To assess the viability evaluation of the used fibroblasts, cells were also seeded onto the standard substrate—tissue-culturing polystyrene (TCPS). At the end of incubation, the number of cells was $21,000 \pm 1000$ per cm^2, which corresponded to good cell viability.

Table 2. Proliferation of fibroblasts on the surface of PAAm gels with different concentrations of Fe_2O_3 or Al_2O_3 nanoparticles.

Nanoparticles Concentration (wt.%)	Cell Monolayer Density (Cells per cm^2)	
	PAAm 1:100 + Al_2O_3	PAAm 1:100 + Fe_2O_3
0.0	400 ± 100	
0.3	900 ± 100 *	1400 ± 400 *
0.6	1900 ± 300 *#	1400 ± 300 *
1.2	2400 ± 300 *#	2500 ± 400 *†

Symbols display significant differences with $p < 0.05$; *: between composite gels and blank gels (PAAm, concentration of NPs = 0.0%); #: between Al_2O_3 composites (NPs = 0.6% or 1.2% and NPs = 0.3%, respectively); †: between Fe_2O_3 composites (NPs = 1.2% and NPs = 0.3% or 0.6%, respectively).

According to the obtained data, the addition of both MNPs and ANPs to the PAAm gel network led to a significant and almost the same increase in the proliferation activity of fibroblasts. Importantly, at the initial NPs concentration of 0.3%, both of the gel-based composites with MNPs and ANPs had similar Young's modules and electrical potentials (see Figures 4 and 5). In general, the effect of NPs concentration on the cell proliferation rate grew gradually in both series of composite gels.

The contribution of specific NPs to cell proliferation rate must be compared in the gel-based substrates with approximately the same mechanical and electrical properties. According to the results presented above (see Figures 4 and 5), the gels filled with ANPs of 0.3% and 0.6% had a similar Young's modulus and electrical potential. In the case of ferrogels, these are substrates with MNP concentrations of 0.6% and 1.2%. One can see in Table 2 that the gain of the increase in cell proliferation rate from 0.3% to 0.6% for gels with Al_2O_3 NPs, and from 0.6% to 1.2% for gels with Fe_2O_3t NPs, was significant and approximately the same.

4. Discussion

In the present study, the biological activity of cells was investigated for two types of gel-based composites: PAAm hydrogels filled with γ-Fe_2O_3 or Al_2O_3 nanoparticles in different concentrations. Despite the difference in the chemical composition of the nanoparticles (alumina or maghemite), these two series of substrates were similar in many senses. Both were based on PAAm gel with the same concentration of monomer (1.6 M) and the same concentration of cross-linker (1:100 molar ratio to monomer). The nanoparticles were synthesized by the same method (LTE) using the same laboratory installation; both types of nanoparticles were spherical, non-agglomerated, and had close characteristic dimensions (10–40 nm) and parameters of PSD. In both series of gel-based substrates, the inner distribution of NPs in the PAAm matrix was approximately the same.

Figure 8 shows typical SEM images of the dried surfaces of MNPs and ANPs based composites. We provide low magnification data in order to emphasize the difficulties of the structural investigation of these types of composites. One can clearly see that the dehydration process resulted in the appearance of strong stresses and surface relief. Even so, these observations confirm the expectation that at a large scale, both MNPs and ANPs were distributed in quite a homogeneous manner without the formation of very large agglomerates.

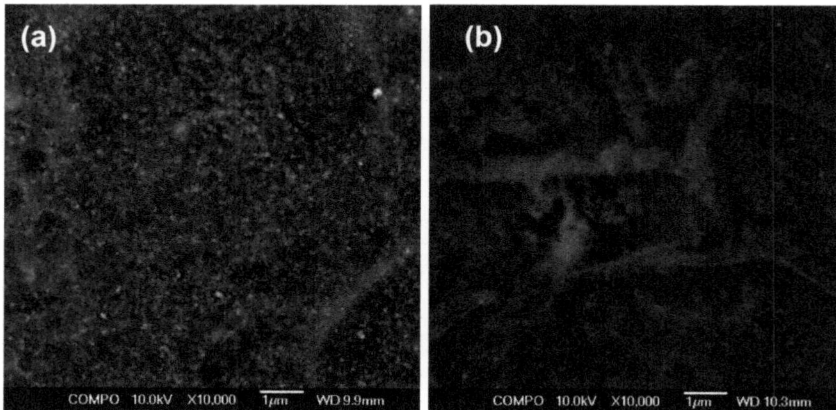

Figure 8. General view of the surface of dry of PAAm gel-base composites with Fe$_2$O$_3$ (**a**) or Al$_2$O$_3$ (**b**) nanoparticles (1.2 wt.%). Scanning electron microscopy.

As discussed in previous work [10,13], as the potential applications of gels and composite gels are under active development, at present, there is a special need for the development of a new technique for gel-based composites characterization. Existing techniques are not adequate for quantitative evaluation of the structural arrangement of the MNPs or ANPs inside the gel. However, understanding the internal structure of filed gels is important, and we used earlier developed techniques for the evaluation of the structure of dried composites with MNPs and AMPs [10,13].

The magnetic characterization revealed that two series of gel-based composite substrates were very different in terms of their magnetic properties. While gels with MNPs showed ferromagnetic behavior with magnetization proportional to the content of iron oxide, gels with ANPs are diamagnetic, it is clear that any difference in the cell cultivation results of gel substrates with embedded ANPs could not be assigned to the difference of magnetic properties of gel composites with different concentration of ANPs.

The possible influence of magnetic interactions between iron oxide nanoparticles could be taken into account for the gel-based composites with embedded MNPs. However, as in the present study, all cell proliferation experiments were performed without the application of the external magnetic field. Although it is unlikely that they might provide a substantial contribution, this point requires further investigation as many other cases involving small magnetic fields [24]. From the low field behavior of magnetization of MNPs based composites, one can estimate the contribution of the stray fields created by the iron oxide MNPs in the field of a few Oe as almost negligible. In the future, it would be interesting to make an evaluation of the cell proliferation rate in the same conditions as in the present case but under application of the external magnetic field of the order of 100 Oe (at least).

The electrical potential and the mechanical properties of composite gels were almost the same at the lowest content of NPs (0.3%), no matter whether NPs were magnetic (maghemite) or diamagnetic (alumina). Meanwhile, different trends in Young's modulus and in the electrical potential were found as the content of NPs increased (see Figures 5 and 6). In the case of gel-based substrates with maghemite, the modulus and the electrical potential enlarged if the content of NPs increased from 0.3% to 1.2%. The opposite trend of the diminishing of the modulus and the electrical potential was found in the case of gel-based substrates with alumina. The decrease in Young's modulus with the concentration of ANPs in gel-based composites cannot be easily explained. Most likely, it might be the result of some specific interaction of ANPs with the PAAm network, or it might stem from the different trends of aggregation of MNPs and ANPs inside the gel structure. A qualitatively similar effect was reported in [25] for the influence of alumina nanoparticles (mean diameter of 30 nm) on the elasticity of composite PAMPS/PAAm gels. For now, we cannot give a plausible explanation for the underlying mechanism but can only state the difference in the mechanical behavior between two series of gel-based composites.

However, despite the marked difference in the magnetic properties, the electrical potential, and the elasticity between magnetic and diamagnetic gel-based substrates, the proliferation of human fibroblasts on these two platforms revealed qualitatively the same result. In both series of gel-based substrates, the increase in NPs content resulted in the enhancement of the proliferation activity of fibroblasts. In meant that the magnetic properties of maghemite NPs did not play a significant role in the determination of the biological activity of fibroblasts on the PAAm gel-based substrates with embedded NPs. In other words, a key factor other than the magnetic properties of the gel-based substrates governs the biocompatibility of nanostructured composites.

Fibroblasts are a type of mechanically sensitive cells, which are able to accept and transform mechanical signals for their vital activity [26–28]. The feasible mechanism of the transduction of mechanical signals into the cell involves heterodimeric trans-membrane receptors, which are referred to as integrins [26,29]. Clusterization of the integrins at the cell membrane in response to the signal triggers the trans-membrane accumulation of the 20 mediators of the signal transduction, including cytoskeleton proteins, which regulate the functioning of the cell by means of the activation of corresponding genes [26,30,31]. It is also assumed that the mechanical signals are transmitted into the cell through the mechanically gated cationic channels through the stretch-activated channels (SAC channels) [32].

In vitro experiments have shown that the surface geometry pattern played a significant role in the realization of the mechanical transduction phenomenon in cells together with chemical composition, wettability, surface charge, elasticity, and other factors [33–35]. Thus, it was shown that the specific geometry pattern of the surface of solid nanocomposite materials (dimples, bumps, their shapes), including the characteristic dimensions of the roughness and its periodicity, could initiate the proliferation and the differentiation of cells [36–38]. Furthermore, it was demonstrated that the specific patterns of the surface of a nanocomposite initiated the corresponding specific signal routes triggering the activity of certain genes [30]. Although the observation with dried composites was quite preliminary and requires further investigation, the difference in the physical properties (for example, magnetostriction) of MNPs and ANPs may be the reason for the fine surface structure formation.

The visualization of the intact surface geometry pattern of gel-based composites is still the unsolved challenge for the conventional microscopic approaches due to the presence of a large amount of solvent in gel interior structure. Such methods of preparation as freeze-drying or vacuum drying strongly disturb the surface of samples. Therefore, the results of microscopic studies like AFM or SEM do not characterize all details of the surface geometry explicitly. Meanwhile, the results of SEM and TEM confirm indirectly the heterogeneity of the surface of nanocomposite gels [39,40]. For instance, such data were reported in our earlier published work [13], and here, we applied this technique for the composites with diamagnetic nanoparticles. One of the consequences of the change of the type of nanoparticles was much lower contrast in SEM studies; as γ-Fe$_2$O$_3$ were characterized by higher "electronic density", i.e., they interacted more actively with the electronic beam, they look much brighter (Figure 8a), and provide much higher contrast in comparison with Al$_2$O$_3$ nanoparticles inside the dried composite (Figure 8b).

Hypothetically, the effect of nanoparticles on the surface geometry can stem both from the absence and from the presence of NPs at the gel interface. For instance, NPs located in the outer layers of the gel could disentangle from the networks and move to the liquid phase, leaving voids in the surface layer. In the case of gel-based composite substrates in the present study, these voids are likely the size of the NPs, which was 10–40 nm, and were separated by approximately 150 nm one from another (evaluation was done for the weight fraction of nanoparticles equal to 1%). Thus, the surface of the gel substrate might be covered with small dimples separated by ca. 0.15 μm.

It is also feasible that NPs, which disentangle from the gel network, do not move to the bulk of the liquid phase but provide the adsorption layer at the surface. There are certain grounds for such supposition. It was shown in previous work [41] that the interaction of polyacrylamide macromolecules with the surface of iron oxide nanoparticles is energetically favorable. Alumina nanoparticles also have high adsorption potential and high catalytic activity at the surface due to a large amount of Al-OH moieties at the surface [42,43]. Hence, it would be reasonable to assume that there are adsorption forces at the gel interface, which can immobilize nanoparticles at the surface. Due to the adsorption of nanoparticles, the roughness of the surface enhances. The exudation of nanoparticles from the gel network and their adsorption on the surface might not be the alternatives, but both could contribute to the surface micro-roughness of gel-based composites. In addition, cell cultivation takes place in the solutions of high ionic strength; the presence of the immobilized at the surface nanoparticles can change the diffusion conditions affecting in this way the cell culture grows.

The mechanism of the bonding of the mechanically sensitive receptors of cells to the heterogeneous surface patterns of nanocomposites is not elucidated yet. Supposedly it may be provided by a certain packing of integrins, which match the characteristic dimensions of nanostructures, like the diameter of nanotubes [44,45]. A model was introduced which described the adhesion of cells to the nanostructured surface based on the energy of deformation of the receptors due to their interaction with the surface geometry pattern [35]. Electrostatic interactions between receptor molecules and nanostructures at the surface might also be feasible [46,47].

Besides, we may suppose that the adhesion of cells to the scaffold can be mediated by proteins, which are the components of the medium for the cell culturing and better adsorb at the rough surface. It was reported that even a small amount of vitronectin and fibronectin at the surface of ferrogel substantially promoted the adhesion of cells due to the presence of a specific amino acid sequence (Arg-Gly-Asp) in their chemical structure, which is favorable for the interaction with membrane receptors of fibroblasts [48,49]. Similar results were obtained for the adhesion and proliferation of human fibroblasts on the surface of Al/Al_2O_3 bi-phasic nanowires (NWs) [50].

Thus, it is reasonable to interpret the results obtained in the present study from the viewpoint of the structuring of the surface of the gel-based composites by the exudation or/and adsorption of nanoparticles. The mechanisms of the surface structuring are likely the same for the composite PAAm gels filled with alumina or magnetite, and therefore the effect of gel-based composites on the adhesion and proliferation of human fibroblasts might be the same as well, i.e., gel-based composites provide a heterogeneous surface geometry pattern, which is favorable for the adhesion of cells.

In general, the obtained results showed that in the elaborated experimental conditions, the proliferation activity of human fibroblasts on the surface of gel-based composites did not depend on the magnetic properties of the embedded nanoparticles. It is worth mentioning that this finding was not influenced by the fact that the gel-based composites of the magnetic and diamagnetic origin significantly differed in their Young's modulus and electrical potential. It may be taken as additional support for the hypothesis that the surface geometry pattern has a key role in the biocompatibility of gel-based composites.

In the meantime, there is quite a lot of studies that address ferrogels as prospective magnetically controlled composites for applications in tissue engineering, regenerative medicine, field-assisted drug delivery, and magnetic biosensing. Various polymers: synthetic, biological, and their blends are used as matrices for these composites. Iron oxide magnetic particles embedded in composite ferrogels also vary in dimensions, chemical compositions, synthetic routes, and other parameters. In general, results obtained in these studies confirm good biocompatibility of ferrogels, and the achieved level of magnetic properties of these materials makes them especially attractive for magnetic field-assisted drug delivery and magnetic biosensing [5,51]. The present combination of the composites based on superparamagnetic and diamagnetic nanoparticles can be of special interest in the understanding of a high-frequency response of polymer matrix with variations of dielectric constants contributions in the formation of magnetoimpedance responses of the gel-based composites [51].

The present study is a special step allowing a comparative evaluation of the contributions of MNPs and ANPs. This comparison finally became possible because of our research works related to the biological activity of cells at hydrogels and ferrogels in similar experimental conditions [8,10–13]. In this series of experimental studies, we used human fibroblasts taken from one patient, magnetic nanoparticles of iron oxide Fe_2O_3 from the same batch, PAAm gel with the same networking, the same procedures, and experimental techniques. According to these studies, the increase in MNPs content in ferrogel always led to the reliable enlargement of the density of cells monolayer at the surface of magnetic composites. Similar results were obtained in the present research, which compared the biocompatibility of gel composites with embedded magnetic iron oxide nanoparticles and embedded non-magnetic alumina nanoparticles, which were very close in terms of their shape and dimensions.

5. Conclusions

Two series of non-agglomerated spherical nanoparticles of 20–40 nm in diameter were fabricated by laser target evaporation technique: superparamagnetic Fe_2O_3 or diamagnetic Al_2O_3 nanoparticles. Composite polyacrylamide gels with Fe_2O_3 or Al_2O_3 embedded nanoparticles were synthesized, aiming to study the magnetic contribution to the ferrogel biocompatibility. The proliferative activity of human dermal fibroblast cell cultures on the surface of these gel-based composites was estimated. The concentration of the fillers in the gel was fixed at 0.0, 0.3, 0.6, or 1.2 wt.%. Mechanical, electrical, and magnetic properties of the composites were characterized by the dependence of Young's modulus, electrical potential, magnetization measurements on the content of embedded nanoparticles. The fibroblast monolayer density on the surface of composite substrates after 96 h of incubation without application of external magnetic field was evaluated for estimation of the composites biocompatibility. It was found that regardless of the nature of the nanoparticles, the increase in their concentration in the composite provided a parallel increase in the cell proliferation on the surface of composite substrates.

The main conclusion of the present study is the statement that the biological activity of cells on the surface of composite gels does not depend on the magnetic properties of nanoparticles, at least in the elaborated experimental conditions. In other words, the obtained results exclude a significant contribution of the magnetic field provided by magnetic nanoparticles inside ferrogel and near its surface on the biocompatibility of magnetic gel composites in the conditions under consideration.

In general, the present study has more methodological than applied value. It is addressed toward the search of a key determinant of FG biocompatibility. In a series of preliminary experiments performed under the same experimental conditions, we tried to exclude systematically the indirect contribution of various factors that can determine cell biological activity at the surface of ferrogels, in particular, the mechanical properties and electrical potential of composites. In this study, we excluded the direct contribution of the FG magnetic effects in very low magnetic fields of the order of terrestrial magnetic field values. Finally, we assumed that the FG biocompatibility is most likely associated with the effect of the particles on the composite surface. At the same time, the results obtained can be useful in the design of gel-based composites for cell technologies. In this context, the application of ferrogels with the use of an external magnetic field is of the greatest interest. From our point of view, research in this direction will make it possible to optimize the principles of controlling biological processes in cell cultures grown onto FG. Special attention should be paid to one very important aspect of this study: iron oxide magnetic nanoparticles from the same batch were used in previous experiments, ensuring a very good basis for comparison of the results obtained for Fe_2O_3 and Al_2O_3 systems fabricated by the same technique and characterized in the same conditions.

Author Contributions: F.A.B., A.P.S. and G.V.K. conceived and designed the experiments; T.F.S., E.B.M., F.A.F., P.A.S., S.F.A., A.P.S. and G.V.K. performed the experiments and prepared graphical materials; F.A.B., E.B.M., T.F.S., A.P.S. and G.V.K. analyzed the data; F.A.B., A.P.S. and G.V.K. wrote the manuscript, prepared illustrations and participated in re-writing of the revised version. F.A.B. and G.V.K.—project administration and funding acquisition. All authors discussed the results and implications and commented on the manuscript at all stages. All authors have read and agreed to the published version of the manuscript.

Funding: The Russian Scientific Foundation (grant 18-19-00090) supported the experimental parts of this study, including the design, performance, and analysis of experiments.

Acknowledgments: We thank I.V. Beketov, A.I. Medvedev, A.M. Murzakaev, Daiana V. Sedneva-Lugovets, I. Orue for special support. Selected measurements (SEM imaging and magnetic measurements) were made using the SGIKER services of UPV/EHU.

Conflicts of Interest: The authors declare no conflict of interest. The funders had no role in the design of the study, in the collection, analyses, or interpretation of data, in the writing of the manuscript, or in the decision to publish the results.

References

1. Burdick, J.A.; Stevence, M.M. Biomedical hydrogels. In *Biomaterials, Artificial Organs, and Tissue Engineering*; Woodhead Publishing: Cambridge, UK, 2005.
2. El-Sherbiny, I.M.; Yacoub, M.H. Hydrogel scaffolds for tissue engineering: Progress and challenges. *Glob. Cardiol. Sci. Pract.* **2013**, *3*, 316–342. [CrossRef]
3. Li, Y.; Huang, G.; Zhang, X.; Li, B.; Chen, Y.; Lu, T.; Lu, T.J.; Xu, F. Magnetic hydrogels and their potential biomedical applications. *Adv. Funct. Matters* **2013**, *3*, 660–672. [CrossRef]
4. Kennedy, S.; Roco, C.; Délérisa, A.; Spoerri, P.; Cezar, C.; Weaver, J.; Vandenburgh, H.; Mooney, D. Improved magnetic regulation of delivery profiles from ferrogels. *Biomaterials* **2018**, *161*, 179–189. [CrossRef] [PubMed]
5. Kurlyandskaya, G.V.; Blyakhman, F.A.; Makarova, E.B.; Buznikov, N.A.; Safronov, A.P.; Fadeyev, F.A.; Shcherbinin, S.V.; Chlenova, A.A. Functional magnetic ferrogels: From biosensors to regenerative medicine. *AIP Adv.* **2020**, *10*, 125128. [CrossRef]
6. Sullivan, K.; Balin, A.K.; Allen, R.G. Effects of static magnetic fields on the growth of various types of human cells. *Bioelectromagnetics* **2011**, *32*, 140–147. [CrossRef] [PubMed]
7. Tang, H.; Wang, P.; Wang, H.; Fang, Z.; Yang, Q.; Ni, W.; Sun, X.; Liu, H.; Wang, L.; Zhao, G.; et al. Effect of static magnetic field on morphology and growth metabolism of Flavobacterium sp. m1-14. *Bioprocess Biosyst. Eng.* **2019**, *42*, 1923–1933. [CrossRef] [PubMed]
8. Stolf, S.; Skorvánek, M.; Stolfa, P.; Rosocha, J.; Vasko, G.; Sabo, J. Effects of static magnetic field and pulsed electromagnetic field on viability of human chondrocytes in vitro. *Physiol. Res.* **2007**, *56*, S45–S49.
9. Lew, W.-Z.; Huang, Y.-C.; Huang, K.-Y.; Lin, C.-T.; Tsai, M.-T.; Huang, H.-M. Static magnetic fields enhance dental pulp stem cell proliferation by activating the p38 mitogen-activated protein kinase pathway as its putative mechanism. *J. Tissue Eng. Regen. Med.* **2018**, *12*, 19–29. [CrossRef]
10. Blyakhman, F.A.; Melnikov, G.Y.; Makarova, E.B.; Fadeyev, F.A.; Sedneva-Lugovets, D.V.; Shabadrov, P.A.; Volchkov, S.O.; Mekhdieva, K.R.; Safronov, A.P.; Fernández Armas, S.; et al. Effects of constant magnetic field to the proliferation rate of human fibroblasts grown onto different substrates: Tissue culture polystyrene, polyacrylamide hydrogel and ferrogels γ-Fe2O3 magnetic nanoparticles. *Nanomaterials* **2020**, *10*, 1697. [CrossRef]
11. Safronov, A.P.; Beketov, I.V.; Komogortsev, S.V.; Kurlyandskaya, G.V.; Medvedev, A.I.; Leiman, D.V.; Larranaga, A.; Bhagat, S.M. Spherical magnetic nanoparticles fabricated by laser target evaporation. *AIP Adv.* **2013**, *3*, 052135. [CrossRef]
12. Blyakhman, F.A.; Safronov, A.P.; Zubarev, A.Y.; Shklyar, T.F.; Makeyev, O.G.; Makarova, E.B.; Melekhin, V.V.; Larrañaga, A.; Kurlyandskaya, G.V. Polyacrylamide ferrogels with embedded maghemite nanoparticles for biomedical engineering. *Results Phys.* **2017**, *6*, 3624–3633. [CrossRef]
13. Blyakhman, F.A.; Makarova, E.B.; Fadeyev, F.A.; Lugovets, D.V.; Safronov, A.P.; Shabadrov, P.A.; Shklyar, T.F.; Melnikov, G.Y.; Orue, I.; Kurlyandskaya, G.V. The contribution of magnetic nanoparticles to ferrogel biophysical properties. *Nanomaterials* **2019**, *9*, 232. [CrossRef] [PubMed]
14. Blyakhman, F.A.; Makarova, E.B.; Shabadrov, P.A.; Fadeyev, F.A.; Shklyar, T.F.; Safronov, A.P.; Komogortsev, S.V.; Kurlyandskaya, G.V. Magnetic nanoparticles as a strong contributor to the biocompatibility of ferrogels. *Phys. Metals Metallogr.* **2020**, *121*, 299–304. [CrossRef]

15. Blyakhman, F.A.; Safronov, A.P.; Makeyev, O.G.; Melekhin, V.V.; Shklyar, T.F.; Zubarev, A.Y.; Makarova, E.B.; Sichkar, D.A.; Rusinova, M.A.; Sokolov, S.Y.; et al. Effect of the polyacrylamide ferrogel elasticity on the cell adhesiveness to magnetic composite. *J. Mech. Med. Biol.* **2018**, *18*, 1850060. [CrossRef]
16. Darton, N.J.; Ionescu, A.; Llandro, J. (Eds.) *Magnetic Nanoparticles in Biosensing and Medicine*; Cambridge University Press: Cambridge, UK, 2019.
17. Scherrer, P. Bestimmung der Grösse und der inneren Struktur von Kolloidteilchen mittels Röntgensrahlen (Determination of the size and internal structure of colloidal particles using X-rays). *Nachr. Ges. Wiss. Göttingen* **1918**, *26*, 98–102.
18. Beketov, I.V.; Safronov, A.P.; Medvedev, A.I.; Alonso, J.; Kurlyandskaya, G.V.; Bhagat, S.M. Iron oxide nanoparticles fabricated by electric explosion of wire: Focus on magnetic nanofluids. *AIP Adv.* **2012**, *2*, 022154. [CrossRef]
19. Hiemenz, P.C.; Rajagopalan, R. *Principles of Colloid and Surface Chemistry*; Marcel Dekker: New York, NY, USA, 1997.
20. Coey, J.M.D. *Magnetism and Magnetic Materials*; Cambridge University Press: New York, NY, USA, 2010.
21. Katchalsky, A.; Lifson, S.; Exsenberg, H. Equation of swelling for polyelectrolyte gels. *J. Polym. Sci.* **1951**, *7*, 571–574. [CrossRef]
22. Rubinstein, M.; Colby, R.H. *Polymer Physics*; Oxford University Press: New York, NY, USA, 2003; p. 59.
23. Benguigui, L.; Boué, F. Homogeneous and inhomogenous polyacrylamide gels as observed by small angle neutron scattering: A connection with elastic properties. *Eur. Phys. J. B* **1999**, *11*, 439–444. [CrossRef]
24. Zhai, Y.; Duan, H.; Meng, X.; Cai, K.; Liu, Y.; Lucia, L. Reinforcement Effects of Inorganic Nanoparticles for Double-Network Hydrogels. *Macromol. Mater. Eng.* **2015**, *12*, 1290–1299. [CrossRef]
25. Glaser, R. *Biophysics*; Springer: Heidelberg, Germany, 1999.
26. Ingber, D.E. Mechanosensation through integrins: Cells act locally but think globally. *Proc. Natl. Acad. Sci. USA* **2003**, *18*, 1472–1474. [CrossRef]
27. Estell, E.G.; Murphya, L.A.; Silversteina, A.M.; Tana, A.R.; Shahb, R.P.; Ateshiana, G.A.; Hunga, C.T. Fibroblast-like synoviocyte mechanosensitivity to fluid shear is modulated by interleukin-1α. *J. Biomech.* **2017**, *60*, 91–99. [CrossRef]
28. Gilles, G.; McCulloch, A.D.; Brakebusch, C.H.; Herum, K.M. Maintaining resting cardiac fibroblasts in vitro by disrupting mechanotransduction. *PLoS ONE* **2020**, *15*, e0241390. [CrossRef] [PubMed]
29. Zheng, H.; Tian, Y.; Gao, Q.; Yu, Y.; Xia, X.; Feng, Z.; Dong, F.; Wu, X.; Sui, L. Hierarchical Micro-Nano Topography Promotes Cell Adhesion and Osteogenic Differentiation via Integrin α2-PI3K-AKT Signaling Axis. *Front. Bioeng. Biotechnol.* **2020**, *8*, 463. [CrossRef] [PubMed]
30. Dalby, M.J.; Andar, A.; Nag, A.; Affrossman, S.; Tare, R.; McFarlane, S.; Oreffo, R.O.C. Genomic expression of mesenchymal stem cells to altered nanoscale topographies. *J. R. Soc. Interface* **2008**, *5*, 1055–1065. [CrossRef] [PubMed]
31. Deguchi, S.; Kato, A.; Wu, P.; Hakamada, M.; Mabuchi, M. Heterogeneous role of integrins in fibroblast response to small cyclic mechanical stimulus generated by a nanoporous gold actuator. *Acta Biomater.* **2021**, *121*, 418–430. [CrossRef] [PubMed]
32. Arnadóttir, J.; Chalfie, M. Eukaryotic Mechanosensitive Channels. *Annu. Rev. Biophys.* **2010**, *39*, 111–137. [CrossRef] [PubMed]
33. Chung, S.H.; Son, S.J.; Min, J. The nanostructure effect on the adhesion and growth rates of epithelial cells with well-defined nanoporous alumina substrates. *Nanotechnology* **2010**, *21*, 125104. [CrossRef] [PubMed]
34. Liu, T.-Y.; Chan, T.-Y.; Wang, K.-S.; Tso, H.-M. Influence of magnetic nanoparticle arrangement in ferrogels for tunable biomolecule diffusion. *RSC Adv.* **2015**, *5*, 90098–90102. [CrossRef]
35. Daniel, M.; Flipi, K.E.; Filová, E.; Fojt, J. Optimal surface topography for cell adhesion is driven by cell membrane mechanics. *arXiv* **2020**, arXiv:2006.16787.
36. Oh, S.; Brammer, K.S.; Li, Y.S.J.; Teng, D.; Engler, A.J.; Chien, S.; Jin, S. Stem cell fate dictated solely by altered nanotube dimension. *Proc. Natl. Acad. Sci. USA* **2009**, *106*, 2130–2135. [CrossRef]
37. Dalby, M.J.; Gadegaard, N.; Oreffo, R.O.C. Harnessing nanotopography and integrin–matrix interactions to influence stem cell fate. *Nat. Mater.* **2014**, *13*, 558–569. [CrossRef] [PubMed]
38. Wang, K.; Bruce, A.; Mezan, R.; Kadiyala, A.; Wang, L.; Dawson, J.; Rojanasakul, Y.; Yang, Y. Nanotopographical modulation of cell function through nuclear deformation. *ACS Appl. Mater. Interfaces* **2016**, *8*, 5082–5092. [CrossRef] [PubMed]
39. Helminger, M.; Wu, B.; Kollmann, T.; Benke, D.; Schwahn, D.; Pipich, V.; Faivre, D.; Zahn, D.; Cölfen, H. Synthesis and characterization of gelatin-based magnetic hydrogels. *Adv. Funct. Mater.* **2014**, *24*, 3187–3196. [CrossRef] [PubMed]
40. Galicia, J.A.; Cousin, F.; Dubois, E.; Sandre, O.; Cabuil, V.; Perzynski, R. Local structure of polymeric ferrogels. *J. Magn. Magn. Mater.* **2011**, *323*, 1211–1215. [CrossRef]
41. Safronov, A.P.; Samatov, O.M.; Tyukova, I.S.; Mikhnevich, E.A.; Beketov, I.V. Heating of polyacrylamide ferrogel by alternating magnetic field. *J. Magn. Magn. Mat.* **2016**, *415*, 24–29. [CrossRef]
42. Zhu, B.; Wei, X.; Song, J.; Zhang, Q.; Jiang, W. Crystalline phase and surface coating of Al_2O_3 nanoparticles and their influence on the integrity and fluidity of model cell membranes. *Chemosphere* **2020**, *247*, 125876. [CrossRef] [PubMed]
43. Ferreira, A.R.; Küçükbenli, E.; De Gironcoli, S.; Souza, W.F.; Chiaro, S.S.X.; Konstantinova, E.; Leitao, A.A. Structural models of activated γ-alumina surfaces revisited: Thermodynamics, NMR and IR spectroscopies from ab initio calculations. *Chem. Phys.* **2013**, *423*, 62–72. [CrossRef]
44. Gongadze, E.; Kabaso, D.; Bauer, S.; Park, J.; Schmuki, P.; Iglič, A. Adhesion of osteoblasts to a vertically aligned TiO_2 nanotube surface. *Mini Rev. Med. Chem.* **2013**, *13*, 194–200. [PubMed]
45. Zhang, Y.; Wang, K.; Dong, K.; Cui, N.; Lu, T.; Han, Y. Enhanced osteogenic differentiation of osteoblasts on $CaTiO_3$ nanotube film. *Colloids Surf. B Biointerfaces* **2020**, *187*, 110773. [CrossRef]

46. Borthakur, P.; Hussain, N.; Darabdhara, G.; Boruah, P.K.; Sharma, B.; Borthakur, P.; Yadav, A.; Das, M.R. Adhesion of gram-negative bacteria onto α-Al$_2$O$_3$ nanoparticles: A study of surface behaviour and interaction mechanism. *J. Environ. Chem. Eng.* **2018**, *6*, 3933–3941. [CrossRef]

47. Zhang, Q.; Wang, H.; Cuicui, G.; Duncan, J.; Kaihong, H.; Adeosun, S.O.; Huaxin, X.; Huiting, P.; Qiao, N. Alumina at 50 and 13 nm nanoparticle sizes have potential genotoxicity. *J. Appl. Toxicol.* **2017**, *37*, 1053–1064. [CrossRef] [PubMed]

48. Machida-Sano, I.; Matsuda, Y.; Namiki, H. In vitro adhesion of human dermal fibroblasts on iron cross-linked alginate films. *Biomed. Mater.* **2009**, *4*, 025008. [CrossRef] [PubMed]

49. Blaeser, A.; Million, N.; Campos, D.F.D.; Lisa Gamrad, L.; Köpf, M.; Rehbock, C.; Nachev, M.; Sures, B.; Barcikowski, S. Laser-based in situ embedding of metal nanoparticles into bioextruded alginate hydrogel tubes enhances human endothelial cell adhesion. *Nano Res.* **2016**, *9*, 3407–3427. [CrossRef]

50. Aktas, O.C.; Sander, M.; Miro, M.M.; Lee, J. Enhanced fibroblast cell adhesion on Al/Al$_2$O$_3$ nanowires. *Appl. Surf. Sci.* **2011**, *257*, 3489–3494. [CrossRef]

51. Blyakhman, F.A.; Buznikov, N.A.; Sklyar, T.F.; Safronov, A.P.; Golubeva, E.V.; Svalov, A.V.; Sokolov, S.Y.; Melnikov, G.Y.; Orue, I.; Kurlyandskaya, G.V. Mechanical, Electrical and Magnetic Properties of Ferrogels with Embedded Iron Oxide Nanoparticles Obtained by Laser Target Evaporation: Focus on Multifunctional Biosensor Applications. *Sensors* **2018**, *18*, 872. [CrossRef]

 *nanomaterials*

Article

Superparamagnetic $ZnFe_2O_4$ Nanoparticles-Reduced Graphene Oxide-Polyurethane Resin Based Nanocomposites for Electromagnetic Interference Shielding Application

Raghvendra Singh Yadav [1,*], Anju [1], Thaiskang Jamatia [1], Ivo Kuřitka [1], Jarmila Vilčáková [1], David Škoda [1], Pavel Urbánek [1], Michal Machovský [1], Milan Masař [1], Michal Urbánek [1], Lukas Kalina [2] and Jaromir Havlica [2]

[1] Centre of Polymer Systems, University Institute, Tomas Bata University in Zlín, Trida Tomase Bati 5678, 760 01 Zlín, Czech Republic; deswal@utb.cz (A.); jamatia@utb.cz (T.J.); kuritka@utb.cz (I.K.); vilcakova@utb.cz (J.V.); dskoda@utb.cz (D.Š.); urbanek@utb.cz (P.U.); machovsky@utb.cz (M.M.); masar@utb.cz (M.M.); murbanek@utb.cz (M.U.)

[2] Materials Research Centre, Brno University of Technology, Purkyňova 464/118, 61200 Brno, Czech Republic; kalina@fch.vut.cz (L.K.); havlica@fch.vut.cz (J.H.)

* Correspondence: yadav@utb.cz; Tel.: +420-576031725

Citation: Yadav, R.S.; Anju; Jamatia, T.; Kuřitka, I.; Vilčáková, J.; Škoda, D.; Urbánek, P.; Machovský, M.; Masař, M.; Urbánek, M.; et al. Superparamagnetic $ZnFe_2O_4$ Nanoparticles-Reduced Graphene Oxide-Polyurethane Resin Based Nanocomposites for Electromagnetic Interference Shielding Application. *Nanomaterials* **2021**, *11*, 1112. https://doi.org/10.3390/nano11051112

Academic Editors: Yurii K. Gun'ko, George C. Hadjipanayis and Cesar De Julian Fernandez

Received: 30 March 2021
Accepted: 23 April 2021
Published: 25 April 2021

Publisher's Note: MDPI stays neutral with regard to jurisdictional claims in published maps and institutional affiliations.

Abstract: Superparamagnetic $ZnFe_2O_4$ spinel ferrite nanoparticles were prepared by the sonochemical synthesis method at different ultra-sonication times of 25 min (ZS25), 50 min (ZS50), and 100 min (ZS100). The structural properties of $ZnFe_2O_4$ spinel ferrite nanoparticles were controlled via sonochemical synthesis time. The average crystallite size increases from 3.0 nm to 4.0 nm with a rise of sonication time from 25 min to 100 min. The change of physical properties of $ZnFe_2O_4$ nanoparticles with the increase of sonication time was observed. The prepared $ZnFe_2O_4$ nanoparticles show superparamagnetic behavior. The prepared $ZnFe_2O_4$ nanoparticles (ZS25, ZS50, and ZS100) and reduced graphene oxide (RGO) were embedded in a polyurethane resin (PUR) matrix as a shield against electromagnetic pollution. The ultra-sonication method has been used for the preparation of nanocomposites. The total shielding effectiveness (SE_T) value for the prepared nanocomposites was studied at a thickness of 1 mm in the range of 8.2–12.4 GHz. The high attenuation constant (α) value of the prepared ZS100-RGO-PUR nanocomposite as compared with other samples recommended high absorption of electromagnetic waves. The existence of electric-magnetic nanofillers in the resin matrix delivered the inclusive acts of magnetic loss, dielectric loss, appropriate attenuation constant, and effective impedance matching. The synergistic effect of $ZnFe_2O_4$ and RGO in the PUR matrix led to high interfacial polarization and, consequently, significant absorption of the electromagnetic waves. The outcomes and methods also assure an inventive and competent approach to develop lightweight and flexible polyurethane resin matrix-based nanocomposites, consisting of superparamagnetic zinc ferrite nanoparticles and reduced graphene oxide as a shield against electromagnetic pollution.

Keywords: sonochemical synthesis; spinel ferrite; nanoparticles; nanocomposites; electromagnetic interference shielding

1. Introduction

Recently, the rapid progress in electronic devices and information technology has endorsed the widespread utilization of high-power electromagnetic waves in scientific, commercial, civil, and military applications [1,2]. The abundant electromagnetic interference (EMI) in the environment has influenced the working of electronic devices [3]. Exposure to electromagnetic radiation has also influenced human health [4]. Electromagnetic shielding or absorption has been demonstrated to be one of the operative approaches to address electromagnetic pollution [5]. Therefore, electromagnetic shielding or absorption material is one of the best procedures of environment and health defense [6]. A lightweight, flexible, and cost-efficient advanced nanocomposite shielding material is required to attenuate the detrimental electromagnetic wave interference [7].

The EMI-shielding mechanism consists of mainly the two-loss factors: (i) reflection loss and (ii) absorption loss. A single component, i.e., an only electric conductive material or magnetic material, has less competence to pay both the mechanism of reflection and absorption. The absorption loss-dominant shielding material is more desirable because it can prevent the second reflection pollution in comparison with the reflection-dominant material. In recent years, researchers and academicians have comprehended that the intention of nanocomposites consisting of both electric and magnetic constituents is a prudent choice to cultivate proficient EMI-shielding material [8]. Such deliberate shielding material is anticipated to close the gap between permittivity and permeability. In recent times, carbon-based materials, including carbon nanotubes, nanofibers, and reduced graphene oxide, have established significant consideration for efficient electromagnetic interference shielding material [9]. Reduced graphene oxide parades a high specific surface area with plenteous functional groups and structural defects on its surface. Therefore, it can deliver copious interfacial polarization and defect dipole polarization [10]. Reduced graphene oxide also exhibits outstanding dielectric characteristics and higher electrical conductivity [11]. In recent years, it is also perceived that the introduction of magnetic nanoparticles, including spinel ferrite in the presence of reduced graphene oxide, can moderate its impedance matching condition and robust attenuation capability of the shielding nanocomposite [12]. The microwave absorption characteristics also depend on the fraction of nanoparticles and reduced graphene oxide [13]. Further, superparamagnetic spinel ferrite nanoparticles have established a broad application due to their competence to hastily respond to an applied magnetic field, which is concomitant with negligible remanence and coercivity [14,15]. For that reason, there is an intensive research subject to develop superparamagnetic spinel ferrite nanoparticles and further their application as electromagnetic interference shielding [16]. A research group, Honglei Yuan et al. [17] reported the superparamagnetic Fe_3O_4/MWCNTs nanocomposites displayed remarkably enhanced microwave absorption characteristics in a high-frequency range (Ku-band). This research group provided an approach to improve the resonance frequency break of the Snoek limit, which considerably boosts the microwave absorption characteristics at high-frequency applications. Traditional shielding materials, such as metals and metallic composites, have drawbacks, including chemical resistance, weak flexibility, corrosion, heavy weight, and hard processibility, etc. Polymer composites as an alternative candidate for EMI shielding can provide lightweight, resistance to corrosion, flexibility, easy availability, processability, and cost-effectiveness. $ZnFe_2O_4$ is one of the most investigated spinel ferrite systems. An interesting characteristic of $ZnFe_2O_4$ is the possibility of controlling the magnetic properties with particle/crystallite size. The variation of magnetic property from paramagnetic to superparamagnetic or ferrimagnetic in $ZnFe_2O_4$ nanoparticles is attributed to cation redistribution at octahedral and tetrahedral sites [18]. The mixed cation distribution in nanosized $ZnFe_2O_4$ spinel ferrite also depends on the synthesis method [19]. Quite a lot of synthesis approaches have been utilized for the preparation of nanoparticles and their nanocomposites [20]. Sonochemistry is an innovative and potent synthesis methodology for the development of nanoparticles and nanocomposites. This technique affords noteworthy rewards, such as well dispersion, size reduction, particle de-agglomeration, homogenization, and emulsification, etc. [21]. The advantage of the sonochemical synthesis approach is cost-effectiveness, strong reaction rate, controllable synthesis, narrow particle size distribution, high purity, and nonpolluting [22,23]. Under extreme sonochemical synthesis conditions, nanomaterials with the required size can be designed at controlled chemical reactions and physical changes [24].

In the present work, structural and physical properties of superparamagnetic $ZnFe_2O_4$ spinel ferrite nanoparticles have been controlled by the sonochemical synthesis approach with increased sonochemical synthesis time. To the best of the authors' knowledge, this is the first report on superparamagnetic $ZnFe_2O_4$ spinel ferrite nanoparticles-reduced graphene oxide (RGO)-polyurethane resin (PUR) nanocomposites as a shield against electromagnetic pollution. RGO exhibited high specific surface area and excellent electrical conductivity. However, non-magnetic RGO exhibited strong dielectric loss and also impedance mismatching issues. Therefore, the combination of magnetic spinel ferrite with the electrically conductive RGO can improve EMI-shielding performance with balanced impedance matching conditions. Additionally, the EMI-shielding performance can be improved with nanoparticles/nanostructures due to its excellent electromagnetic properties and high surface area. The structural and electromagnetic properties/parameters of developed superparamagnetic $ZnFe_2O_4$ nanoparticles and RGO embedded PUR-based nanocomposites were examined in detail. The confirmation of structural formation of prepared nanoparticles and nanocomposites was carried out by X-ray diffractometry (XRD), infrared spectrometry (FTIR), and Raman spectrometry. Microstructural features were studied by transmission electron microscopy (TEM) and field emission-scanning electron spectroscopy (FE-SEM). The sonochemical synthesis approach can also be used for the industrial-scale formation of various spinel ferrite nanoparticles. The developed lightweight and flexible EMI-shielding nanocomposite can find potential application in the field of portable electronics and wearable devices that need to be lightweight, thin, and flexible material.

2. Materials and Methods

2.1. Materials

Sodium hydroxide, zinc nitrate, and iron nitrate were acquired from Alfa Aesar GmbH & Co KG, Germany. Potassium permanganate and graphite flakes were procured from Sigma-Aldrich, Germany. Further, sodium nitrate was procured from Lach-Ner, the Czech Republic. The reducing agent, Vitamin C (Livsane), for the reduction of graphene oxide, was obtained from Dr. Kleine Pharma GmbH, Bielefeld, Germany. Polyurethane resin (PUR) was selected as a polymer elastomeric casting matrix for very low shrinkage and flexibility. PUR matrix was prepared by using Biresin® U1404 elastomeric casting resin for mold-making from Sika Advanced Resins GmbH, Bad Urach, Germany, which has a description as a basis: two-component PUR system; Component A: Biresin® U1404, isocyanate prepolymer, colorless-transparent, unfilled; Component B: Biresin® U1404, amine, reddish-transparent, unfilled; Component B: Biresin® U1434, amine, beige, filled. It provides product benefits, such as insensitive to moisture, very soft, high elongation at break, good tensile strength, and elasticity, with component B Biresin® U1404 for shore hardness of A 40, with component B Biresin® U1434 for shore hardness of A 55, and very low shrinkage. The PUR matrix for $ZnFe_2O_4$ and RGO as nanofillers was obtained in liquid form with isocyanate prepolymer (component A, Biresin® U1404) and amine (component B, Biresin® U1434) as a curing agent (hardener) with a density of 1.3 g/cm^3; viscosity of ~3700 mPa-s; shore hardness of 55A; tear strength of 9 N/mm; tensile strength of 4 MPa; elongation at break >600%; linear shrinkage internal <0.02%.

2.2. Sonochemical Preparation of $ZnFe_2O_4$ Nanoparticles

$ZnFe_2O_4$ nanoparticles were synthesized by the sonochemical synthesis technique. First, 2.63 g $Zn(NO_3)_2 \cdot 6H_2O$ and 7.56 g $Fe(NO_3)_3 \cdot 9H_2O$ were mixed in 60 mL of deionized water. Additionally, aqueous 1.6 M sodium hydroxide (NaOH) was mixed in the above set solution with continuous stirring by a magnetic stirrer. Moreover, the attained mixed solution was placed to high-intensity ultrasonic waves for 25 min, 50 min, and 100 min with the use of the UZ SONOPULS HD 2070 Ultrasonic homogenizer (Berlin, Germany) at 70 W power and 20 kHz frequency. The achieved product was washed by utilizing deionized water and ethanol to eliminate unwanted chemical impurity. Finally, these washed nanoparticles were dried at 60 °C for 18 h. The prepared $ZnFe_2O_4$ nanoparticles were designated as ZS25, ZS50, and ZS100 according to their sonochemical synthesis time of 25 min, 50 min, and 100 min, respectively.

2.3. Preparation of Reduced Graphene Oxide (RGO)

First, graphene oxide (GO) was prepared from natural graphite powders by utilizing the Hummers' method. Typically, the mixture comprised of graphite powder (1.5 g) and sodium nitrate (1.5 g) was added slowly to concentrated H_2SO_4 (75 mL). The beaker, which contains the above reaction mixture, was placed on an ice bath, and potassium permanganate (9 g) was added gradually within 20 min, and this mixed solution was stirred further for 30 min under an ice bath (temperature in the range of 0–5 °C). This solution was stirred for an additional 48 h at room temperature. Afterward, 138 mL of deionized water was gradually added into the above mixture and stirred for 10 min. In this achieved mixture, 420 mL warm deionized water was then poured and kept under a continuous strong stirring. Moreover, H_2O_2 (30 mL) was mixed into the above reaction mixture to eliminate the remaining $KMnO_4$ and stirred until the color of the product transformed into bright yellow, which signaled the formation of graphite oxide from graphite. Finally, the obtained bright yellow product suspension was centrifuged and washed with ethanol and deionized water until the pH ~7 was reached. The obtained product was annealed at 60 °C in a vacuum oven for 24 h.

Vitamin C (10 g) was utilized as a reducing agent in the development of reduced graphene oxide (RGO) from 3 g of graphene oxide (GO). For this, the above-prepared GO was mixed in deionized water and additionally placed to high-intensity ultrasonic waves for 15 min with the use of UZ SONOPULS HD 2070 Ultrasonic homogenizer. Then, in this attained solution, vitamin C was added gradually, and then the obtained suspension was continuously stirred for the time of 3 h at temperature 90 °C. Additionally, the achieved product suspension was centrifuged and washed with ethanol and deionized water. Finally, the washed product was dried in a vacuum oven at 60 °C for 15 h.

2.4. Ultrasonic Preparation of Nanocomposites

Nanocomposites of polyurethane resin (PUR) (50 wt.%) with nanofillers (40 wt.% zinc ferrite nanoparticles and 10 wt.% RGO) were prepared. For the PUR matrix, component A and component B were used in the ratio of 100–50. For the preparation of nanocomposites, in a 25-mL beaker, isocyanate prepolymer (component A, Biresin U1404) were mixed with nanofillers ($ZnFe_2O_4$ (90%) + RGO (10%)) by using a EURO-ST-D mechanical stirrer for 30 min and then sonicated by using a UP 400S ultra probe (Hielscher Ultrasonics GmbH, Teltow, Germany) (frequency: 24 kHz, power: 400 W) for 30 min in an ice bath. Further, amine (component B) as a curing agent was mixed to the above mixture and then sonicated at another 10 min by using a UP 400S ultra probe (frequency: 24 kHz, power: 400 W). Finally, the prepared sample was closed and retained in a drying oven, where the composite material was cured at 25 °C for 5 days. Three PUR-based nanocomposite using zinc ferrite nanoparticles (ZS25, ZS50, or ZS100) and RGO as nanofillers, namely, (i) ZS25-RGO-PUR, (ii) ZS50-RGO-PUR, and (iii) ZS100-RGO-PUR, were prepared. Additionally, rectangle-shaped samples $22.86 \times 10.16 \times 1 \text{ mm}^3$ were produced by cast molding.

2.5. Characterization Techniques

The sonochemically prepared zinc ferrite nanoparticles X-ray Diffraction (XRD) was performed using an X-ray powder diffraction from Rigaku Corporation, Tokyo, Japan. The Raman spectroscopy of PUR-based nanocomposites was performed on a Raman spectrometer of Thermo Fisher Scientific, Waltham, MA, USA. XPS study of graphene oxide and reduced graphene oxide was performed on an X-ray photoelectron spectroscope of Kratos Analytical Ltd. (Manchester, UK). The FTIR spectroscopy of $ZnFe_2O_4$ nanoparticles and PUR-based nanocomposites was performed on Nicolet 6700 (Thermo Scientific, Waltham, MA, USA). The high-resolution transmission electron microscope (JEOL JEM 2100) (JEOL, Peabody, MA, USA) was utilized to investigate the morphology and lattice fringes of $ZnFe_2O_4$ nanoparticles. The surface morphology and structure of the PUR nanocomposites were investigated with an FE-SEM of FEI NanoSEM450 (The Netherland, FEI Company). Magnetic hysteresis curves of sonochemically prepared superparamagnetic

ZnFe$_2$O$_4$ nanoparticles were studied utilizing a VSM 7407, Lake Shore, Westerville, OH, USA. ZFC and FC temperature-dependent magnetization study of the sonochemically prepared zinc ferrite nanoparticles were investigated using a SQUID magnetometer of Quantum Design MPMS XL-7. The electromagnetic interference shielding effectiveness of the developed PUR-based nanocomposite with zinc ferrite nanoparticles and RGO as nanofillers was studied by using a vector network analyzer (Agilent N5230A, Agilent Technologies, Santa Clara, CA, USA) in 8.2–12.4 GHz (X band).

3. Results

3.1. X-ray Diffraction Study

The X-ray diffraction pattern of the sonochemically synthesized ZnFe$_2$O$_4$ spinel ferrite nanoparticles at sonication times of 25 min, 50 min, and 100 min is displayed in Figure 1. The observed diffraction peaks correspond to the reflection of (220), (311), (222), (400), (331), (422), (511), (440), (531), and (442) planes of an Fd$\bar{3}$m spinel crystal structure. Additionally, there is no presence of an impurity peak, which designates the high purity of spinel ferrite material. It is worth noting that as the sonication synthesis time increased, the intensity of diffraction peaks increased, and the width of the diffraction peak decreased, which suggests grain growth with an increase of sonication time. The average crystallite size of synthesized ZnFe$_2$O$_4$ nanoparticles was studied by utilizing the Debye–Scherrer equation [25]:

$$D = \frac{(0.9)\,\lambda}{\beta\,\cos\theta} \tag{1}$$

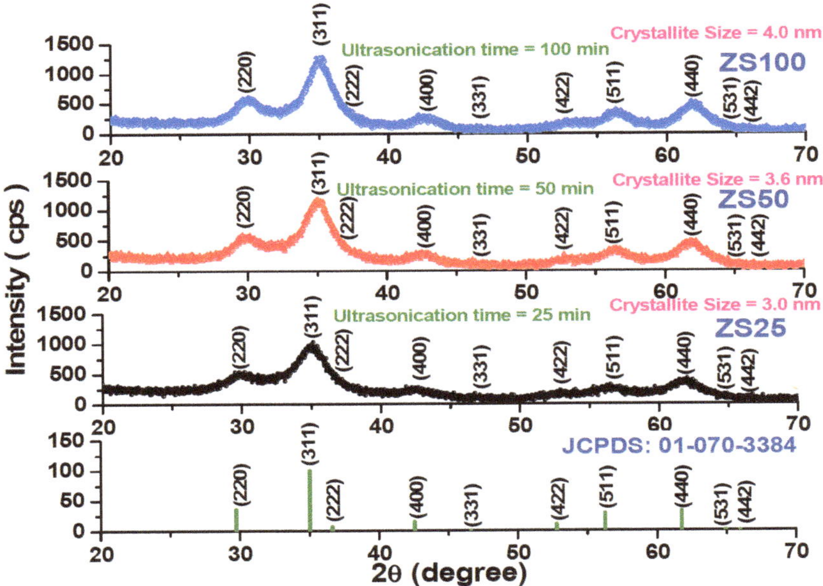

Figure 1. X-ray diffraction pattern of sonochemically prepared ZnFe$_2$O$_4$ spinel ferrite nanoparticles.

Herein, λ, β, and θ are the wavelength of X-ray, the full-width at half maximum (FWHM), and the Bragg angle, respectively. The average crystallite size increases from 3.0 nm to 4.0 nm with an increase in sonication time, as shown in Table 1. The growth of spinel ferrite nanocrystals was associated with an increase in ultrasonic time [26].

Table 1. Crystallite size, Lattice Parameter, X-ray Density, and Ionic Radii (r_A, r_B) for the prepared $ZnFe_2O_4$ spinel ferrite nanoparticles by the sonochemical synthesis approach.

Sample	Crystallite Size (nm)	Lattice Parameter, a (Å)	X-ray Density d_x (g/cm^3)	Ionic Radii r_A (Å)	Ionic Radii r_B (Å)
ZS 25	3.0	7.219	8.51	0.2757	1.4026
ZS 50	3.6	7.245	8.42	0.2816	1.4087
ZS 100	4.0	7.248	8.41	0.2822	1.4093

The lattice parameter was determined by utilizing the following relation [25]:

$$a^2 = \frac{\lambda^2 \left(h^2 + k^2 + l^2\right)^{1/2}}{4\sin^2\theta} \tag{2}$$

Herein, θ is the Bragg angle, and (hkl) are the Miller indices of the planes. The lattice parameter increases from 7.219 Å to 7.248 Å with an increase in sonication time from 25 min to 100 min, as shown in Table 1. The observed increase in the lattice constant with sonication time follows Vegard's law [27]. Generally, the lattice constant in the case of spinel ferrite correlates with microstructure, ordering/reordering of cations, valence states, and defects, etc. [28]. In the present work, the variation in the lattice constant can be attributed to changes in microstructure and ultrasonic-activated ordering/reordering of cations in $ZnFe_2O_4$ spinel ferrite nanoparticles.

The X-ray density (d_x) of prepared spinel ferrite nanoparticles is evaluated by the following relation [25]:

$$d_x = \frac{ZM}{NV} \tag{3}$$

Herein, Z, M, N, and V are the number of the nearest neighbor, the molecular weight, the Avogadro number, and the volume of the unit cell ($V = a^3$), respectively. The evaluated value of the X-ray density was 8.51 g/cm^3, 8.42 g/cm^3, and 8.41 g/cm^3 for ZS25, ZS50, and ZS100 samples, respectively (Table 1). Thus, an increase of sonication time to 25 min, 50 min, and 100 min decreased the density of the prepared spinel ferrite nanoparticles.

Additionally, structural parameters, such as ionic radii, hopping length for the octahedral and tetrahedral site, tetrahedral and octahedral bond length, tetrahedral edge, and the shared and unshared octahedral edge, for prepared $ZnFe_2O_4$ nanoparticles were assessed [29,30]. The variation in these parameters with sonication times of 25 min, 50 min, and 100 min was noticed, as mentioned in Tables 1 and 2. The increase in ionic radii, hopping length for the octahedral and tetrahedral site, tetrahedral and octahedral bond length, tetrahedral edge, and the shared and unshared octahedral edge for prepared $ZnFe_2O_4$ nanoparticles with an increase of sonication time was noticed. Microstructure and ultrasonic-activated ordering/reordering of cations in $ZnFe_2O_4$ nanoparticles was associated with an increase in sonication time, which can affect the physical properties of the material [31].

Table 2. Structural parameters for prepared $ZnFe_2O_4$ nanoparticles synthesized by sonochemical approach: hopping length for the octahedral and tetrahedral site, tetrahedral and octahedral bond length, tetrahedral edge, and the shared and unshared octahedral edge.

Sample	Hopping Length for Tetrahedral Site d_A (Å)	Hopping Length for Octahedral Site d_B (Å)	Tetrahedral Bond Length, d_{Ax} (Å)	Octahedral Bond Length, d_{Bx} (Å)	Tetrahedral Edge, d_{AxE} (Å)	Shared Octahedral Edge, d_{BxE} (Å)	Unshared Octahedral Edge, d_{BxEU} (Å)
ZS 25	3.1263	2.5526	1.6757	1.7424	2.7364	2.3688	2.5559
ZS 50	3.1374	2.5617	1.6816	1.7486	2.7461	2.3772	2.5650
ZS 100	3.1384	2.5625	1.6822	1.7491	2.7470	2.3780	2.5658

3.2. TEM Study

TEM measurements were carried out to investigate the structural features of prepared $ZnFe_2O_4$ spinel ferrite nanoparticles. Figure 2 represents TEM and HRTEM images of prepared nanoparticles, namely ZS25, ZS50, and ZS100. The TEM image of ZS25 is depicted in Figure 2a, which shows particles in the range of 2–4.5 nm (Figure S1 in supplementary material). The HRTEM image of ZS25 is shown in Figure 2b, which displays the lattice of (220) planes (d spacing 0.29 nm), (311) planes (d spacing 0.25 nm), and (400) planes (d spacing 0.21 nm) of $ZnFe_2O_4$ spinel ferrite [32]. Further, Figure 2c depicts a low-resolution TEM image of the ZS50 sample, which illustrated that the product consisted of particles with sizes of 2.5–5 nm. Figure 2d shows lattice fringes with an interplanar spacing of 0.29 nm, which is consistent with (220) planes of spinel ferrite. Additionally, the TEM image of ZS100 is depicted in Figure 2e, which demonstrated that the prepared nanoparticles exhibited size 3–12 nm. Figure 2f depicts the HRTEM image of ZS100. The investigation of the HRTEM image depicts the interplanar spacing of 0.25 nm, 0.21 nm, and 0.17 nm of lattice fringes corresponding to (311), (400), and (422) plane of $ZnFe_2O_4$ spinel ferrite.

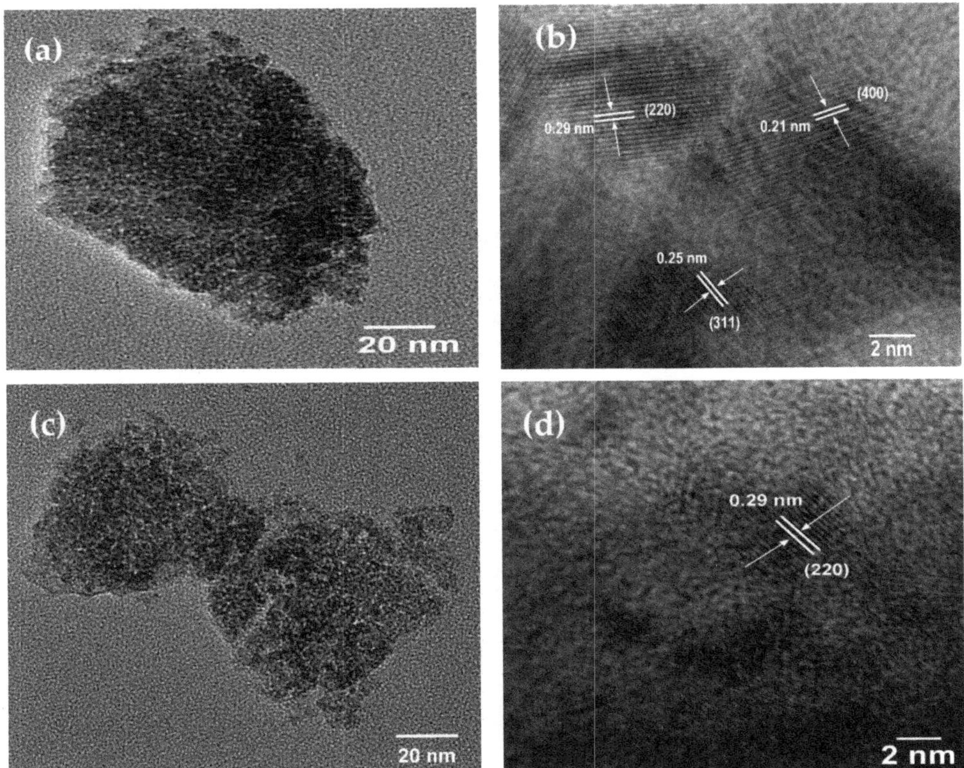

Figure 2. *Cont.*

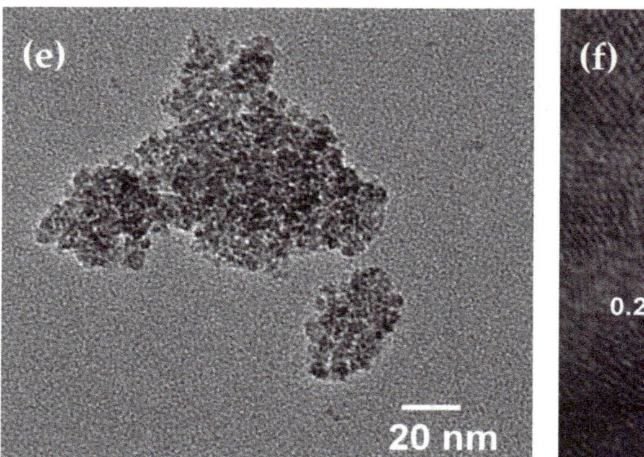
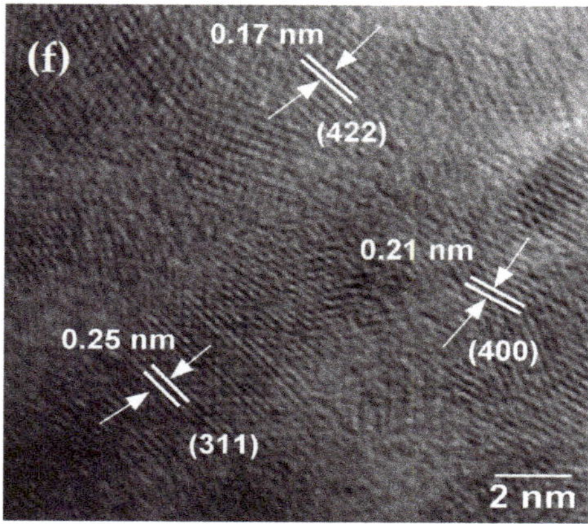

Figure 2. (**a**) TEM image of ZS25, (**b**) HRTEM image of ZS25, (**c**) TEM image of ZS50, (**d**) HRTEM image of ZS50, (**e**) TEM image of ZS100, and (**f**) HRTEM image of ZS100.

3.3. FE-SEM Study

Figure 3 depicts the typical SEM image of RGO and prepared polyurethane resin-based nanocomposites. Wrinkled and curled graphene sheets can be noticed in Figure 3a. Further, the presence of RGO and prepared $ZnFe_2O_4$ nanoparticles in polyurethane resin can be noticed in SEM images of the surfaces of the PUR-based nanocomposites, as shown in Figure 3b–d. The increase in the thickness of RGO may be due to the agglomeration of RGO during the processing and formation of polymer nanocomposite [33].

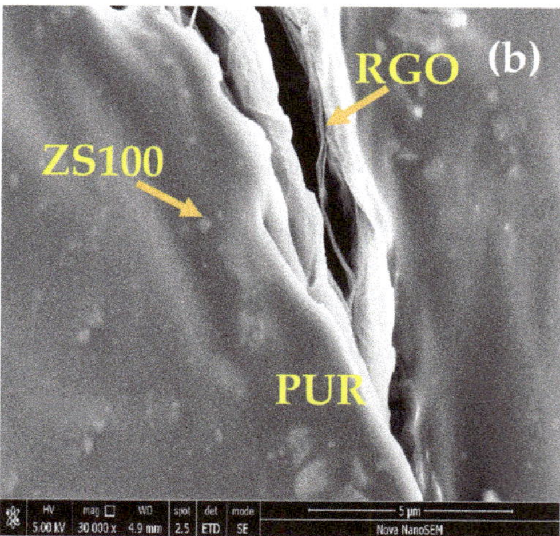

Figure 3. *Cont.*

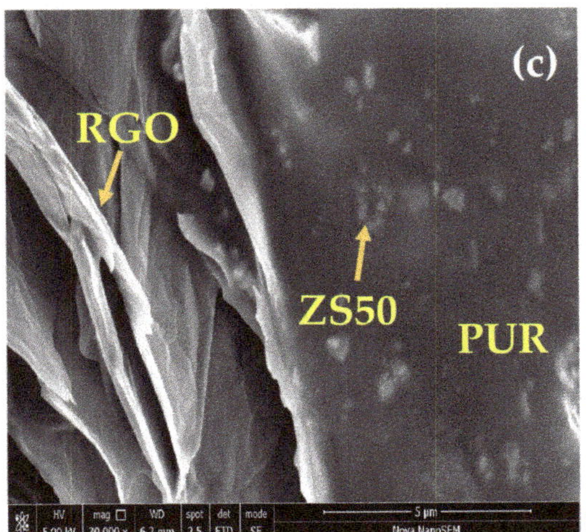

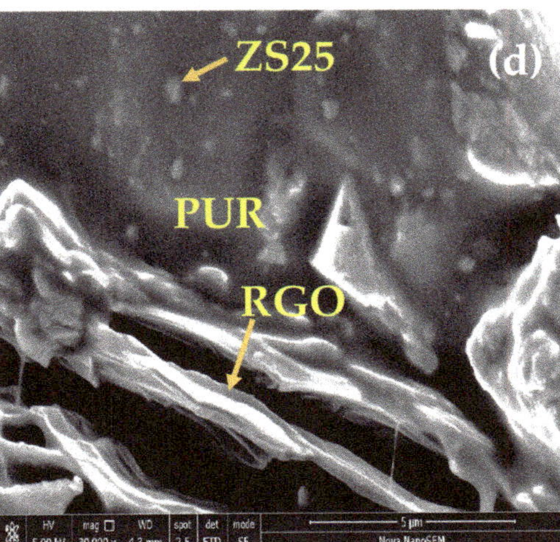

Figure 3. FE-SEM image of RGO (**a**), and FE-SEM image of the fracture surface of ZS100-RGO-PUR (**b**), ZS50-RGO-PUR (**c**), and ZS25-RGO-PUR (**d**).

3.4. X-ray Photoelectron Spectroscopy

The prepared GO and RGO were examined by X-ray photoelectron spectroscopy (XPS). Figure 4 shows the XPS spectra of prepared graphene oxide (GO) and reduced graphene oxide (RGO). Figure 4a,c signifies the survey scan spectra of GO and RGO, which display the existence of carbon and oxygen. Figure 4b depicts the high-resolution XPS spectra of the C 1s region for GO. The deconvoluted C 1s peak displays the peak binding energy of 284.1 eV, 284.7 eV, 286.5 eV, 288.4 eV, and 290.0 eV, which resembles C=C (sp^2 carbon), C-C (sp^3 carbon), C-O, C=O, and O-C=O bonds, respectively [34]. Additionally, Figure 4d denotes the high-resolution XPS spectra of C 1s for RGO. It displays the peak binding energy of 284.4 eV, 285.9 eV, 287.7 eV, 289.1 eV, and 290.6 eV related to C=C, C-OH, C=O, O-C=O, and π-π* satellite bonds, respectively [35]. The XPS investigation demonstrated that after reduction treatment, the functional group of GO is reduced, and the sp^3 carbon is altered to sp^2 carbon.

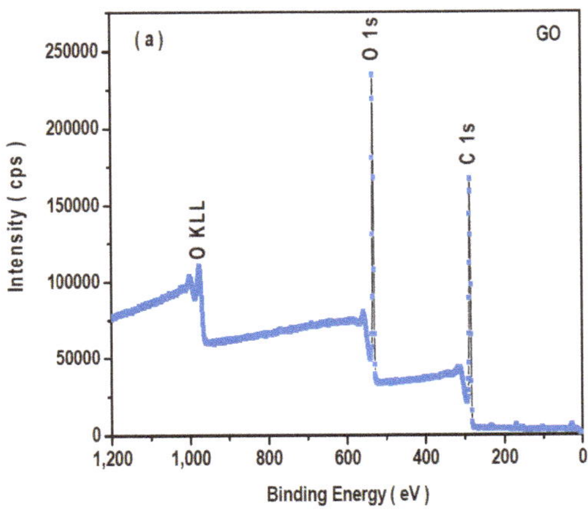

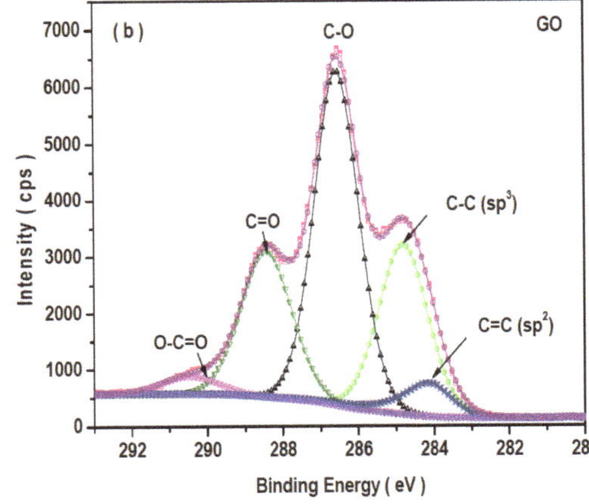

Figure 4. *Cont.*

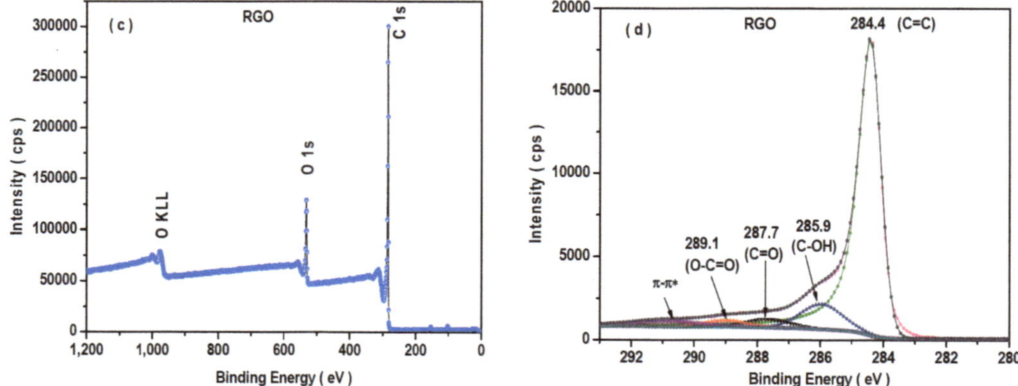

Figure 4. XPS spectra of GO and RGO: (**a**) survey spectrum of GO, (**b**) C1s spectrum of GO, (**c**) survey spectrum of RGO, and (**d**) C1s spectrum of RGO.

3.5. Raman Spectroscopy

The Raman spectroscopy examined the structural properties of synthesized nanoparticles and PUR-based nanocomposites. Figure 5 shows the Raman spectra of prepared nanoparticles, namely ZS25, ZS50, ZS100, and nanocomposites designated as ZS25-RGO-PUR, ZS50-RGO-PUR, and ZS100-RGO-PUR. Two characteristics feature Raman peaks of reduced graphene oxide, located at 1345 cm^{-1} (D band) and 1556 cm^{-1} (G band). The observed D band at 1345 cm^{-1} is ascribed to the existence of sp^3 defects within the reduced graphene oxide sheets, whereas the G band at 1556 cm^{-1} is attributed to the E$_{2g}$ phonon mode of in-plane sp^2 carbon atoms [36]. Further, the observed Raman bands at ~240 cm^{-1}, ~330–340 cm^{-1}, ~470–490 cm^{-1}, ~570 cm^{-1}, and ~650 cm^{-1} were ascribed to T$_{2g}$(1), E$_g$, T$_{2g}$(2), T$_{2g}$(3), and A$_{1g}$ modes for spinel ferrite ZnFe$_2$O$_4$ nanoparticles [37].

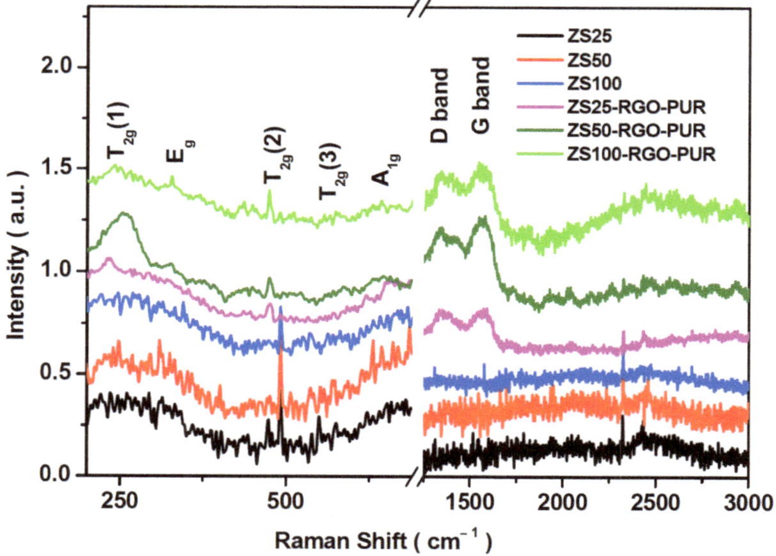

Figure 5. Raman spectra of ZS25, ZS50, ZS100, ZS25-RGO-PUR, ZS50-RGO-PUR, and ZS100-RGO-PUR samples.

3.6. FTIR Spectroscopy

FTIR spectroscopy is an outstanding complementary characterization tool for Raman spectroscopy characterization of nanocomposites. Figure 6a represents the FTIR spectra of prepared $ZnFe_2O_4$ nanoparticles. Two absorption bands at ~545 cm^{-1} and ~385 cm^{-1} were noted. The absorption band ~545 cm^{-1} can be ascribed to the tetrahedral Zn^{2+} (Zn-O) stretching vibration for the $ZnFe_2O_4$ spinel ferrite crystal structure. The band ~385 cm^{-1} can be assigned to the octahedral Fe^{3+} (Fe-O) stretching vibration [38]. Additionally, Figure 6b displays the FTIR spectra of pure polyurethane resin (PUR) and its nanocomposites. The presence of an absorption band ~545 cm^{-1} confirms the existence of $ZnFe_2O_4$ in the developed nanocomposite. Furthermore, Figure 6b demonstrated other absorption bands related to main characteristics peaks for polyurethane, which was observed at ~3310 cm^{-1}, 2966 cm^{-1}, 2868 cm^{-1}, 1728 cm^{-1}, 1640 cm^{-1}, 1534 cm^{-1}, 1223 cm^{-1}, and 1094 cm^{-1}. The absorption band ~3310 cm^{-1} can be ascribed to the stretching vibration of the N-H group [39]. The bands at 2966 cm^{-1} and 2868 cm^{-1} can be attributed to the non-symmetric and symmetric stretching vibration of CH_2. Further, the absorption bands at 1728 cm^{-1}, 1223 cm^{-1}, and 1094 cm^{-1} can be attributed to carbonyl (C=O), aromatic C-O stretching vibration, and C-O-C non-symmetric stretching vibration. Additionally, the band ~1640 cm^{-1} can be ascribed as the abundance of amide I bands [40]. In addition, the absorption band at ~1534 cm^{-1} can be associated with the N-H bonds of the urethane group [41]. In combination with Raman and FTIR spectroscopy results, the existence of spinel ferrite $ZnFe_2O_4$ nanoparticles and reduced graphene oxide in the polyurethane matrix were verified.

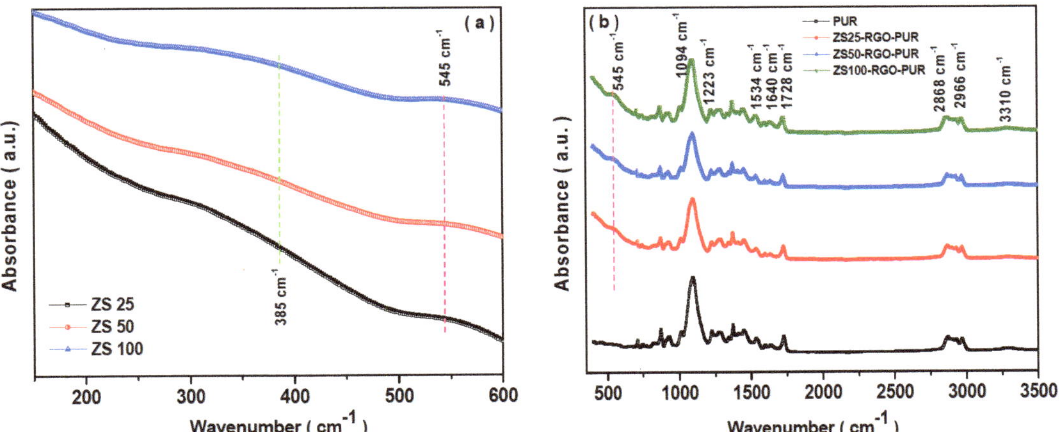

Figure 6. (**a**) FTIR spectra of ZS25, ZS50, and ZS100; (**b**) FTIR spectra of PUR, ZS25-RGO-PUR, ZS50-RGO-PUR, and ZS100-RGO-PUR samples.

3.7. Magnetic Properties

The magnetic characteristics of sonochemically synthesized spinel ferrite $ZnFe_2O_4$ nanoparticles were examined. Figure 7a depicts magnetic hysteresis curves of synthesized $ZnFe_2O_4$ nanoparticles at different sonication times of 25 min, 50 min, and 100 min. The prepared spinel ferrite nanoparticles exhibit zero remanent and zero coercivity, which is associated with superparamagnetic characteristics. The saturation magnetization value at the applied magnetic field of 800 kA/m was 0.78 Am^2/kg, 1.05 Am^2/kg, and 1.33 Am^2/kg for ZS25, ZS50, and ZS100, respectively [42]. The reduced saturation magnetization of ZS25 is associated with the existence of a magnetically dead or anti-ferromagnetic layer on the border of the nanoparticle [43]. The observation of superparamagnetism or ferrimagnetism characteristics at room temperature in nanosized $ZnFe_2O_4$ spinel ferrite has been attributed to cation redistribution between Zn^{2+} ions at the tetrahedral sites and Fe^{3+} ions at the octahedral sites [44,45]. Further, Figure 7b depicts magnetic hysteresis curves of prepared nanoparticle sample ZS25 at various temperatures 2 K, 77 K, and 300 K. At room temperature (300 K) and 77 K, the coercivity and remanent are zero for the ZS25 sample; however, it exhibits coercivity (67 kA/m) and remanent value (5 Am^2/kg) at 2 K, which indicates superparamagnetic behavior of prepared ZS25 sample [46,47].

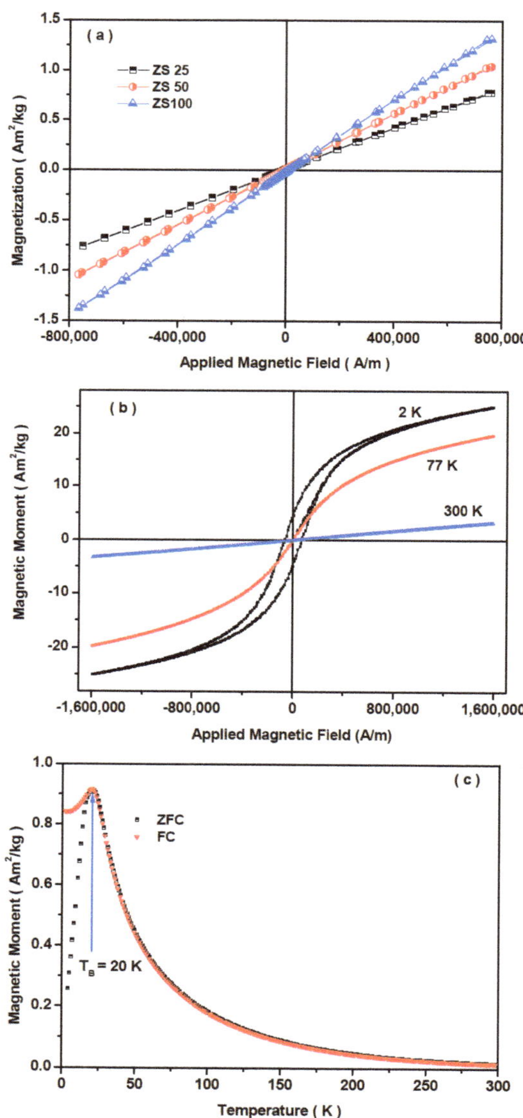

Figure 7. (**a**) M-H plots obtained at 300 K for ZS25, ZS50, and ZS100 samples; (**b**) M-H plots obtained at 2 K, 77 K, and 300 K for the ZS25 sample; (**c**) FC-ZFC curves for the ZS25 sample.

The temperature dependences of zero-field cooled (ZFC) and field-cooled (FC) of prepared nanoparticle sample ZS25 were also measured in wide temperature interval

2–300 K for magnetic field 7.96 kA/m. The ZFC and FC investigation is utilized to define the blocking temperature. Figure 7c displays the irreversibility of ZFC and FC curves and the occurrence of a maximum in ZFC curves. ZFC-FC curve display irreversibility characteristic due to the blocking/freezing and unblocking mechanism of the magnetic moment of magnetic nanoparticles [48]. The rise of ZFC magnetization with temperature is associated with an unblocking progression [49]. The ZFC and FC measurements for the ZS25 sample display the blocking temperature of 20 K. The prepared ZS25 sample displays ferromagnetic behavior below 20 K and superparamagnetic above 20 K. Above the blocking temperature, the magnetic moment of nanoparticles freely fluctuates in the applied magnetic field, which leads to superparamagnetism.

Nevertheless, below the blocking temperature, the magnetic moment of each nanoparticle is blocked/freeze in the applied magnetic field direction, and the magnetic hysteresis with coercivity value 67 kA/m was noticed, as presented in Figure 7b. The blocking temperature depends on various factors such as effective anisotropy constant, magnetic coupling, particle size, applied magnetic field, etc. [50,51]. A research group, Qi Chen et al. [52], observed that the blocking temperature increases with the increase in particle size of $MgFe_2O_4$ nanoparticles. Further, Chao Liu et al. [53] reported the increase in the blocking temperature from 20 to 250 K with the increase in the size of the $MnFe_2O_4$ nanoparticles from 4.4 to 13.5 nm.

It is well-known that the saturation magnetization (M_s) and coercivity (H_c) of the electromagnetic wave absorber material are the most important factor to influence the magnetic loss of the electromagnetic wave shielding material [54]. In general, for the application of electromagnetic interference shielding, an initial permeability (μ_i) signifies strong magnetic loss capacity of electromagnetic wave absorber material, which can be expressed as [55]:

$$\mu_i = \frac{M_S^2}{akH_cM_S + b\lambda\xi} \tag{4}$$

where a and b are constants, which have a dependence on the material. In the above relation, λ, ξ, and k are the magnetostriction constant, elastic strain parameter of the crystal, and proportionality coefficient, respectively [56]. The above equation signifies that the lower H_c and higher M_s are supportive of the value of μ_i increasing and, consequently, the performance of electromagnetic wave absorption enhancing [57]. Superparamagnetic nanoparticles that exhibited zero coercivity and higher saturation magnetization could be high-performance electromagnetic interference shielding material. In the VSM study, it is noticed the ZS100 sample exhibited a high value of magnetization as compared to ZS25 and ZS100; therefore, the permeability of nanocomposites containing ZS100 would be higher. Further, this suggests that a nanocomposite based on ZS100 could have higher electromagnetic shielding performance as compared to ZS25 and ZS50.

3.8. Electromagnetic Interference Shielding Effectiveness (EMI SE) Study

The total electromagnetic interference (EMI) shielding effectiveness (SE_T) can be ascribed by the logarithmic ratio between the incoming power (P_{in}) and outgoing power (P_{out}) of electromagnetic radiation [58]:

$$SE_T = 10\log(P_{in}/P_{out}) = SE_A + SE_R + SE_{MR} \tag{5}$$

Herein, SE_A, SE_R, SE_{MR}, and SE_T are the absorption, reflection, multiple reflections, and total EMI shielding, respectively. In general, SE_{MR} can be ignored when SE_T is larger than 10 dB [58]. Hence, SE_T can be expressed as:

$$SE_T = SE_A + SE_R \tag{6}$$

SE_A in dB can be expressed as [4]:

$$SE_A = -8.68t \frac{\sqrt{(f\mu_r \sigma_T)}}{2} \tag{7}$$

where f is the frequency of the electromagnetic wave; σ_T is the total conductivity (S/cm) of shielding material; (μ_r) is the complex permeability of shielding material. It can be noticed that SE_A is directly proportional to the conductivity (σ_T) and permeability (μ_r) of the shielding material. Further, SE_R can be expressed as [4]:

$$SE_R = -10\log_{10}\left(\frac{\sigma_T}{\mu_r}\right) \tag{8}$$

It can be noticed from the above expression that SE_R is the function of the ratio of σ_T and μ_r. Furthermore, SE_{MR} can be expressed as [4]:

$$SE_{MR} = 20\log_{10}\left(1 - 10^{\frac{SE_A}{10}}\right) \tag{9}$$

SE_{MR} can be neglected at $SE_A \geq 10$ dB.

Figure 8 depicts the EMI-shielding performance of prepared nanocomposites based on the polyurethane matrix with sonochemically prepared superparamagnetic $ZnFe_2O_4$ nanoparticles and RGO as nanofillers. The maximum total shielding effectiveness (SE_T) value for prepared nanocomposites of thickness 1 mm in the frequency range of 8.2–12.4 GHz was 12.7 dB, 13.8 dB, and 16.7 dB, for ZS25-RGO-PUR, ZS50-RGO-PUR, and ZS100-RGO-PUR, respectively. The EMI SE_T value increases with an increase in the size of utilized superparamagnetic $ZnFe_2O_4$ spinel ferrite nanoparticles in developed nanocomposites. The maximum SE_A value was 3.6 dB, 5.9 dB, and 10.2 dB for prepared nanocomposites ZS25-RGO-PUR, ZS50-RGO-PUR, and ZS100-RGO-PUR, respectively. The higher SE_A value of the ZS100-RGO-PUR sample indicates that this nanocomposite exhibits higher electrical conductivity and higher magnetic permeability. Additionally, the maximum SE_R value was noticed to be 10.7 dB, 8.7 dB, and 6.7 dB for developed composite material ZS25-RGO-PUR, ZS50-RGO-PUR, and ZS100-RGO-PUR, respectively.

In recent years, a research group, Xiaogu Huang et al. [59], reported the minimum reflection loss of −3.8 dB, −5 dB, and −8.1 dB for $ZnFe_2O_4$ nanoparticles, nanorods, nanofibers, respectively. Another research group, Chang Sun et al. [60], noticed a minimum reflection loss value of −10.4 dB at 17.7 GHz with a matching thickness of 16.7 mm for $LiFeO_2/ZnFe_2O_4$ composite. Further, Ruiwen Shu et al. [61] noticed the value of the minimum reflection loss of −12.0 dB with a thickness of 4.5 mm for hybrid nanocomposites of reduced graphene oxide/zinc ferrite fabricated by a facile one-pot hydrothermal strategy. Furthermore, Junsheng Xue et al. [62] reported microwave absorption characteristics of $ZnFe_2O_4$ nanoparticles of 130 nm and noticed minimum reflection loss of $ZnFe_2O_4$/paraffin of −23.4 dB at 17.9 GHz for higher thickness around 9 mm. Moreover, Wei Ma et al. [63] reported that the reduced graphene oxide/zinc ferrite/nickel nanohybrids exhibited the reflection loss (RL) of −22.57 dB at 4.21 GHz with a thickness of 2.5 mm. The electromagnetic interference shielding or microwave absorption characteristics of $ZnFe_2O_4$ nanoparticles depend on various factors, including size, morphology, compositions, the thickness of the sample, and composite matrix, etc. [64–66].

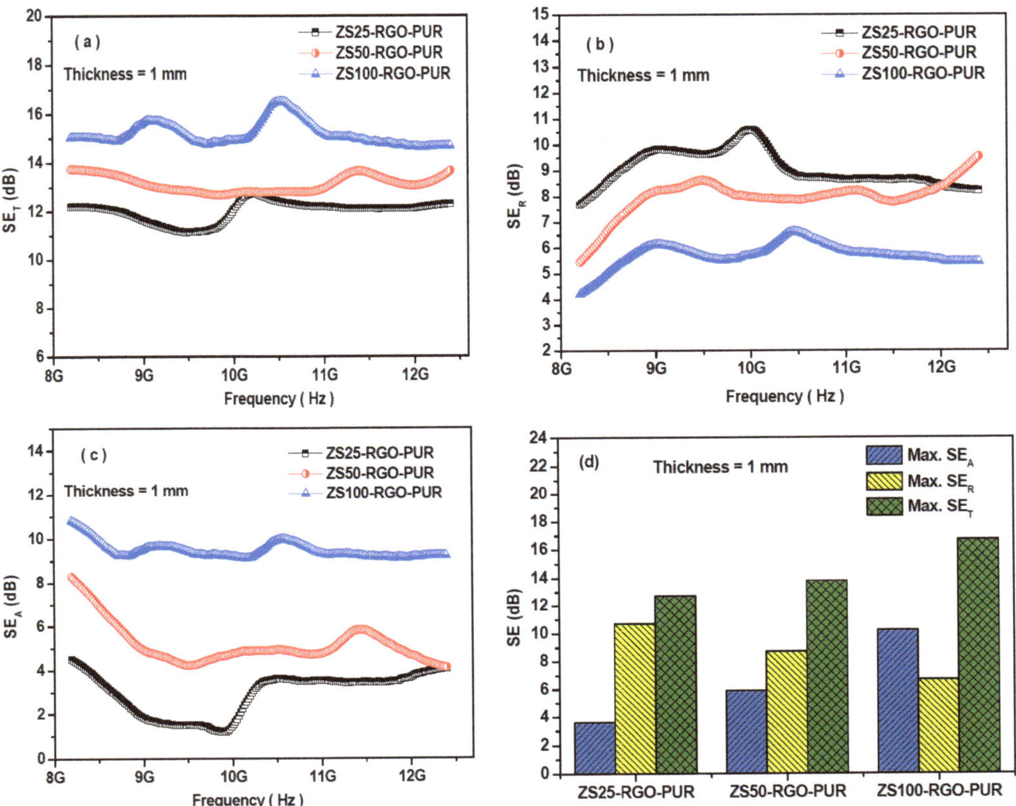

Figure 8. The variation of EMI-shielding effectiveness (**a**) SE_T, (**b**) SE_R, and (**c**) SE_A of superparamagnetic $ZnFe_2O_4$ and RGO-based PUR epoxy nanocomposites at the thickness of 1 mm; (**d**) a comparison of the maximum values of SE_T, SE_R, and SE_A.

3.9. Electromagnetic Properties and Parameters

To elaborate more on the shielding characteristics of prepared superparamagnetic $ZnFe_2O_4$ and RGO-based PUR nanocomposites, complex permittivity and permeability of the nanocomposites were investigated. The real permittivity (ε') implies the storage ability of electrical energy, while the imaginary permittivity (ε'') signifies energy dissipation. Figure 9a represents the frequency dependence real permittivity for developed nanocomposites in the 8.2–12.4 GHz frequency range. The value of ε' is in the ranges of 6.8 to 7.3, 7.5 to 7.9, 8.4 to 9.1 for developed nanocomposites ZS25-RGO-PUR, ZS50-RGO-PUR, and ZS100-RGO-PUR, respectively. Figure 9b represents the frequency dependence imaginary permittivity for nanocomposites in the 8.2–12.4 GHz frequency range. The value of ε'' is in the range of 0.35 to 0.74, 0.55 to 0.88, 0.65 to 1.31 for ZS25-RGO-PUR, ZS50-RGO-PUR, and ZS100-RGO-PUR, respectively. The complex permittivity is the result of the polarizability of the nanocomposite material associated with the dipolar and electric polarization, initiated by an EM wave [5,67]. The input to the space charge polarization acts because of the heterogeneity of the nanocomposite material. In heterogeneous dielectric materials, there is an accumulation of virtual charges on the interfaces of two mediums with different dielectric constants and conductivities, which lead to interfacial polarization and is called Maxwell–Wagner polarization [68,69]. It can be observed that the values of ε' and ε'' are increased with the increase of the size of superparamagnetic $ZnFe_2O_4$ nanoparticles in prepared nanocomposites.

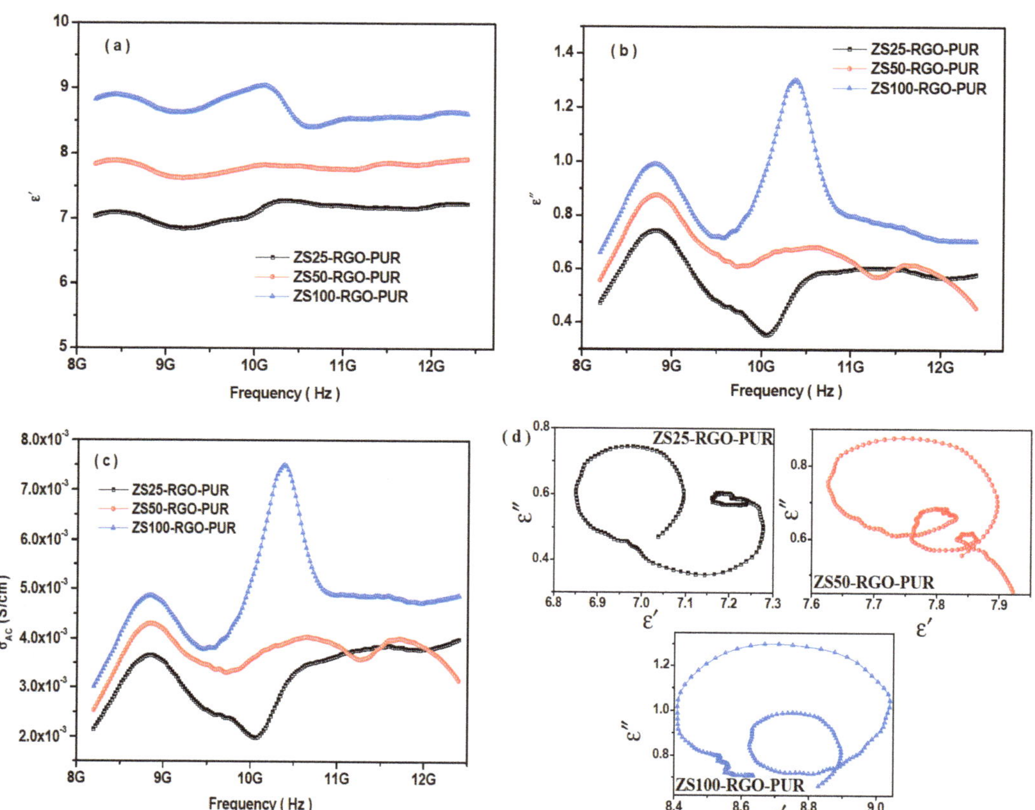

Figure 9. (**a**) Frequency dependence of the real permittivity (ε'), (**b**) frequency dependence of the imaginary permittivity (ε''), (**c**) frequency dependence of the ac conductivity, and (**d**) Cole–Cole plots of nanocomposites.

The relation between electrical conductivity (σ_{AC}) and imaginary permittivity (ε'') can be stated as [70]:

$$\sigma_{AC} = \varepsilon_0 \varepsilon'' 2\pi f \tag{10}$$

Herein, ε_0 is the dielectric constant of free space; f is the frequency of the electromagnetic wave. The above relation signifies that the electrical conductivity will increase with an increase in the value of imaginary permittivity. Therefore, the enhanced value of the complex permittivity can be associated with the increase in the electrical conductivity of the prepared nanocomposites with an increase in the size of embedded superparamagnetic ferrite nanoparticles. Figure 9c represents the change in electrical conductivity with the frequency of prepared nanocomposites. The electrical conductivity is in the range of 1.9×10^{-3} to 3.9×10^{-3} S/cm, 2.5×10^{-3} to 4.3×10^{-3} S/cm, 2.9×10^{-3} to 7.5×10^{-3} S/cm for ZS25-RGO-PUR, ZS50-RGO-PUR, and ZS100-RGO-PUR, respectively. Further, in reported literature by other researchers, the Debye theory is generally utilized to clarify the relaxation process of dipoles [71,72]. According to the Debye theory for dielectric loss characteristics, the real permittivity (ε') and imaginary permittivity (ε'') can be written as [73]:

$$\varepsilon' = \varepsilon_\infty + \frac{\varepsilon_s - \varepsilon_\infty}{1+(\omega\tau)^2}$$
$$\varepsilon'' = \varepsilon''_{relax} + \varepsilon''_\sigma = \frac{\varepsilon_s - \varepsilon_\infty}{1+(\omega\tau)^2}\omega\tau + \frac{\sigma}{\omega\varepsilon_o} \tag{11}$$

Herein, ε_s and ε_∞ are the static and infinite permittivity; $\omega = 2\pi f$ is the angular frequency; τ is the relaxation time; σ is the conductivity. It can be seen from the above relation that the ε' and ε'' are the functions of $\omega\tau$. Hence, both the ε' and ε'' are mutually dependent on one another. A relationship between ε' and ε'' can be inferred after ignoring the contribution of σ and by eliminating $\omega\tau$ [74]:

$$\left(\varepsilon' - \frac{\varepsilon_s + \varepsilon_\infty}{2}\right)^2 + (\varepsilon'')^2 = \left(\frac{\varepsilon_s - \varepsilon_\infty}{2}\right)^2 \tag{12}$$

From the above relation, it is easy to recognize that the curves of ε' and ε'' would be a semi-circle, which is known as the Cole–Cole semicircle [75].

Figure 9d depicts the Cole–Cole plots for the developed PUR-based nanocomposites. In general, the relaxation is associated with a delay in polarization concerning the change in the electrical field. Some obvious Cole–Cole semicircles can be noticed in Figure 9d, which signifies that the relaxation contributed to the dielectric loss. Additionally, one Cole–Cole semicircle represents a Debye dipolar relaxation, and the existence of more semicircles is attributed to multiple relaxation processes [76]. These other semicircles are associated with Maxwell–Wagner relaxation, electron/ion polarization, and interfacial polarization [77]. The multiple dielectric losses were responsible for the improvement of the absorption characteristics of PUR-based nanocomposites.

It is well-known that the real permeability (μ') represents the storage ability of magnetic energy, and the imaginary permeability (μ'') signifies the magnetic loss. Figure 10a represents the frequency dependence of the real permeability (μ') of PUR-based nanocomposites. The μ' is in the range of 0.86 to 0.96, 0.91 to 0.99, and 0.90 to 1.09 for nanocomposites ZS25-RGO-PUR, ZS50-RGO-PUR, and ZS100-RGO-PUR, respectively. The value of real permeability (μ') was increased with an increase of grain size of utilized superparamagnetic $ZnFe_2O_4$ spinel ferrite nanoparticles. Further, the value of μ'' is in the range of -0.06 to 0.03, -0.01 to 0.07, and 0.03 to 0.19 for the prepared composites ZS25-RGO-PUR, ZS50-RGO-PUR, and ZS100-RGO-PUR, respectively, as shown in Figure 10b. Remarkably, it is noticed that the μ'' exhibited negative value also for some PUR-based nanocomposites, which is associated with the motion of charges [78].

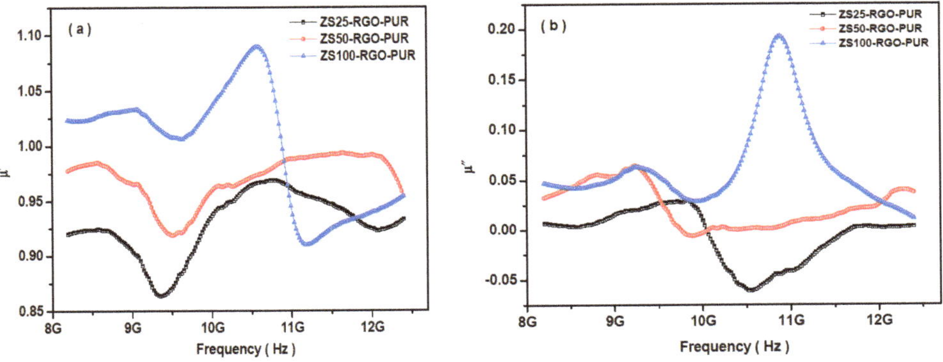

Figure 10. *Cont.*

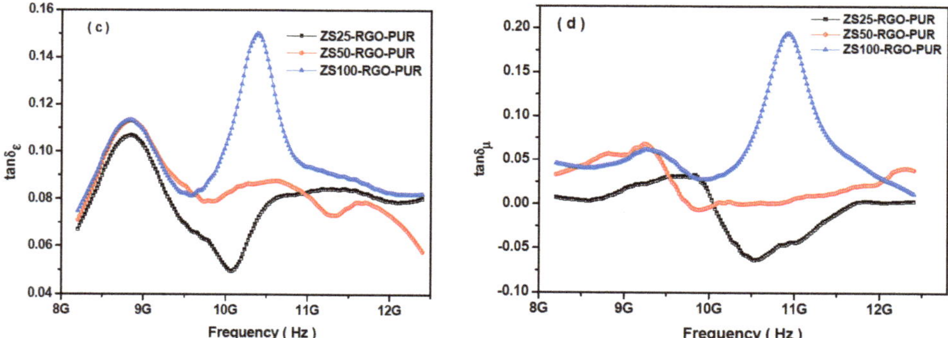

Figure 10. (**a**) Frequency dependence of the real permeability (μ'), (**b**) frequency dependence of the imaginary permeability (μ''), (**c**) dielectric loss tangent, and (**d**) magnetic loss tangent of nanocomposites.

Additionally, the Globus equation is expressed as [79]:

$$\mu \propto \left(M_s^2 D / K_1 \right)^{1/2} \tag{13}$$

This equation signifies that to get a higher complex permeability, a higher saturation magnetization (M_S), larger grain size (D), and smaller magnetocrystalline anisotropy constant (K_1) are needed. The increased magnetization and larger grain size of the ZS100 sample may add to the larger permeability for prepared ZS100-RGO-PUR nanocomposites, as compared with ZS25-RGO-PUR and ZS50-RGO-PUR nanocomposites.

Further, based on the following relations [80]:

$$\tan \delta_\varepsilon = \frac{\varepsilon_r''}{\varepsilon_r'}$$
$$\tan \delta_\mu = \frac{\mu''}{\mu'} \tag{14}$$

and utilizing electromagnetic parameters for ZS25-RGO-PUR, ZS50-RGO-PUR, and ZS100-RGO-PUR nanocomposites, the dielectric loss tangent ($\tan\delta_\varepsilon$) and magnetic loss tangent ($\tan\delta_\mu$) were evaluated. Figure 10c represents dielectric loss tangent vs. frequency curves for prepared PUR-based nanocomposites. The dielectric loss tangent ($\tan\delta_\varepsilon$) of samples ZS25-RGO-PUR, ZS50-RGO-PUR, and ZS100-RGO-PUR, fluctuated with an increase of frequency of electromagnetic wave between 0.05 to 0.10, 0.06 to 0.11, and 0.07 to 0.15, respectively. Additionally, the dielectric loss is related to dipole polarization and interfacial polarization at higher frequencies [81]. It can be also noticed that the dielectric loss (ε'') value of the ZS100-RGO-PUR sample is much higher than the other two samples (i.e., ZS25-RGO-PUR, and ZS50-RGO-PUR). The higher dielectric loss in the ZS100-RGO-PUR sample is associated with enhanced electrical conductivity and dielectric constant induced by micro-currents and polarization in nanocomposites [82].

The magnetic loss tangent variation with the frequency of an electromagnetic wave of prepared PUR-based nanocomposites is presented in Figure 10d. It can be perceived that the magnetic loss tangent fluctuated between -0.06 to 0.03, -0.01 to 0.07, and 0.03 to 0.19 for samples ZS25-RGO-PUR, ZS50-RGO-PUR, and ZS100-RGO-PUR, respectively.

It is well-known that natural resonance, exchange resonance, and eddy current are the main contributors to the magnetic loss of nanoparticles [83]. The eddy current loss can be stated by the following relation when the size of magnetic nanoparticle (D) is smaller than the skin depth (δ) [84]:

$$\frac{\mu''}{\mu'} \propto \frac{\mu' f D}{\rho} \tag{15}$$

where f is the electromagnetic wave frequency; ρ is the electric resistivity of the nanoparticles. Based on this above relation, $C_o = f^{-1}(\mu')^{-2}\mu''$ should be constant, if the magnetic loss is mainly contributed from the eddy current loss. It can be seen in Figure 11a that the value C_o is not constant for all the prepared PUR-based nanocomposites. It signifies that the eddy current loss would not be a dominant contributor to magnetic loss.

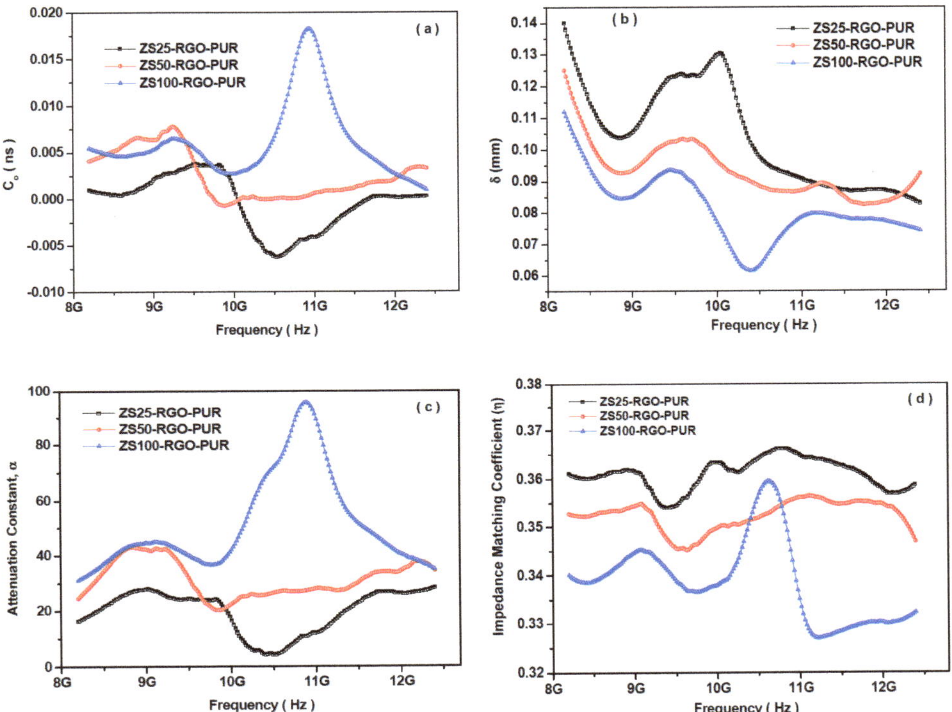

Figure 11. Frequency dependence of (**a**) the eddy current loss, (**b**) skin depth, (**c**) attenuation constant, and (**d**) impedance matching coefficient for prepared PUR-based nanocomposites.

Besides dielectric and magnetic losses, skin depth (δ) is another important factor that stimulates the absorption of electromagnetic waves. Skin depth states the distance at which the field drops to 1/e of the incident value and stated as [85]:

$$\delta = 1/\sqrt{\pi f \mu \sigma} \tag{16}$$

Herein, f is the frequency; σ is the electrical conductivity; μ is the permeability. This relation signifies that skin depth reduces with an increase in frequency, permeability, and conductivity. Figure 11b depicts the frequency dependence variation of skin depth of the prepared PUR-based nanocomposites. A smaller skin depth states a stronger absorption capacity [86]. The skin depth of samples ZS25-RGO-PUR, ZS50-RGO-PUR, and ZS100-RGO-PUR fluctuated with an increase in the frequency of electromagnetic waves between 0.08 to 0.14 mm, 0.08 to 0.12 mm, and 0.06 to 0.11 mm, respectively. The prepared ZS100-RGO-PUR nanocomposite exhibits smaller skin depth and, therefore, stronger absorption.

The attenuation constant (α) is an important factor that governs the electromagnetic wave absorption capabilities of shielding nanocomposites. It can be assessed by the following relation [87]:

$$\alpha = \frac{\sqrt{2}\pi f}{c}\sqrt{(\mu''\varepsilon'' - \mu'\varepsilon') + \sqrt{(\mu'\varepsilon'' + \mu''\varepsilon')^2 + (\mu''\varepsilon'' - \mu'\varepsilon')^2}} \tag{17}$$

Figure 11c displays the frequency dependence variation of the attenuation constant (α) of prepared PUR-based nanocomposites. The high attenuation constant (α) value of the prepared ZS100-RGO-PUR nanocomposite compared with other samples demonstrated high absorption of electromagnetic waves. Another key factor that governs electromagnetic wave absorption is impedance matching. It is stated by the modulus of the normalized characteristic impedance (Z), which can be calculated by utilizing the following relation [88]:

$$Z = |Z_1/Z_o| \tag{18}$$

where $Z_1 = Z_o\sqrt{\mu_r/\varepsilon_r}$; Zo is the impedance in free space; ε_r is the value of complex permittivity; μ_r is the value of complex permeability. Figure 11d shows the variation of the impedance matching coefficient for prepared PUR-based nanocomposites with frequency. The values of Z are below one, and the ZS100-RGO-PUR nanocomposite has a lower Z value compared to other samples. A high attenuation constant and moderate Z value of ZS100-RGO-PUR nanocomposite provided the high value of EMI-shielding effectiveness [89]. The schematic illustration of the electromagnetic interference shielding mechanism in the prepared nanocomposite is shown in Figure 12. When electromagnetic waves interact at the surface of the prepared nanocomposite, a part of it is reflected, another part is absorbed, and the remaining part has multiple reflections and scattering [90]. The reflection is associated with moving charge carriers interacted with electromagnetic waves [91]. The absorption signifies the dissipation of energy of the electromagnetic waves due to the interaction of electromagnetic waves with the electric and magnetic dipoles [92]. The multiple reflections are reflections at different surfaces or interfaces present due to inhomogeneity within the prepared nanocomposite. The prepared nanocomposites consisted of conductive RGO sheets and $ZnFe_2O_4$ magnetic nanoparticles not only improve impedance matching but also creates a micro-current network and attains interfacial polarization [93]. The appropriate conductivity of RGO sheets gifted the prepared nanocomposites exhibited a moderate conductivity loss. Residual functional groups and defects in RGO sheets and $ZnFe_2O_4$ originated dipole polarization and defect polarization [94]. The interaction of electromagnetic waves also creates hopping and migrating electrons across the defects of RGO sheets.

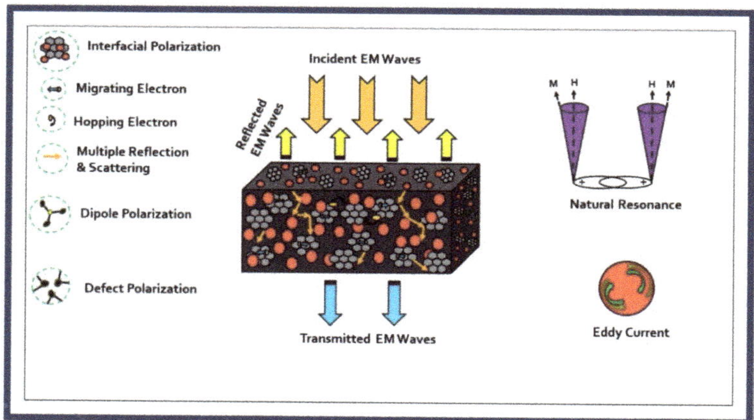

Figure 12. Schematic illustration of the electromagnetic interference shielding mechanism in prepared nanocomposites.

The magnetic characteristics of the superparamagnetic $ZnFe_2O_4$ nanoparticles component in the developed nanocomposites provided a degree of magnetic loss such as natural resonance and eddy current loss. It improves the impedance matching between complex permittivity and permeability, which provides well absorption condition for electromagnetic waves [95]. The improved electromagnetic wave shielding characteristics of the ZS100-RGO-PUR nanocomposite can be primarily attributed to the inclusive acts of magnetic loss, dielectric loss, and appropriate attenuation constant derived from various nanofillers in the matrix.

4. Conclusions

In summary, superparamagnetic $ZnFe_2O_4$ spinel ferrite nanoparticles were prepared successfully by the sonochemical synthesis approach at various ultra-sonication times of 25 min (ZS25), 50 min (ZS50), and 100 min (ZS100). The average crystallite size increased from 3.0 nm to 4.0 nm with an increase in sonication time. The lattice parameter increased from 7.219 Å to 7.248 Å with an increase in sonication time from 25 min to 100 min. The increase in ionic radii, hopping length for the octahedral and tetrahedral site, tetrahedral and octahedral bond length, tetrahedral edge, and shared and unshared octahedral edge for prepared $ZnFe_2O_4$ nanoparticles with an increase in sonication time is associated with cation redistribution in $ZnFe_2O_4$ nanoparticles with an increase in sonication time. The prepared spinel ferrite nanoparticles exhibited zero remanent and zero coercivity, which is associated with superparamagnetic characteristics. The prepared magnetic $ZnFe_2O_4$ nanoparticles (ZS25, ZS50, and ZS100) and electrically conductive reduced graphene oxide (RGO) were embedded in a polyurethane resin (PUR) matrix to develop lightweight and flexible nanocomposites for electromagnetic interference shielding application. The maximum total shielding effectiveness (SE_T) value for developed nanocomposites of thickness 1 mm in the range of 8.2–12.4 GHz frequency was 12.7 dB, 13.8 dB, and 16.7 dB, for ZS25-RGO-PUR, ZS50-RGO-PUR, and ZS100-RGO-PUR, respectively. The higher attenuation constant (α) value of in prepared ZS100-RGO-PUR nanocomposite as compared with other samples demonstrated high absorption of electromagnetic waves. This work demonstrated an ingenious and effective strategy to develop polyurethane resin-based nanocomposites consisting of superparamagnetic $ZnFe_2O_4$ spinel ferrite nanoparticles with RGO for shielding electromagnetic pollution.

Supplementary Materials: The following are available online at https://www.mdpi.com/article/10.3390/nano11051112/s1, Figure S1: (a) TEM image of ZS25, (b) size distribution of ZS25, (c) TEM image of ZS50, (d) size distribution of ZS50, (e) TEM image of ZS100, (f) size distribution of ZS100.

Author Contributions: A. and T.J. performed the experiments; D.Š., P.U., M.M. (Michal Machovský), M.M. (Milan Masař), M.U., and L.K. performed the characterizations; R.S.Y., I.K., J.V. and J.H. analyzed the data and wrote the manuscript. All authors have read and agreed to the published version of the manuscript.

Funding: We thank the financial support of the Czech Science Foundation (GA19-23647S) project at the Centre of Polymer Systems, Tomas Bata University in Zlin, Czech Republic. One author, Anju, also acknowledges the financial support by the internal grant no. IGA/CPS/2020/003 for specific research from Tomas Bata University in Zlín.

Institutional Review Board Statement: Not applicable.

Informed Consent Statement: Not applicable.

Data Availability Statement: Data can be available upon request from the corresponding author.

Acknowledgments: Authors acknowledge Oldrich Schneeweiss, Institute of Physics of Materials, Academy of Sciences of the Czech Republic, Zizkova 22, 616 62 Brno, Czech Republic, for the zero-field-cooled (ZFC) and field-cooled (FC) temperature-dependent magnetization measurement.

Conflicts of Interest: The authors declare no conflict of interest.

References

1. Abbasi, H.; Antunes, M.; Velasco, J.I. Recent advances in carbon-based polymer nanocomposites for electromagnetic interference shielding. *Prog. Mater. Sci.* **2019**, *103*, 319–373. [CrossRef]
2. Gupta, S.; Tai, N.-H. Carbon materials and their composites for electromagnetic interference shielding effectiveness in X-band. *Carbon* **2019**, *152*, 159–187. [CrossRef]

3. Wang, C.; Murugadoss, V.; Kong, J.; He, Z.; Mai, X.; Shao, Q.; Chen, Y.; Guo, L.; Liu, C.; Angaiah, S.; et al. Overview of carbon nanostructures and nanocomposites for electromagnetic wave shielding. *Carbon* **2018**, *140*, 696–733. [CrossRef]

4. Sankaran, S.; Deshmukh, K.; Ahamed, M.B.; Pasha, S.K.K. Recent advances in electromagnetic interference shielding properties of metal and carbon filler reinforced flexible polymer composites: A review. *Compos. Part A* **2018**, *114*, 49–71. [CrossRef]

5. Sushmita, K.; Madras, G.; Bose, S. Polymer Nanocomposites Containing Semiconductors as Advanced Materials for EMI Shielding. *ACS Omega* **2020**, *5*, 4705–4718. [CrossRef] [PubMed]

6. Li, S.; Li, J.; Ma, N.; Liu, D.; Sui, G. Super-Compression-Resistant Multiwalled Carbon Nanotube/Nickel Coated Carbonized Loofah Fiber/Polyether Ether Ketone Composite with Excellent Electromagnetic Shielding Performance. *ACS Sustain. Chem. Eng.* **2019**, *7*, 13970–13980. [CrossRef]

7. Kumaran, R.; Kumar, S.D.; Balasubramanian, N.; Alagar, M.; Subramanian, V.; Dinakaran, K. Enhanced Electromagnetic Interference Shielding in a Au-MWCNT Composite Nanostructure Dispersed PVDF Thin Films. *J. Phys. Chem. C* **2016**, *120*, 13771–13778. [CrossRef]

8. Chhetri, S.; Adak, N.C.; Samanta, P.; Murmu, N.C.; Srivastava, S.K.; Kuila, T. Synergistic effect of Fe_3O_4 anchored N-doped rGO hybrid on mechanical, thermal and electromagnetic shielding properties of epoxy composites. *Compos. Part B* **2019**, *166*, 371–381. [CrossRef]

9. Song, W.-L.; Gong, C.; Li, H.; Cheng, X.-D.; Chen, M.; Yuan, X.; Chen, H.; Yang, Y.; Fang, D. Graphene-Based Sandwich Structures for Frequency Selectable Electromagnetic Shielding. *ACS Appl. Mater. Interfaces* **2017**, *9*, 36119–36129. [CrossRef] [PubMed]

10. Li, M.; Yang, K.; Zhu, W.; Shen, J.; Rollinson, J.; Hella, M.; Lian, J. Copper-Coated Reduced Graphene Oxide Fiber Mesh-Polymer Composite Films for Electromagnetic Interference Shielding. *ACS Appl. Nano Mater.* **2020**, *3*, 5565–5574. [CrossRef]

11. Jia, Z.; Zhang, M.; Liu, B.; Wang, F.; Wei, G.; Su, Z. Graphene Foams for Electromagnetic Interference Shielding: A Review. *ACS Appl. Nano Mater.* **2020**, *3*, 6140–6155. [CrossRef]

12. Prasad, J.; Singh, A.K.; Haldar, K.K.; Tomar, M.; Gupta, V.; Singh, K. $CoFe_2O_4$ nanoparticles decorated MoS2-reduced graphene oxide nanocomposite for improved microwave absorption and shielding performance. *RSC Adv.* **2019**, *9*, 21881. [CrossRef]

13. Liang, C.; Qiu, H.; Han, Y.; Gu, H.; Song, P.; Wang, L.; Kong, J.; Cao, D.; Gu, J. Superior electromagnetic interference shielding 3D graphene nanoplatelets/reduced graphene oxide foam/epoxy nanocomposites with high thermal conductivity. *J. Mater. Chem. C* **2019**, *7*, 2725. [CrossRef]

14. Xiao, Y.; Du, J. Superparamagnetic nanoparticles for biomedical applications. *J. Mater. Chem. B* **2020**, *8*, 354–367. [CrossRef]

15. Li, B.; Cao, H.; Shao, J.; Qu, M.; Warner, J.H. Superparamagnetic Fe_3O_4 nanocrystals@graphene composites for energy storage devices. *J. Mater. Chem.* **2011**, *21*, 5069. [CrossRef]

16. Liu, P.; Huang, Y.; Zhang, X. Superparamagnetic $NiFe_2O_4$ particles on poly(3,4-ethylenedioxythiophene)-graphene: Synthesis, characterization and their excellent microwave absorption properties. *Compos. Sci. Technol.* **2014**, *95*, 107–113. [CrossRef]

17. Yuan, H.; Xu, Y.; Jia, H.; Zhou, S. Superparamagnetic Fe_3O_4/MWCNTs heterostructures for high frequency microwave absorption. *RSC Adv.* **2016**, *6*, 67218. [CrossRef]

18. Mozaffari, M.; Arani, M.E.; Amighian, J. The Effect of Cation Distribution on Magnetization of $ZnFe_2O_4$ Nanoparticles. *J. Magn. Magn. Mater.* **2010**, *322*, 3240–3244. [CrossRef]

19. Ammar, S.; Jouini, N.; Fievet, F.; Stephan, O.; Marhic, C.; Richard, M.; Villain, F.; Moulin, C.C.D.; Brice, S.; Sainctavit, P. Influence of the Synthesis Parameters on the Cationic Distribution of $ZnFe_2O_4$ Nanoparticles Obtained by Forced Hydrolysis in Polyol Medium. *J. Non-Cryst. Solids* **2004**, *345–346*, 658–662. [CrossRef]

20. Selvarajan, S.; Suganthi, A.; Rajarajan, M. Fabrication of g-C3N4/NiO heterostructured nanocomposite modified glassy carbon electrode for quercetin biosensor. *Ultrason. Sonochem.* **2018**, *41*, 651–660. [CrossRef]

21. Rani, K.K.; Karuppiah, C.; Wang, S.-F.; Alaswad, S.O.; Sireesha, P.; Devasenathipathy, R.; Jose, R.; Yang, C.-C. Direct pyrolysis and ultrasound assisted preparation of N, S co-doped graphene/Fe3C nanocomposite as an efficient electrocatalyst for oxygen reduction and oxygen evolution reactions. *Ultrason. Sonochem.* **2020**, *66*, 105111. [CrossRef] [PubMed]

22. Zhang, R.; Wang, Y.; Ma, D.; Ahmed, S.; Qin, W.; Liu, Y. Effects of ultrasonication duration and graphene oxide and nano-zinc oxide contents on the properties of polyvinyl alcohol nanocomposites. *Ultrason. Sonochem.* **2019**, *59*, 104731. [CrossRef] [PubMed]

23. Mirzajani, R.; Pourreza, N.; Burromandpiroze, J. Fabrication of magnetic Fe_3O_4@nSiO_2@mSiO_2-NH_2 core—Shell mesoporous nanocomposite and its application for highly efficient ultrasound assisted dispersive µSPE-spectrofluorimetric detection of ofloxacin in urine and plasma samples. *Ultrason. Sonochem.* **2018**, *40*, 101–112. [CrossRef]

24. Bhanvase, B.A.; Veer, A.; Shirsath, S.R.; Sonawane, S.H. Ultrasound assisted preparation, characterization and adsorption study of ternary chitosan-ZnO-TiO_2 nanocomposite: Advantage over conventional method. *Ultrason. Sonochem.* **2019**, *52*, 120–130. [CrossRef]

25. Cullity, B.D.; Stock, S.R. *Elements of X-ray Diffraction*, 3rd ed.; Prentice-Hall: New York, NY, USA, 2001.

26. Yadav, R.S.; Mishra, P.; Pandey, A.C. Growth mechanism and optical property of ZnOnanoparticles synthesized by sonochemical method. *Ultrason. Sonochem.* **2008**, *15*, 863–868. [CrossRef]

27. Yadav, R.S.; Kuřitka, I.; Vilcakova, J.; Havlica, J.; Masilko, J.; Kalina, L.; Tkacz, J.; Švec, J.; Enev, V.; Hajdúchová, M. Impact of grain size and structural changes on magnetic, dielectric, electrical, impedance and modulus spectroscopic characteristics of $CoFe_2O_4$ nanoparticles synthesized by honey mediated sol-gel combustion method. *Adv. Nat. Sci. Nanosci. Nanotechnol.* **2017**, *8*, 045002. [CrossRef]

28. Yadav, R.S.; Havlica, J.; Masilko, J.; Kalina, L.; Wasserbauer, J.; Hajdúchová, M.; Enev, V.; Kuřitka, I.; Kožáková, Z. Effects of annealing temperature variation on the evolution of structural and magnetic properties of $NiFe_2O_4$ nanoparticles synthesized by starch-assisted sol-gel auto-combustion method. *J. Magn. Magn. Mater.* **2015**, *394*, 439–447. [CrossRef]

29. Sinha, A.; Dutta, A. Structural, optical, and electrical transport properties of some rare-earth-doped nickel ferrites: A study on effect of ionic radii of dopants. *J. Phys. Chem. Solids* **2020**, *145*, 109534. [CrossRef]

30. Gul, S.; Yousuf, M.A.; Anwar, A.; Warsi, M.F.; Agboola, P.O.; Shakir, I.; Shahid, M. Al-substituted zinc spinel ferrite nanoparticles: Preparation and evaluation of structural, electrical, magnetic and photocatalytic properties. *Ceram. Int.* **2020**, *46*, 14195–14205. [CrossRef]

31. Patange, S.M.; Shirsath, S.E.; Jangam, G.S.; Lohar, K.S.; Jadhav, S.S.; Jadhav, K.M. Rietveld structure refinement, cation distribution and magnetic properties of Al^{3+} substituted $NiFe_2O_4$ nanoparticles. *J. Appl. Phys.* **2011**, *109*, 053909. [CrossRef]

32. Zhou, X.; Li, X.; Sun, H.; Sun, P.; Liang, X.; Liu, F.; Hu, X.; Lu, G. Nanosheet-Assembled $ZnFe_2O_4$ Hollow Microspheres for High-Sensitive Acetone Sensor. *ACS Appl. Mater. Interfaces* **2015**, *7*, 15414–15421. [CrossRef] [PubMed]

33. Mallik, A.K.; Habib, M.L.; Robel, F.N.; Shahruzzaman, M.; Haque, P.; Rahman, M.M.; Devanath, V.; Martin, D.J.; Nanjundan, A.K.; Yamauchi, Y.; et al. Reduced Graphene Oxide (rGO) Prepared by Metal-Induced Reduction of Graphite Oxide: Improved Conductive Behavior of a Poly(methyl methacrylate) (PMMA)/rGO Composite. *ChemistrySelect* **2019**, *4*, 7954–7958. [CrossRef]

34. Genorio, B.; Harrison, K.L.; Connell, J.G.; Dražić, G.; Zavadil, K.R.; Markovic, N.M.; Strmcnik, D. Tuning the Selectivity and Activity of Electrochemical Interfaces with Defective Graphene Oxide and Reduced Graphene Oxide. *ACS Appl. Mater. Interfaces* **2019**, *11*, 34517–34525. [CrossRef] [PubMed]

35. Ossonon, B.D.; Belanger, D. Synthesis and characterization of sulfophenyl functionalized reduced graphene oxide sheets. *RSC Adv.* **2017**, *7*, 27224. [CrossRef]

36. Liu, H.; Dong, M.; Huang, W.; Gao, J.; Dai, K.; Guo, J.; Zheng, G.; Liu, C.; Shen, C.; Guo, Z. Lightweight conductive graphene/thermoplastic polyurethane foams with ultrahigh compressibility for piezoresistive sensing. *J. Mater. Chem. C* **2017**, *5*, 73. [CrossRef]

37. Dolcet, P.; Kirchberg, K.; Antonello, A.; Suchomski, C.; Marschall, R.; Diodati, S.; Muñoz-Espí, R.M.; Landfester, K.; Gross, S. Exploring wet chemistry approaches to $ZnFe_2O_4$ spinel ferrite nanoparticles with different inversion degrees: A comparative study. *Inorg. Chem. Front.* **2019**, *6*, 1527. [CrossRef]

38. Sun, Q.; Wu, K.; Zhang, J.; Sheng, J. Construction of $ZnFe_2O_4$/rGO composites as selective magnetically recyclable photocatalysts under visible light irradiation. *Nanotechnology* **2019**, *30*, 315706. [CrossRef]

39. Gavgani, J.N.; Adelnia, H.; Zaarei, D.; Gudarzi, M.M. Lightweight flexible polyurethane/reduced ultralarge graphene oxide composite foams for electromagnetic interference shielding. *RSC Adv.* **2016**, *6*, 27517–27527. [CrossRef]

40. Barman, S.; Parasar, B.; Kundu, P.; Roy, S. A copper based catalyst for poly-urethane synthesis from discarded motherboard. *RSC Adv.* **2016**, *6*, 75749. [CrossRef]

41. Xiao, Y.; Huang, H.; Peng, X. Synthesis of self-healing waterborne polyurethanes containing sulphonate groups. *RSC Adv.* **2017**, *7*, 20093. [CrossRef]

42. Zhang, W.; Wang, M.; Zhao, W.; Wang, B. Magnetic composite photocatalyst $ZnFe_2O_4$/$BiVO_4$: Synthesis, characterization, and visible-light photocatalytic activity. *Dalton Trans.* **2013**, *42*, 15464–15474. [CrossRef]

43. Xuan, S.; Wang, F.; Wang, Y.-X.J.; Yu, J.C.; Leung, K.C.-F. Facile synthesis of size-controllable monodispersed ferrite nanospheres. *J. Mater. Chem.* **2010**, *20*, 5086–5094. [CrossRef]

44. Cobos, M.A.; de la Presa, P.; Llorente, I.; Alonso, J.M.; García-Escorial, A.; Marín, P.; Hernando, A.; Jiménez, J.A. Magnetic Phase Diagram of Nanostructured Zinc Ferrite as a Function of Inversion Degree δ. *J. Phys. Chem. C* **2019**, *123*, 17472–17482. [CrossRef]

45. Lopez-Maldonado, K.L.; de la Presa, P.; Betancourt, I.; Farias Mancilla, J.R.; Matutes Aquino, J.A.; Hernando, A.; Elizalde Galindo, J.T. Superparamagnetic response of zinc ferrite incrusted nanoparticles. *J. Alloys Compd.* **2015**, *637*, 443–448. [CrossRef]

46. Zhang, J.; Zhang, Y.; Wu, X.; Ma, Y.; Chien, S.-Y.; Guan, R.; Zhang, D.; Yang, B.; Yan, B.; Yang, J. Correlation between Structural Changes and Electrical Transport Properties of Spinel $ZnFe_2O_4$ Nanoparticles under High Pressure. *ACS Appl. Mater. Interfaces* **2018**, *10*, 42856–42864. [CrossRef]

47. Ivashchenko, O.; Peplińska, B.; Gapiński, J.; Flak, D.; Jarek, M.; Załęski, K.; Nowaczyk, G.; Pietralik, Z.; Jurga, S. Silver and ultrasmall iron oxides nanoparticles in hydrocolloids: Effect of magnetic field and temperature on self-organization. *Sci. Rep.* **2018**, *8*, 4041. [CrossRef] [PubMed]

48. Singh, S.; Tovstolytkin, A.; Lotey, G.S. Magnetic properties of superparamagnetic β-$NaFeO_2$ nanoparticles. *J. Magn. Magn. Mater.* **2018**, *458*, 62–65. [CrossRef]

49. Stoner, E.C.; Wohlfarth, E.P. A mechanism of magnetic hysteresis in heterogeneous alloys. *Philos. Trans. R. Soc. A* **1948**, *240*, 599–642. [CrossRef]

50. Presa, P.D.L.; Luengo, Y.; Multigner, M.; Costo, R.; Morales, M.P.; Rivero, G.; Hernando, A. Study of Heating Efficiency as a Function of Concentration, Size, and Applied Field in γ-Fe_2O_3 Nanoparticles. *J. Phys. Chem. C* **2012**, *116*, 25602–25610. [CrossRef]

51. Zeleňáková, A.; Kováč, J.; Zeleňák, V. Magnetic properties of Fe_2O_3 nanoparticles embedded in hollows of periodic nanoporous silica. *J. Appl. Phys.* **2010**, *108*, 034323. [CrossRef]

52. Chen, Q.; Zhang, Z.J. Size-dependent superparamagnetic properties of $MgFe_2O_4$ spinel ferrite nanocrystallites. *Appl. Phys. Lett.* **1998**, *73*, 3156. [CrossRef]

53. Liu, C.; Zhang, Z.J. Size-Dependent Superparamagnetic Properties of Mn Spinel Ferrite Nanoparticles Synthesized from Reverse Micelles. *Chem. Mater.* **2001**, *13*, 2092–2096. [CrossRef]

54. Zhu, X.; Qiu, H.; Chen, P.; Chen, G.; Min, W. Anemone-shaped ZIF-67@CNTs as effective electromagnetic absorbent covered the whole X-band. *Carbon* **2021**, *173*, 1–10. [CrossRef]

55. Wang, C.; Han, X.; Xu, P.; Wang, J.; Du, Y.; Wang, X.; Qin, W.; Zhang, T. Controlled Synthesis of Hierarchical Nickel and Morphology-Dependent Electromagnetic Properties. *J. Phys. Chem. C* **2010**, *114*, 3196–3203. [CrossRef]

56. Zhang, W.; Zhang, X.; Zheng, Y.; Guo, C.; Yang, M.; Li, Z.; Wu, H.; Qiu, H.; Yan, H.; Qi, S. Preparation of Polyaniline@MoS$_2$@Fe$_3$O$_4$ Nanowires with a Wide Band and Small Thickness toward Enhancement in Microwave Absorption. *ACS Appl. Nano Mater.* **2018**, *1*, 5865–5875. [CrossRef]

57. Wei, H.; Zhang, Z.; Hussain, G.; Zhou, L.; Li, Q.; Ostrikov, K.K. Techniques to enhance magnetic permeability in microwave absorbing materials. *Appl. Mater. Today* **2020**, *19*, 100596. [CrossRef]

58. Cheng, K.; Li, H.; Zhu, M.; Qiu, H.; Yang, J. In situ polymerization of graphenepolyaniline@polyimide composite films with high EMI shielding and electrical properties. *RSC Adv.* **2020**, *10*, 2368. [CrossRef]

59. Huang, X.; Qin, Y.; Ma, Y.; Chen, Y. Preparation and electromagnetic properties of nanosized ZnFe$_2$O$_4$ with various shapes. *Ceram. Int.* **2019**, *45*, 18389–18397. [CrossRef]

60. Sun, C.; Cheng, C.; Sun, M.; Zhang, Z. Facile synthesis and microwave absorbing properties of LiFeO$_2$/ZnFe$_2$O$_4$ composite. *J. Magn. Magn. Mater.* **2019**, *482*, 79–83. [CrossRef]

61. Shu, R.; Li, W.; Zhou, X.; Tian, D.; Zhang, G.; Gan, Y.; Shi, J.; He, J. Facile preparation and microwave absorption properties of RGO/MWCNTs/ZnFe$_2$O$_4$ hybrid nanocomposites. *J. Alloys Compd.* **2018**, *743*, 163–174. [CrossRef]

62. Xue, J.; Zhang, H.; Zhao, J.; Ou, X.; Ling, Y. Characterization and microwave absorption of spinel MFe$_2$O$_4$ (M=Mg, Mn, Zn) nanoparticles prepared by a facile oxidation-precipitation process. *J. Magn. Magn. Mater.* **2020**, *514*, 167168. [CrossRef]

63. Ma, W.; Yang, R.; Yang, Z.; Duan, C.; Wang, T. Synthesis of reduced graphene oxide/zinc ferrite/nickel nanohybrids: As a lightweight and high-performance microwave absorber in the low frequency. *J. Mater. Sci. Mater. Electron.* **2019**, *30*, 18496–18505. [CrossRef]

64. Ge, Y.; Li, C.; Waterhouse, G.I.N.; Zhang, Z.; Yu, L. ZnFe$_2$O$_4$@SiO$_2$@Polypyrrole nanocomposites with efficient electromagnetic wave absorption properties in the K and Ka band regions. *Ceram. Int.* **2021**, *47*, 1728–1739. [CrossRef]

65. Di, X.; Wang, Y.; Fu, Y.; Wu, X.; Wang, P. Wheat flour-derived nanoporous carbon@ZnFe$_2$O$_4$ hierarchical composite as an outstanding microwave absorber. *Carbon* **2021**, *173*, 174–184. [CrossRef]

66. Chai, L.; Wang, Y.; Zhou, N.; Du, Y.; Zeng, X.; Zhou, S.; He, Q.; Wu, G. In-situ growth of core-shell ZnFe$_2$O$_4$ @ porous hollow carbon microspheres as an efficient microwave absorber. *J. Colloid Interface Sci.* **2021**, *581*, 475–484. [CrossRef] [PubMed]

67. Arjmand, M.; Moud, A.A.; Li, Y.; Sundararaj, U. Outstanding electromagnetic interference shielding of silver nanowires: Comparison with carbon nanotubes. *RSC Adv.* **2015**, *5*, 56590. [CrossRef]

68. Chen, N.; Jiang, J.-T.; Xu, C.-Y.; Yan, S.-J.; Zhen, L. Rational Construction of Uniform CoNi-Based Core-Shell Microspheres with Tunable Electromagnetic Wave Absorption Properties. *Sci. Rep.* **2018**, *8*, 3196. [CrossRef] [PubMed]

69. Singh, A.K.; Kumar, A.; Srivastava, A.; Yadav, A.N.; Haldar, K.; Gupta, V.; Singh, K. Lightweight reduced graphene oxide-ZnO nanocomposite for enhanced dielectric loss and excellent electromagnetic interference shielding. *Compos. Part B* **2019**, *172*, 234–242. [CrossRef]

70. Liu, W.; Liu, L.; Ji, G.; Li, D.; Zhang, Y.; Ma, J.; Du, Y. Composition Design and Structural Characterization of MOF-Derived Composites with Controllable Electromagnetic Properties. *ACS Sustain. Chem. Eng.* **2017**, *5*, 7961–7971. [CrossRef]

71. Zhao, Y.; Wang, W.; Wang, J.; Zhai, J.; Lei, X.; Zhao, W.; Li, J.; Yang, H.; Tian, J.; Yan, J. Constructing multiple heterogeneous interfaces in the composite of bimetallic MOF-derivatives and rGO for excellent microwave absorption performance. *Carbon* **2021**, *173*, 1059–1072. [CrossRef]

72. Cui, Y.; Yang, K.; Wang, J.; Shah, T.; Zhang, Q.; Zhang, B. Preparation of pleated RGO/MXene/Fe$_3$O$_4$ microsphere and its absorption properties for electromagnetic wave. *Carbon* **2021**, *172*, 1–14. [CrossRef]

73. Hou, Y.; Cheng, L.; Zhang, Y.; Du, X.; Zhao, Y.; Yang, Z. High temperature electromagnetic interference shielding of lightweight and flexible ZrC/SiC nanofiber mats. *Chem. Eng. J.* **2021**, *404*, 126521. [CrossRef]

74. Zhang, Y.; Qiu, M.; Yu, Y.; Wen, B.; Cheng, L. A Novel Polyaniline-Coated Bagasse Fiber Composite with Core-Shell Heterostructure Provides Effective Electromagnetic Shielding Performance. *ACS Appl. Mater. Interfaces* **2017**, *9*, 809–818. [CrossRef]

75. Hou, Y.; Xiao, B.; Yang, G.; Sun, Z.; Yang, W.; Wu, S.; Huang, X.; Wen, G. Enhanced electromagnetic wave absorption performance of novel carbon-coated Fe$_3$Si nanoparticles in an amorphous Si CO ceramic matrix. *J. Mater. Chem. C* **2018**, *6*, 7661. [CrossRef]

76. Wu, S.; Zou, M.; Li, Z.; Chen, D.; Zhang, H.; Yuan, Y.; Pei, Y.; Cao, A. Robust and Stable Cu Nanowire@Graphene Core–Shell Aerogels for Ultraeffective Electromagnetic Interference Shielding. *Small* **2018**, *14*, 1800634. [CrossRef]

77. Guo, Z.; Huang, H.; Xie, D.; Xia, H. Microwave properties of the single layer periodic structure composites composed of ethylene-vinyl acetate and polycrystalline iron fibers. *Sci. Rep.* **2017**, *7*, 11331. [CrossRef] [PubMed]

78. Sun, X.; He, J.; Li, G.; Tang, J.; Wang, T.; Guo, Y.; Xue, H. Laminated magnetic graphene with enhanced electromagnetic wave absorption properties. *J. Mater. Chem. C* **2013**, *1*, 765. [CrossRef]

79. Yang, Z.; Li, Z.; Yang, Y.; Xu, Z.J. Optimization of Zn$_x$Fe$_{3-x}$O$_4$ Hollow Spheres for Enhanced Microwave Attenuation. *ACS Appl. Mater. Interfaces* **2014**, *6*, 21911–21915. [CrossRef]

80. Yadav, R.S.; Kuřitka, I.; Vilcakova, J.; Skoda, D.; Urbánek, P.; Machovsky, M.; Masař, M.; Kalina, L.; Havlica, J. Lightweight NiFe$_2$O$_4$-Reduced Graphene Oxide-Elastomer Nanocompositeflexible sheet for electromagnetic interference shielding application. *Compos. Part B* **2019**, *166*, 95–111. [CrossRef]
81. Zheng, X.; Feng, J.; Zong, Y.; Miao, H.; Hu, X.; Bai, J.; Li, X. Hydrophobic graphene nanosheets decorated by monodispersed superparamagnetic Fe$_3$O$_4$ nanocrystals as synergistic electromagnetic wave absorbers. *J. Mater. Chem. C* **2015**, *3*, 4452–4463. [CrossRef]
82. Wang, Y.; Guan, H.; Dong, C.; Xiao, X.; Du, S.; Wang, Y. Reduced graphene oxide (RGO)/Mn$_3$O$_4$ nanocomposites for dielectric loss properties and electromagnetic interference shielding effectiveness at high frequency. *Ceram. Int.* **2016**, *42*, 936–942. [CrossRef]
83. Wang, X.; Pan, F.; Xiang, Z.; Zeng, Q.; Pei, K.; Che, R.; Lu, W. Magnetic vortex core-shell Fe$_3$O$_4$@C nanorings with enhanced microwave absorption performance. *Carbon* **2020**, *157*, 130–139. [CrossRef]
84. Wu, M.; Zhang, Y.D.; Hui, S.; Xiao, T.D.; Ge, S.; Hines, W.A.; Budnick, J.I.; Taylor, G.W. Microwave magnetic properties of Co$_{50}$(SiO$_2$)$_{50}$ nanoparticles. *Appl. Phys. Lett.* **2002**, *80*, 4404. [CrossRef]
85. Zhang, Y.; Yang, Z.; Wen, B. An Ingenious Strategy to Construct Helical Structure with Excellent Electromagnetic Shielding Performance. *Adv. Mater. Interfaces* **2019**, *6*, 1900375. [CrossRef]
86. Teotia, S.; Singh, B.P.; Elizabeth, I.; Singh, V.N.; Ravikumar, R.; Singh, A.P.; Gopukumar, S.; Dhawan, S.K.; Mathur, R.B. Multifunctional, Robust, Light Weight, Free Standing MWCNT/Phenolic Composite Paper as Anode for Lithium Ion Batteries and EMI Shielding Material. *RSC Adv.* **2014**, *4*, 33168–33174. [CrossRef]
87. He, N.; He, Z.; Liu, L.; Lu, Y.; Wang, F.; Wu, W.; Tong, G. Ni^{2+} guided phase/structure evolution and ultra-wide band width microwave absorption of Co$_x$Ni$_{1-x}$ alloy hollow microspheres. *Chem. Eng. J.* **2020**, *381*, 1227432. [CrossRef]
88. Ghosh, S.; Remanan, S.; Mondal, S.; Ganguly, S.; Das, P.; Singha, N.; Das, N.C. An approach to prepare mechanically robust full IPN strengthened conductive cotton fabric for high strain tolerant electromagnetic interference shielding. *Chem. Eng. J.* **2018**, *344*, 138–154. [CrossRef]
89. Quan, B.; Liang, X.; Ji, G.; Lv, J.; Dai, S.S.; Xu, G.; Du, Y. Laminated graphene oxide-supported high-efficiency microwave absorber fabricated by an in situ growth approach. *Carbon* **2018**, *129*, 310–320. [CrossRef]
90. Anand, S.; Pauline, S. Electromagnetic Interference Shielding Properties of BaCo$_2$Fe$_{16}$O$_{27}$ Nanoplatelets and RGO Reinforced PVDF Polymer Composite Flexible Films. *Adv. Mater. Interfaces* **2021**, *8*, 2001810. [CrossRef]
91. Zhang, M.; Zhang, J.; Lin, H.; Wang, T.; Ding, S.; Li, Z.; Wang, J.; Meng, A.; Li, Q.; Lin, Y. Designable synthesis of reduced graphene oxide modified using CoFe$_2$O$_4$ nanospheres with tunable enhanced microwave absorption performances between the whole X and Ku bands. *Compos. Part B* **2020**, *190*, 107902. [CrossRef]
92. Wang, X.; Lu, Y.; Zhu, T.; Chang, S.; Wang, W. CoFe$_2$O$_4$/N-doped reduced graphene oxide aerogels for high-performance microwave absorption. *Chem. Eng. J.* **2020**, *388*, 124317. [CrossRef]
93. Peng, J.; Peng, Z.; Zhu, Z.; Augustine, R.; Mahmoud, M.M.; Tang, H.; Rao, M.; Zhang, Y.; Li, G.; Jiang, T. Achieving ultra-high electromagnetic wave absorption by anchoring Co$_{0.33}$Ni$_{0.33}$Mn$_{0.33}$Fe$_2$O$_4$ nanoparticles on graphene sheets using microwave assisted polyol method. *Ceram. Int.* **2018**, *44*, 21015–21026. [CrossRef]
94. Xu, F.; Chen, R.; Lin, Z.; Qin, Y.; Yuan, Y.; Li, Y.; Zhao, X.; Yang, M.; Sun, X.; Wang, S.; et al. Super flexible Interconnected Graphene Network Nanocomposites for High-Performance Electromagnetic Interference Shielding. *ACS Omega* **2018**, *3*, 3599–3607. [CrossRef] [PubMed]
95. Fu, P.; Huan, X.; Luo, J.; Ren, S.; Jia, X.; Yang, X. Magnetically Aligned Fe$_3$O$_4$ Nanowires-Reduced Graphene Oxide for Gas Barrier, Microwave Absorption, and EMI Shielding. *ACS Appl. Nano Mater.* **2020**, *3*, 9340–9355. [CrossRef]

nanomaterials

Article

Thiolation of Chitosan Loaded over Super-Magnetic Halloysite Nanotubes for Enhanced Laccase Immobilization

Avinash A. Kadam [1], Bharat Sharma [2], Surendra K. Shinde [3], Gajanan S. Ghodake [3], Ganesh D. Saratale [4], Rijuta G. Saratale [1], Do-Yeong Kim [1,*] and Jung-Suk Sung [5,*]

[1] Research Institute of Biotechnology and Medical Converged Science, Dongguk University-Seoul, Ilsandong-gu, Goyang-si, Gyeonggi-do, Seoul 10326, Korea; kadamavinash@dongguk.edu (A.A.K.); rijutaganesh@gmail.com (R.G.S.)
[2] Department of Materials Science and Engineering, Incheon National University, Academy Road Yeonsu, Incheon, Seoul 22012, Korea; bharatsharma796@gmail.com
[3] Department of Biological and Environmental Science, Dongguk University-Seoul, Ilsandong-gu, Goyang-si, Gyonggido, Seoul 10326, Korea; surendrashinde.phy@gmail.com (S.K.S.); ghodakegs@gmail.com (G.S.G.)
[4] Department of Food Science and Biotechnology, Dongguk University-Seoul, Ilsandong-gu, Goyang-si, Gyeonggi-do, Seoul 10326, Korea; gdsaratale@gmail.com
[5] Department of Life Science, College of Life Science and Biotechnology, Dongguk University-Seoul, Ilsandong-gu, Goyang-si, Gyonggido, Seoul 10326, Korea
[*] Correspondence: dykimm@dongguk.edu (D.-Y.K.); sungjs@dongguk.edu (J.-S.S.); Tel.: +82-31-961-5624 (D.-Y.K.); +82-31-961-5132 (J.-S.S.)

Received: 1 December 2020; Accepted: 17 December 2020; Published: 20 December 2020

Abstract: This study focuses on the development of a nanosupport based on halloysite nanotubes (HNTs), Fe_3O_4 nanoparticles (NPs), and thiolated chitosan (CTs) for laccase immobilization. First, HNTs were modified with Fe_3O_4 NPs (HNTs-Fe_3O_4) by the coprecipitation method. Then, the HNTs-Fe_3O_4 surface was tuned with the CTs (HNTs-Fe_3O_4-CTs) by a simple refluxing method. Finally, the HNTs-Fe_3O_4-CTs surface was thiolated (-SH) (denoted as; HNTs-Fe_3O_4-CTs-SH) by using the reactive NHS-ester reaction. The thiol-modified HNTs (HNTs-Fe_3O_4-CTs-SH) were characterized by FE-SEM, HR-TEM, XPS, XRD, FT-IR, and VSM analyses. The HNTs-Fe_3O_4-CTs-SH was applied for the laccase immobilization. It gave excellent immobilization of laccase with 100% activity recovery and 144 mg/g laccase loading capacity. The immobilized laccase on HNTs-Fe_3O_4-CTs-SH (HNTs-Fe_3O_4-CTs-S-S-Laccase) exhibited enhanced biocatalytic performance with improved thermal, storage, and pH stabilities. HNTs-Fe_3O_4-CTs-S-S-Laccase gave outstanding repeated cycle capability, at the end of the 15th cycle, it kept 61% of the laccase activity. Furthermore, HNTs-Fe_3O_4-CTs-S-S-Laccase was applied for redox-mediated removal of textile dye DR80 and pharmaceutical compound ampicillin. The obtained result marked the potential of the HNTs-Fe_3O_4-CTs-S-S-Laccase for the removal of hazardous pollutants. This nanosupport is based on clay mineral HNTs, made from low-cost biopolymer CTs, super-magnetic in nature, and can be applied in laccase-based decontamination of environmental pollutants. This study also gave excellent material HNTs-Fe_3O_4-CTs-SH for other enzyme immobilization processes.

Keywords: laccase; immobilization; thiolated chitosan; super-magnetic nanosupports for enzyme immobilization; ampicillin; Direct Red 80

1. Introduction

Environmental pollution is a daunting challenge to face in the modern world [1]. Pollutants from land and water are foremost in reducing the quality of life and human health [2]. Textile dyes released from textile industries into the water lead to several hazards to the environment, mainly due to their recalcitrant and toxic nature [3]. In recent decades, emerging pharmaceutical contaminants released in very small concentrations caused a significant challenge to the environment, mainly due to their bioactive nature and the development of antibiotic resistance in the microbes [4]. The pharmaceutical compounds caused the development of antibiotic resistance in the microbes, hence taking the entire world into a serious health challenge [5]. Various methods and combinations of treatment were employed to face this challenge [6]. However, enzymatic methods unveiled a new outlook to treat waste streams to get rid of intractable organic pollutants [7]. Laccase is one of the most important enzymes in terms of waste pollutant removal. Laccase catalyzes one-electron oxidation of the pollutant using molecular oxygen as the electron donor [8]. However, free laccase has several disadvantages: its proteinaceous nature, it is more prone to denaturation, stability issues, and higher cost due to use in a single reaction [9]. The immobilization approach will cover all of these disadvantages and provide a more suitable and accessible biocatalyst. Therefore, suitable, novel, and efficient nanosupports for the immobilization of laccase are required for its potential applications in the remediation of environmental pollutants [1].

Several supports have been reported for laccase immobilization [10]. Of the nanosupports studied for immobilization, the naturally available clay mineral halloysite nanotube (HNT) is catching significant attention. This is mainly due to a number of important properties of HNTs [11]. These include: easy and low-cost availability, structural morphology in the form of nanotubes, nanotubular lumen, and an extremely modifiable surface [12]. The surface modifications of HNTs are the main key to developing them into a comprehensive nanosupport for laccase immobilization [13,14]. The surface modification of HNTs with Fe_3O_4 nanoparticles (NPs) makes HNTs super-magnetic. This adds the important parameter of magnetic separation [15]. Being HNT-immobilized, laccase separation from solution is an extremely important parameter for laccase recovery and enhanced laccase applications [16]. Furthermore, the surface of the super-magnetic HNTs can be modified with a biopolymer like chitosan (CTs) [17]. As CTs is a ubiquitous biopolymer used for enzyme immobilization due to having the surface amino ($-NH_2$) functional group. In our previous research, modification of the HNT surface with Fe_3O_4 and CTs proved to be excellent for laccase immobilization [13,14,18]. In these studies, the immobilization of "laccase" and "CTs and Fe_3O_4 modified HNTs" was carried out by glutaraldehyde cross-linking. Still, there is a huge scope to adopt new immobilization strategies. One such important strategy is thiolation ($-SH$) of the amino ($-NH_2$) group of CTs loaded on magnetic-HNTs [19–21]. Therefore, this study mainly emphasized the thiolation of the amino ($-NH_2$) group present of CTs-modified magnetic HNTs.

Thiolation of chitosan has attracted significant attention in recent years for many desirable applications such as enzyme immobilization, drug delivery, cosmetics, and tissue engineering [19,20]. Thiolation of the CTs involves modification of the surface amino ($-NH_2$) group to the thiol ($-SH$) group by NHS-ester reaction [22]. The enzymes possess the $-SH$ group due to the presence of amino acids such as cysteine and methionine [7]. The $-SH$ group from the thiolated supports and enzyme react at pH 5 to form a very stable ($-S-S-$) disulfide bond for immobilization [23]. Thus, thiolation of the CTs loaded over magnetized HNTs can provide highly applicable nanosupports for enzyme immobilization applications. As per the author's literature survey, thiolated CTs, Fe_3O_4, and HNT nanocomposite has not yet been reported for enzyme immobilization.

Thus, in summary, this study attempts the synthesis of nanocomposites mainly containing HNTs, Fe_3O_4 NPs, and thiolated CTs for laccase immobilization and application of the newly synthesized immobilization system for the biocatalytic degradation of the environmental pollutants Direct Red 80 (DR80) and the pharmaceutical compound ampicillin. The structural and morphological details of the nanocomposites were assessed with various techniques such as; FE-SEM, HR-TEM, XPS, FT-IR, and XRD analyses. The biocatalysis of immobilized laccase was carried out in detail. The degradation of DR80 and ampicillin were checked by the developed immobilized system.

2. Materials and Methods

2.1. Materials

Halloysite nanotubes (nanopowder), ammonium hydroxide (ACS Grade Solvents, 25%, NH_4OH), chitosan (low molecular weight), N-(3-dimethylaminopropyl)-N'-ethylcarbodiimide hydrochloride ($\geq$98% (titration), EDAC.HCl), N-hydroxysuccinimide ($\geq$97.0% (titration) NHS), thioglycolic acid ($\geq$98%), N,N-dimethylformamide (99.8% DMF), ampicillin (anhydrous), Direct Red 80 (powder), syringaldehyde (SA, assay- $\geq$ 98%), guaiacol (GUA, assay- $\geq$ 98%), p-Coumaric acid (CA, $\geq$98.0% high-performance liquid chromatography (HPLC)), 1-Hydroxybenzotriazole hydrate (HBT, $\geq$97.0% (T)), 2,2-Azino-bis(3-ethylbenzothiazoline-6-sulfonic acid) diammonium (ABTS, Liquid Substrate System), methanol (HPLC grade), acetonitrile (HPLC grade), water (HPLC grade) and laccase from *Trametes Versicolor* (powder) were obtained from the Sigma Aldrich, St. Louis, MO, USA. $FeCl_3 \cdot 6H_2O$ and $FeCl_2 \cdot 4H_2O$ were received from JUNSEI (Kyoto) Japan.

2.2. Synthesis of HNTs-Fe$_3$O$_4$-CTs-SH

The HNTs-Fe$_3$O$_4$-CTs-SH was synthesized in three steps. First, the HNTs were tuned with Fe$_3$O$_4$ NPs by the coprecipitation method [13]. Second, Fe$_3$O$_4$ NP-modified HNTs (HNTs-Fe$_3$O$_4$) were functionalized with the chitosan (CTs) (HNTs-Fe$_3$O$_4$-CTs) [14]. Finally, in the third step, NH$_2$ functional groups of HNTs-Fe$_3$O$_4$-CTs were thiolated (-SH) by NHS-ester reaction (HNTs-Fe$_3$O$_4$-CTs). In the first step, 0.2–2 g of HNTs was added to 200 mL of deionized water. The mixture was ultrasonicated by Sonics Vibra-Cell VC130 Ultrasonic Processor, power—130 W, frequency—20 kHz, and amplitude—60 µM (Sonics & Materials, Inc., Newtown, CT, USA) for 1 h. This mixture was added to 100 mL of 2% FeCl$_3$·6H$_2$O, and 100 mL of 1% FeCl$_2$·4H$_2$O (dropwise) in the presence of N$_2$ gas at 60 °C under constant stirring. Then, 30 mL of 25% NH$_4$OH solution was added to the mixture. The obtained black colored precipitate of nanocomposite was stirred at 60 °C for 1 h. Further, it was separated magnetically, washed thoroughly with water, ethanol, and methanol, and dried in an oven at 60 °C for 48 h. The obtained sample of HNTs-Fe$_3$O$_4$ was powdered in a mortar and pestle for further use.

In the second step, the CTs modification of HNTs-Fe$_3$O$_4$ was done. In a typical reaction, the CTs (1 g) was taken in 100 mL of 1% acetic acid and vortexed for 4 h on the magnetic stirrer adjusted to the 50 °C. The CTs solution was prepared in 1% acetic acid in warm conditions to obtain a clear solution of completely dissolved CTs. Then 1 g of HNTs-Fe$_3$O$_4$ was taken in 100 mL distilled water and ultrasonicated with Sonics Vibra-Cell VC130 Ultrasonic Processor, power—130 W, frequency—20 kHz, and amplitude—60 µM (Sonics & Materials, Inc., Newtown, CT, USA) for 1 h. Next, both the CTs solution and ultrasonicated HNTs-Fe$_3$O$_4$ solution were poured into a 500 mL beaker and stirred for 15 min. At this point, 2 mL of glutaraldehyde solution (2.5%) was added to the mixture. Further, the beaker was covered with the aluminum foil and continued the stirring for 8 h on the magnetic stirrer adjusted to 50 °C. The obtained HNTs-Fe$_3$O$_4$-CTs nanocomposite was separated magnetically, washed thoroughly with distilled water, ethanol, and methanol, dried in an oven at 60 °C for 48 h, and finally powdered for further use.

Further, the thiolation of the HNTs-Fe$_3$O$_4$-CTs was done with the NHS-ester reaction. In a typical reaction, EDAC (6.08 mM), NHS (5.79 mM), and 5 mL of TGA were added to the 10 mL of the DMF. The mixture was kept in shaking conditions of 200 rpm for 24 h at 25 °C to form a reactive NHS-ester. Then, the HNTs-Fe$_3$O$_4$-CTs and NHS-ester reaction were carried out to form a thiol functionalized HNTs-Fe$_3$O$_4$-CTs (HNTs-Fe$_3$O$_4$-CTs-SH). In this typical reaction, HNTs-Fe$_3$O$_4$-CTs (6.6 g/L) was taken in pH 5.0 buffer (Na-acetate buffer 100 mM) and further added with the NHS-ester (0.67 mL/L) in dark conditions. The mixture was vortexed rapidly at 200 rpm and 25 °C for 4 h in dark conditions. Once the reaction was completed, the materials were immediately removed magnetically and washed thoroughly with pH 5.0 buffer (Na-acetate buffer 100 mM). The obtained materials were dried in a freeze dryer and powdered for further use for enzyme immobilization study.

2.3. Laccase Immobilization Experiment

The laccase immobilization experiment was carried out in a 20 mL glass tube. The materials HNTs-Fe$_3$O$_4$-CTs and HNTs-Fe$_3$O$_4$-CTs-SH (1 g/L) were taken in 20 mL glass tubes separately. Each of these tubes was added with 10 mL of laccase solution (1.5 g/L) prepared in the 100 mM Na-acetate buffer (pH-4). These mixtures were immediately kept shaking (200 rpm) and at a temperature of 25 °C for 24 h. After completion of the immobilization process, the materials were magnetically removed and washed thoroughly with the 100 mM Na-acetate buffer (pH-4) to remove all unbound laccase. The laccase immobilized materials were then tested for activity recovery (%) and laccase-loading capacities. Laccase activity was determined in the 2 mL reaction mixture having 0.9 mL of sodium acetate buffer (100 mM, pH 4), 1 mL of ABTS solution (90 µM), and free laccase (0.1 mL from 1.5 mg/mL of laccase solution) or immobilized laccase (0.1 mL solution containing 1 mg of HNTs-Fe$_3$O$_4$-CTs-S-S-Laccase). The reaction mixtures of both free laccase and immobilized laccase were kept gently mixing for 30 min. ABTS oxidation was quantified by the absorption at 420 nm for ABTS (molar extinction coefficient [ε420], 0.0360 µM^{-1} cm^{-1}). The 1 U of laccase activity was measured as the ABTS (µM) oxidation per min of incubation time and per mL of solution. The detailed formulae of the measurements of activity recovery (%) are given by Equation (1) [13].

$$Activity\ recovery\ (\%)\ =\ A_{IL}/A_{FL} \times 100 \tag{1}$$

where A_{IL} is the immobilized laccase activity, and A_{FL} is the free laccase activity before the immobilization procedure. The laccase loading capacity (mg/g) was determined by the Bradford method (Add reference) using the Pierce Coomassie (Bradford) Protein Assay Kit, Thermo Scientific, Massachusetts, MA, USA. The detailed formulae of the measurements of laccase loading capacity are given by Equation (2) [13].

$$Laccase\ loading\ (mg/g)\ =\ (C_{bi} - C_{ai})V/W \tag{2}$$

where C_{bi} is laccase concentration before immobilization (mg/L), C_{ai} is retained laccase concentration in solution after immobilization (mg/L), V is the volume of the solution in liters (L), and W is the weight of the nanocomposites in grams (g). Further, the immobilized laccase was analyzed for biocatalytic properties.

2.4. Characterizations

The modified materials were morphologically characterized by scanning electron microscopy (SEM, FC-SM10, Hitachi S-4800, Ibaraki, Japan) and high-resolution transmission electron microscopy (HR-TEM, Tecnai G2 transmission electron microscope, Hillsboro, OR, USA). The samples for SEM and TEM were prepared in 10 mL distilled water containing 10 mg of each sample. The samples were ultrasonicated (Sonics Vibra-Cell VC130 Ultrasonic Processor (Sonics & Materials, Inc., Newtown, CT, USA) for 1 h. The well-dispersed samples were drop coated on TEM carbon grid and silicon wafer for SEM and TEM analysis, respectively. The crystalline purity of the samples (in powder form) were analyzed by X-ray powder diffractometer (XRD, Cu-Kα radiation (λ = 1.5418 Å), Ultima IV/Rigaku, Tokyo, Japan). The magnetic properties of samples (in powder form) were analyzed by vibrating sample magnetometer (VSM, Lakeshore, Model: 7407, LA, USA). The functional group profile of samples (in powder form) were analyzed by Fourier transform infrared spectroscopy (FT-IR, Spectrum 100, PerkinElmer, Waltham, MA, USA). The surface elemental profile of the samples (in powder form) were analyzed by X-ray photoelectron spectroscopy (XPS, Theta Probe AR-XPS System, Thermo Fisher Scientific, Dartford, UK).

2.5. Biocatalysis of Immobilized Laccase

The effect of the initial laccase concentration on laccase loading on HNTs-Fe$_3$O$_4$-CTs-SH was assessed by taking laccase concentrations of 0.5, 0.75, 1.0, 1.25, 1.5, and 1.75 mg/mL. The laccase loading was evaluated as per the Equation (2) from Section 2.3. The temperature stability was analyzed by incubating the free laccase and immobilized laccase in acetate buffer (100 mM, pH 4.2) for 200 min at 60 °C. The samples were withdrawn after each 40 min interval and analyzed for relative activity (%). The relative activity was calculated by considering the initial activity before the stability experiment as 100% [13]. The calculation of relative activity is given by the following formulae Equation (3).

$$Relative\ activity\ (\%) \ = \ A_e/A_i \times 100 \tag{3}$$

where A_e is the activity after the stability experiment, and A_i is the initial activity before the stability experiment. Further, the storage stability was evaluated by incubating free and immobilized laccase acetate buffer (100 mM, pH 4.2) for 30 days at 4 °C. The sample was tested for relative activity (%) after every 5 days. The pH stability was also evaluated by incubating the free and immobilized laccase at various pH 1–9 for 1 h and at a temperature of 20 °C. The samples were measured for the relative activity (%) after the incubation process. The reusability experiments of immobilized laccase were carried out by magnetically removing the immobilized laccase after completion of the first reaction cycle, and this was followed by the addition of the new reaction mixture for the next cycle. This was continued for 15 cycles of reactions by the same immobilized laccase.

2.6. Immobilized Laccase Mediated Degradation of Environmental Pollutants

The target pollutants used were DR80 and ampicillin. Their degradation experiment was carried out in a typical reaction mixture that included acetate buffer (100 mM, pH 4.2), pollutant (DR80 (15 ppm)/ampicillin (25 ppm)), 1 mM redox mediators, and immobilized laccase/free laccase (0.1 mL) for 4 h, with a shaking condition of 200 rpm and temperature of 20 °C. The DR80 degradation was calculated by performing UV–vis spectroscopic analysis. The high-performance liquid chromatography (HPLC, Shimadzu LC-20AD, Kyoto, Japan) analysis was carried out for assessment of the ampicillin degradation. The HPLC analysis used the following parameters such as detection wavelength (230 nm), a mobile phase of [A:B:C:D] [A] acetonitrile, [B] water, [C] 1M potassium phosphate monobasic in water, [D] 1N acetic acid in water (80:909:10:1), the flow rate of 0.6 mL/min and the C18 column (Ascentis Express 90 Å, C18 10 cm × 4.6 mm, 5 µm, Sigma-Aldrich, St. Louis, MO, USA). The repeated cycle degradation of DR80 was carried out as follows, after the completion of the first degradation cycle as mentioned above, the immobilized laccase was magnetically removed, washed with sodium acetate buffer pH 4.2. This washed immobilized laccase was added to the fresh reaction mixture which included fresh acetate buffer (100 mM, pH 4.2), 1 mM redox mediator, and DR80 (15 ppm) for the next cycle. Similarly, 10 cycles were carried out to assess the reusability potential.

3. Results and Discussion

3.1. Synthesis

The synthesis of HNTs-Fe$_3$O$_4$-CTs, reactive NHS ester, and HNTs-Fe$_3$O$_4$-CTs-S-S-Laccase are presented in Figure 1A–C. The pristine HNTs were first tuned with Fe$_3$O$_4$ by the coprecipitation method (Figure 1A). Further, the HNTs-Fe$_3$O$_4$ was modified with the CTs (Figure 1A) in a simple reaction process. It is well known that CTs exhibits ubiquitous NH$_2$ groups on its surface, and this can be essential for thiolation. The process of thiolation involves modification of the materials with the thio (-SH) group by using the reactive NHS ester. The detailed synthesis of the reactive NHS ester was presented in Figure 1B. In the reaction, the thioglycolic acid reacted with EDAC.HCl to form an unstable reactive o-acylisourea ester. The NHS attached to the unstable o-acylisourea ester and formed the reactive NHS ester. In the next reaction of NHS ester with HNTs-Fe$_3$O$_4$-CTs, NHS ester attaches to the amino (NH$_2$) functional group present on the surface of the HNTs-Fe$_3$O$_4$-CTs. This transforms the thiol moiety on the surface of the HNTs-Fe$_3$O$_4$-CTs (denoted as; HNTs-Fe$_3$O$_4$-CTs-SH). The reaction at pH 5 is very important for the reaction to occur. A similar mechanism for thiolated chitosan was explained by Hanif et al., 2015 [21]. The thiolated and CTs/Fe$_3$O$_4$ modified HNTs were applied for the laccase immobilization. The thiol group from the nanocomposite reacted with the thiol group from the laccase to form the strong disulfide (-S-S-) bond for immobilization. The thiol group displayed excellent immobilization performance for the enzyme laccase [7].

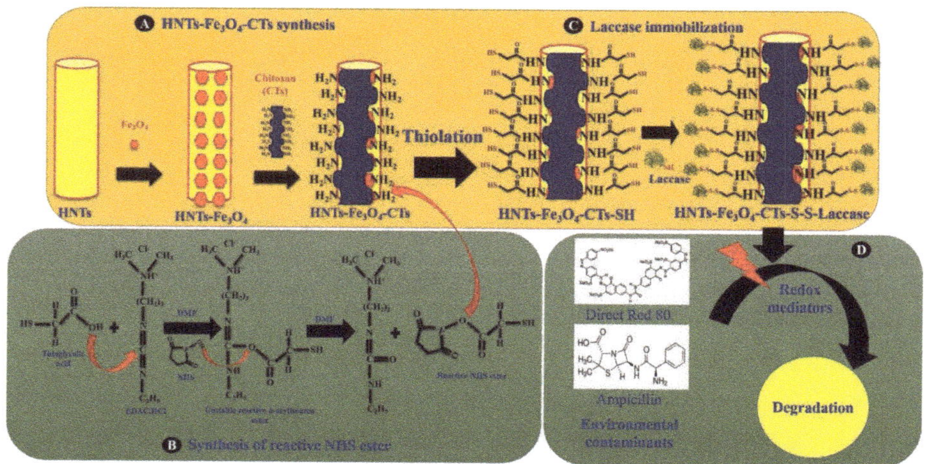

Figure 1. The schematic presentation of the synthesis of (**A**) HNTs-Fe$_3$O$_4$-CTs, (**B**) reactive NHS ester, and (**C**) HNTs-Fe$_3$O$_4$-CTs-S-S-Laccase, and (**D**) degradation of the environmental contaminants Direct Red 80 and ampicillin by the immobilized laccase.

The laccase immobilized HNTs-Fe$_3$O$_4$-CTs-SH is denoted as HNTs-Fe$_3$O$_4$-CTs-S-S-Laccase (Figure 1C). Finally, after laccase immobilization, the HNTs-Fe$_3$O$_4$-CTs-S-S-Laccase was applied for the redox-mediated degradation of the textile dye Direct Red 80 (DR80) and pharmaceutical compound ampicillin (Figure 1D).

3.2. Characterizations

3.2.1. SEM and HR-TEM Analysis

The morphological observations of the modified material were done by SEM analysis as shown in Figure 2A,B. Figure 2A shows the unmodified HNTs. The image represents the diverse sized nanotubes with the plane and unmodified surface. The ends of the tubes were found open. Figure 2B shows the modified form of the HNTs, i.e., HNTs-Fe$_3$O$_4$-CTs-SH. A close look reveals that the surface of the HNTs was heavily modified with the subsequent modifications done on the pristine HNTs; such as Fe$_3$O$_4$, CTs, and thiolation. The tubes seemed to be broadened, possibly due to the coatings of the CTs. Similar observations for chitosan modified HNTs was seen in an earlier report [13,14]. The SEM morphologies gave an idea of the significant modification of the plane surface of the HNTs. Furthermore, it was very important to analyze the morphological observations in more detail. To achieve this, the HR-TEM analysis of the HNTs-Fe$_3$O$_4$-CTs-SH was carried out. Figure 2C shows the HR-TEM image of the HNTs-Fe$_3$O$_4$-CTs-SH. The image represents the surface of a single nanotube. The surface of the tube showed the Fe$_3$O$_4$ NPs decorated over the tubular surface. The Fe$_3$O$_4$ NPs were found to be 5–10 nm in size. The shape of the Fe$_3$O$_4$ NPs was found to be circular and quasi-polyhedral. The selected area (electron) diffraction (abbreviated as SAED) analysis of HR-TEM image of HNTs-Fe$_3$O$_4$-CTs-SH is shown in Figure 2C(i). The SAED pattern revealed the polycrystalline nature resulting from HNTs and Fe$_3$O$_4$ NPs. Moreover, it was important to understand the subsequent modification on the nanotube with the evident elemental distribution. To observe this, high-angle annular dark-field imaging

(HAADF) S-TEM analysis was carried out in Figure 2D,E. The combination of all the elements was revealed in Figure 2E.

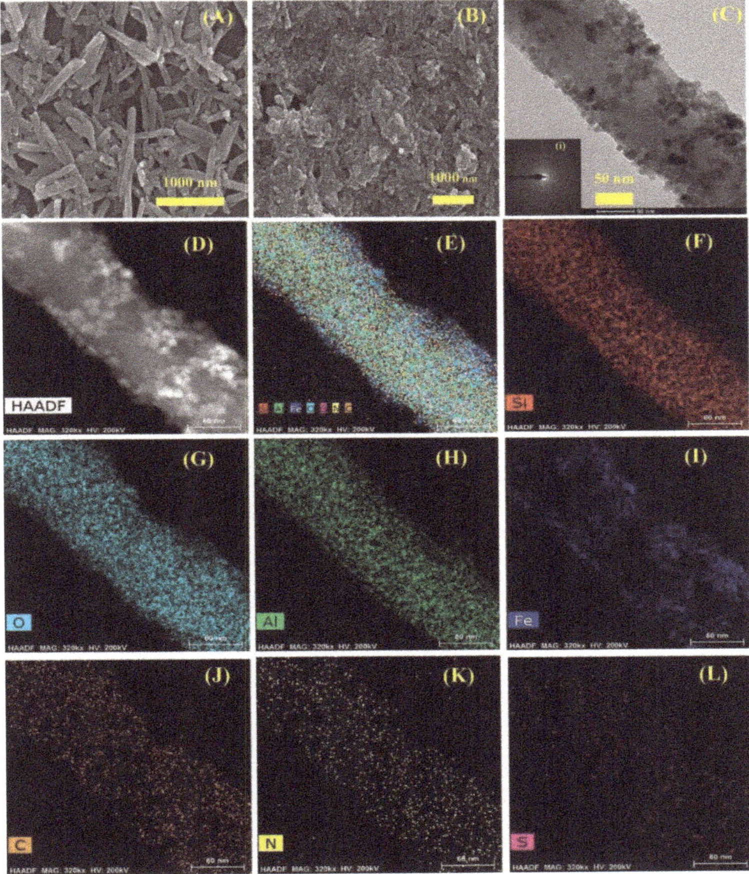

Figure 2. SEM analysis of (**A**) HNTs, and (**B**) HNTs-Fe$_3$O$_4$-CTs-SH. HR-TEM analysis of the HNTs-Fe$_3$O$_4$-CTs-SH (**C**) inset (i) SAED pattern of the HNTs-Fe$_3$O$_4$-CTs-SH. The STEM HAADF image of HNTs-Fe$_3$O$_4$-CTs-SH (**D**), elemental mapping of HNTs-Fe$_3$O$_4$-CTs-SH (**E**), and corresponding elemental distribution of Si (**F**), O (**G**), Al (**H**), Fe (**I**), C (**J**), N (**K**) and S (**L**).

The element maps of silicon (Si), oxygen (O), aluminum (Al), iron (Fe), carbon (C), nitrogen (N), and sulfur (S) are shown in Figure 2F–L, respectively. The dense presence of Si, Al, and O elements outlined the basic backbone structure of the HNTs (Figure 2F–H). The presence of Fe represented the Fe$_3$O$_4$ NPs decorated over the tubular surface (Figure 2I). The presence of C and N designated the CTs modification (Figure 2J,K). Finally, the presence of the S over the nanotube confirmed the thiolation of the HNTs-Fe$_3$O$_4$-CTs (Figure 2L). Likewise, the elemental distribution was also confirmed by the HR-TEM energy-dispersive X-ray spectroscopy (EDS) analysis carried out to confirm the elemental distributions (Figure 3A). The presence of Fe, C, N, and S corroborated the successful modification of the HNT surface with the Fe$_3$O$_4$ NPs, CTs, and thiolation. The overall SEM and HR-TEM analysis gave the idea about the surface morphology, polycrystalline nature, and elemental distributions of the modified HNT materials.

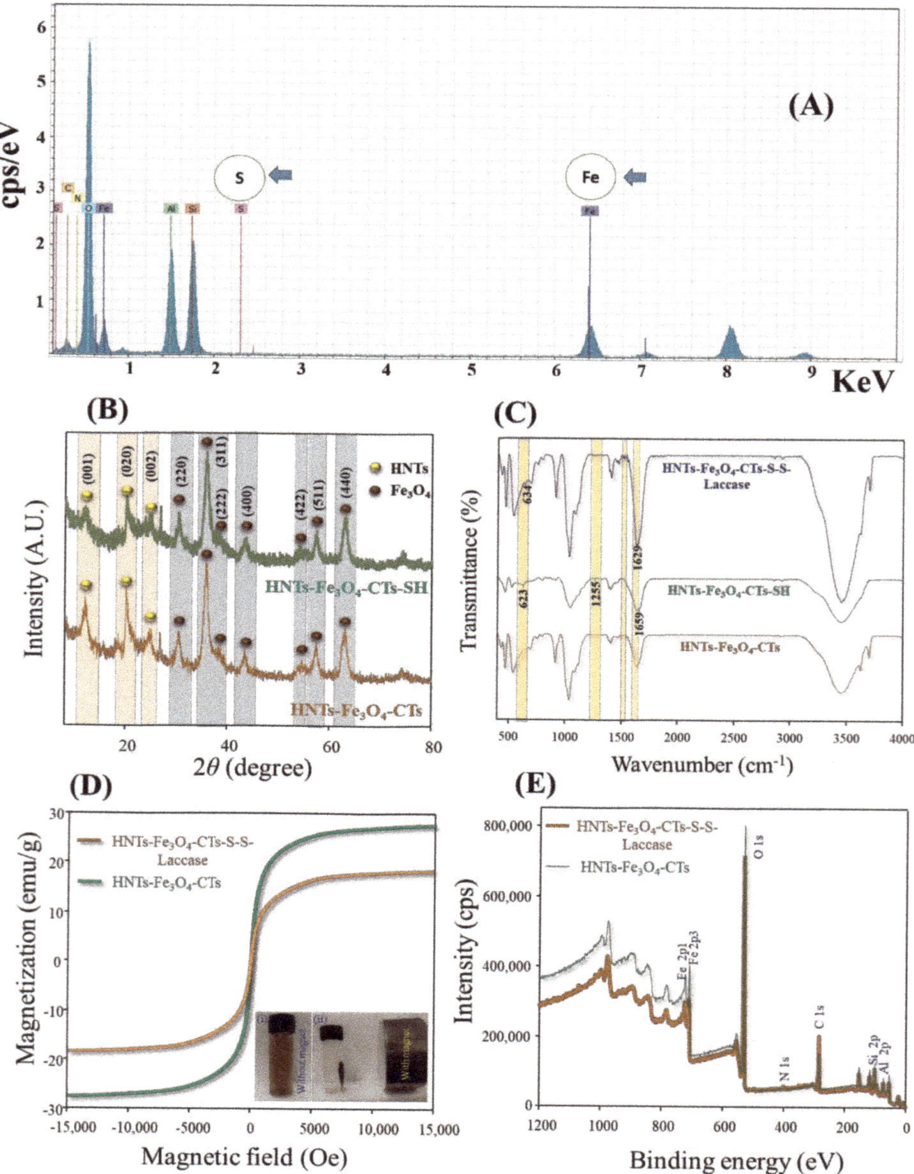

Figure 3. (**A**) HR-TEM EDS analysis of HNTs-Fe$_3$O$_4$-CTs-SH, (**B**) XRD analysis of HNTs-Fe$_3$O$_4$-CTs and HNTs-Fe$_3$O$_4$-CTs-SH, (**C**) FT-IR analysis of HNTs-Fe$_3$O$_4$-CTs, HNTs-Fe$_3$O$_4$-CTs-SH, and HNTs-Fe$_3$O$_4$-CTs-S-S-Laccase (**D**) VSM analysis of HNTs-Fe$_3$O$_4$-CTs and HNTs-Fe$_3$O$_4$-CTs-S-S-Laccase inset (**i**) HNTs-Fe$_3$O$_4$-CTs-S-S-Laccase without magnet, and (**ii**) HNTs-Fe$_3$O$_4$-CTs-S-S-Laccase with magnet, (**E**) XPS analysis of HNTs-Fe$_3$O$_4$-CTs and HNTs-Fe$_3$O$_4$-CTs-S-S-Laccase.

3.2.2. XRD, FT-IR, VSM, and XPS Analysis

The crystalline nature of HNTs-Fe$_3$O$_4$-CTs before and after thiolation was assessed using the XRD analysis, Figure 3B. The HNTs-Fe$_3$O$_4$-CTs and HNTs-Fe$_3$O$_4$-CTs-SH showed similar diffraction peak profiling. This revealed the intact crystalline pattern after the thiolation process. The typical diffraction peak patterning of HNTs, i.e., 11.94°, 20.33°, and 24.36° 2θ angles corresponding to the crystalline planes of (001), (020), and (002) was obtained in both the samples. Similar profiling of the HNTs peaks after modifications with Fe$_3$O$_4$ NPs and amino-salinization was reported [16]. The Fe$_3$O$_4$ NPs diffraction peaks were seen in both the samples at 30.35, 35.65, 37.46, 43.45, 54.54, 57.36, and 62.92° 2θ angles corresponding to the crystalline planes of the (220), (311), (222), (400), (422), (511), and (440), respectively, as per the JCPDS cards 75-0033 data [24]. Thus, the XRD analysis confirmed the Fe$_3$O$_4$ NPs modification over the HNTs.

The functional group profile of the HNTs-Fe$_3$O$_4$-CTs, HNTs-Fe$_3$O$_4$-CTs-SH, and HNTs-Fe$_3$O$_4$-CTs-S-S-Laccase was analyzed by the FT-IR analysis, shown in Figure 3C. The absorption peaks obtained in all the samples at 1028 and 910 cm^{-1} corresponded to siloxane vibration and silanol vibration, respectively [16]. The absorption peaks obtained in all the samples at 472 cm^{-1} corresponded to the Fe-O bond from Fe$_3$O$_4$ NPs [9]. The absorption peaks of 3448, 1550, 1443, 1322, and 1119 cm^{-1} obtained in all the samples, corresponded to the O–H stretching, N–H deformation, C–H deformation, C–N vibrations, and C–OH stretching of the CTs [9]. Hence, all the samples gave a similar peak profile corresponding to the peaks of HNTs, Fe$_3$O$_4$, and CTs. This corroborates the successful synthesis of the HNTs-Fe$_3$O$_4$-CTs. However, the additional peaks of 1659, 1255, and 623 cm^{-1} were observed in the HNTs-Fe$_3$O$_4$-CTs-SH. These peaks mainly corresponded to the C=O stretching amides, C–SH stretching, and C–S stretching, respectively [7,25]. The appearance of these peaks successfully corroborated the thiolation of the chitosan. However, in the spectra of HNTs-Fe$_3$O$_4$-CTs-SH, the S-H stretch in the region of 2600–2550 cm^{-1} was absent. A similar observation for the absence of S-H stretch in the region of 2600–2550 cm^{-1} for thiolated chitosan was observed by an earlier report [7]. This might be due to selective thiolation rather than thiolation of most of the -NH$_2$ group of the chitosan. Moreover, the FT-IR analysis of the HNTs-Fe$_3$O$_4$-CTs-S-S-Laccase is shown in Figure 3C. The peak corresponding to the disulfide bond was obtained at the 634 cm^{-1} [26,27]. The presence of the absorption peak at 1629 cm^{-1} corresponds to the amide II from proteins [28]. These all observations obtained in the FT-IR spectra of the HNTs-Fe$_3$O$_4$-CTs-S-S-Laccase indicated the successful loading of the laccase over the Fe$_3$O$_4$-CTs-SH through the disulfide covalent bond. The obtained results of the FT-IR are in strong agreement with the HAADF STEM elemental mapping and HR-TEM EDS analyses.

The magnetic potential of the materials played a crucial role in the separation mechanism. Hence, it was very important to know the magnetic properties of the synthesized material HNTs-Fe$_3$O$_4$-CTs and immobilized laccase HNTs-Fe$_3$O$_4$-CTs-S-S-Laccase. The magnetic properties of the HNTs-Fe$_3$O$_4$-CTs and HNTs-Fe$_3$O$_4$-CTs-S-S-Laccase were assessed by the VSM analysis (Figure 3D). The VSM analysis of both the samples exhibited the typical hysteresis curve of the magnetization with coercivity and remanence values to be zero. The magnetic potential value was observed to be 18.38 and 27.53 emu/g. All these observations were designated the superparamagnetic nature of both the samples. Besides, the magnetization potential of HNTs-Fe$_3$O$_4$-CTs-S-S-Laccase decreased with 9.15 emu/g in comparison with the HNTs-Fe$_3$O$_4$-CTs. This result confirmed the enhanced laccase loading over the surface of the modified material. Furthermore, the Figure 3D insets (i) and (ii) show the external magnet mediated separation of the HNTs-Fe$_3$O$_4$-CTs-S-S-Laccase from solution. All the VSM analyses confirmed the magnetic potential and laccase immobilization on materials.

Furthermore, the surface elemental profile of the HNTs-Fe$_3$O$_4$-CTs and HNTs-Fe$_3$O$_4$-CTs-S-S-Laccase were examined by XPS analysis, Figure 3E. The HNTs-Fe$_3$O$_4$-CTs and HNTs-Fe$_3$O$_4$-CTs-S-S-Laccase showed the peaks Al 2p, Si 2p, C 1s, N 1s, O 1s, and Fe 2p at the binding energies of 74.88, 103.28, 284.60, 400.07, 532.41, and 710.47 eV, respectively. All the obtained elements confirmed the presence of Fe$_3$O$_4$ and CTs. However, the peaks of C1s from

HNTs-Fe_3O_4-CTs-S-S-Laccase were found significantly higher in intensity than HNTs-Fe_3O_4-CTs. This corroborated the successful loading of the laccase. Additionally, the high-resolution spectra of Fe 2p for both the samples were examined with the curve fitting analysis (Figure 4). The close look at the curve fittings of Fe 2p showed two main peaks of Fe 2p1/2 and Fe 2p3/2 at the binding energies of the 710.42 and 723.80 eV, respectively. The spin energy separation of Fe $2p_{1/2}$ and Fe $2p_{3/2}$ was found to be 13.38 eV, which matches the standard data for Fe_3O_4 [29]. Iron in Fe_3O_4 exists in Fe^{2+} and Fe^{3+} oxidation states [30]. Hence, the Fe 2p curve fitting analysis showed both the states of Fe^{2+} and Fe^{3+} [31]. The spectra of both the samples were found to be split into Fe $2p_{3/2}$ and Fe $2p_{1/2}$ peaks at 710.4 and 724.0 eV corresponding to Fe^{2+}, similarly Fe $2p_{3/2}$ and Fe $2p_{1/2}$ peaks of Fe^{3+} at 712.81 and 728.5 eV. However, the characteristic Fe satellite peak was observed in the HNTs-Fe_3O_4-CTs-S-S-Laccase. Similar Fe 2p peak profiling for Fe_3O_4-rGO was reported [32]. Thus, the overall XPS spectrum analysis corroborated the successful synthesis of modified HNTs and laccase immobilization.

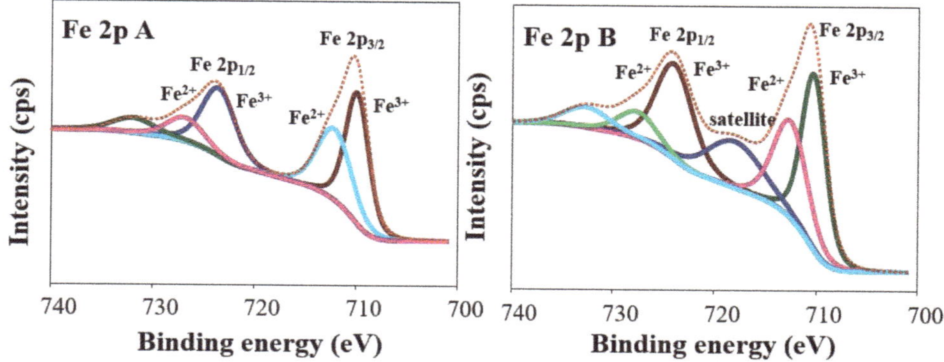

Figure 4. The high-resolution XPS spectrum of the Fe 2p peaks of (**A**) HNTs-Fe_3O_4-CTs, and (**B**) HNTs-Fe_3O_4-CTs-S-S-Laccase.

3.3. Laccase Immobilization on HNTs-Fe_3O_4-CTs-SH

After structural and morphological confirmation of the HNTs-Fe_3O_4-CTs-SH synthesis, the HNTs-Fe_3O_4-CTs-SH was applied for immobilization of the laccase. The activity of free laccase, laccase immobilized on the HNTs-Fe_3O_4-CTs, and HNTs-Fe_3O_4-CTs-S-S-Laccase are shown in Figure 5A. The measure of 1 U of laccase activity is given as μM of ABTS oxidized/min/mL. The free laccase oxidized gave 10.60 U of the activity (Figure 5A). The laccase immobilized on HNTs-Fe_3O_4-CTs exhibited 3.3 U of the laccase activity (Figure 5A). This might be due to the adsorption of the laccase on HNTs-Fe_3O_4-CTs. However, the immobilization of laccase over HNTs-Fe_3O_4-CTs-SH (HNTs-Fe_3O_4-CTs-S-S-Laccase) possessed 10.64 U of the activity. Hence, HNTs-Fe_3O_4-CTs-S-S-Laccase exhibited 100% activity recovery (Figure 5A). Figure 5B shows the ABTS oxidized potential of free laccase and HNTs-Fe_3O_4-CTs-S-S-Laccase in photographic form. These results confirmed the significant enhancement in laccase activity after the thiolation of the support HNTs-Fe_3O_4-CTs. The thiol (-SH) group modification of chitosan provided unique covalent binding sites for the laccase immobilization. The thiol (-SH) group from the laccase covalently bound to the (-SH) group over the HNTs-Fe_3O_4-CTs-SH, to form the disulfide bond (-S-S-). This led to a significant enhancement of laccase loading over the HNTs-Fe_3O_4-CTs-SH. A similar mechanism of the laccase immobilization over the (-SH) modified supports was reported earlier [7,19,33,34].

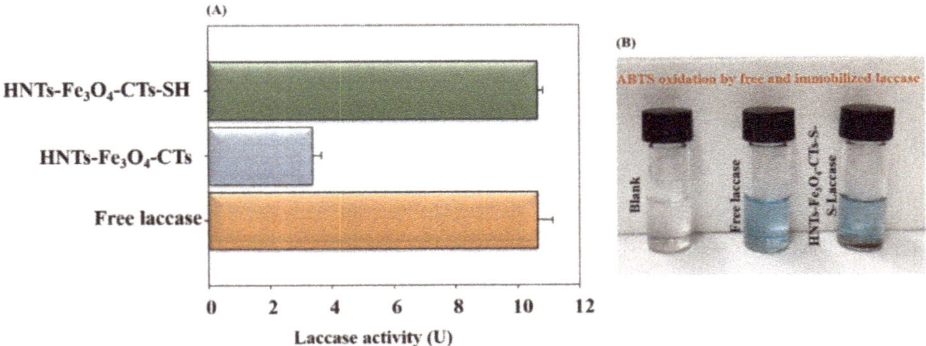

Figure 5. (**A**) Unit activity of free laccase (0.1 mL from 1.5 mg/mL laccase solution in sodium acetate buffer (100 mM, pH 4)), laccase immobilized on HNTs-Fe$_3$O$_4$-CTs (0.1 mL solution containing 1 mg of laccase immobilized HNTs-Fe$_3$O$_4$-CTs in sodium acetate buffer (100 mM, pH 4)), and HNTs-Fe$_3$O$_4$-CTs-S-S-Laccase (0.1 mL solution containing 1 mg of HNTs-Fe$_3$O$_4$-CTs-S-S-Laccase in sodium acetate buffer (100 mM, pH 4)). (**B**) Photographic image representing the color change after the oxidation of ABTS by free laccase and HNTs-Fe$_3$O$_4$-CTs-S-S-Laccase.

Furthermore, the effect of initial laccase concentrations on laccase loading over the surface of HNTs-Fe$_3$O$_4$-CTs-SH was assessed (Figure 6A). In the presence of the initial laccase concentrations of 0.5, 0.75, 1.0, 1.25, 1.5 and 1.75 mg/mL, HNTs-Fe$_3$O$_4$-CTs-SH displayed 45, 64, 84, 108, 145, and 144 mg/g of laccase loading. The laccase loading on HNTs-Fe$_3$O$_4$-CTs-SH was found to increase with a subsequent increase in the initial laccase concentrations from 0.5 to 1.5 mg/mL. The optimum laccase loading (144 mg/g) was reached 1.5 mg/mL and remained stable with a further increase in the laccase concentrations. This result indicated that the optimum occupation of the immobilization sites on HNTs-Fe$_3$O$_4$-CTs-SH was reached at the 1.5 mg/mL initial laccase concentration. The obtained value of the laccase loading was found to be excellent in comparison to the most recent reports (Table 1). The obtained laccase loading in this study was found to be higher in comparison with the recently reported nanosupports. This indicates that thiol functioned CTs over clay mineral HNTs with magnetic properties can be excellent supports for the enzyme immobilization.

Table 1. Laccase loading capacity (mg/g) comparison of the recently reported nanosupports.

Used Material	Laccase Loading (mg/g)	Reference
HNTs-Fe$_3$O$_4$-CTs-SH	144	This study
Magnetic biochar (L-MBC)	27	[35]
MACS-NIL-Cu-Laccase	47	[36]
Polyacrylamide-alginate cryogel	68	[8]
LA-Au/PDA@SiO$_2$-MEPCM	50	[37]
ZrO$_2$–SiO$_2$	86	[38]
Fe$_3$O$_4$@Chitosan	32	[39]
ZrO$_2$–SiO$_2$/Cu^{2+}	94	[38]
HNTs-M-chitosan (1%)	100	[14]
Aminosilanized magnetic HNTs	84	[16]
Fe$_3$O$_4$-NIL-DAS@lac	60	[40]
Magnetized chitosan modified α-Cellulose	73	[9]
PD-GMA-Ca@ABTS beads	8	[41]
Magnetized chitosan modified HNTs	92	[13]
Chitosan microspheres	8	[42]
Sepharose-linked antibody	33	[43]

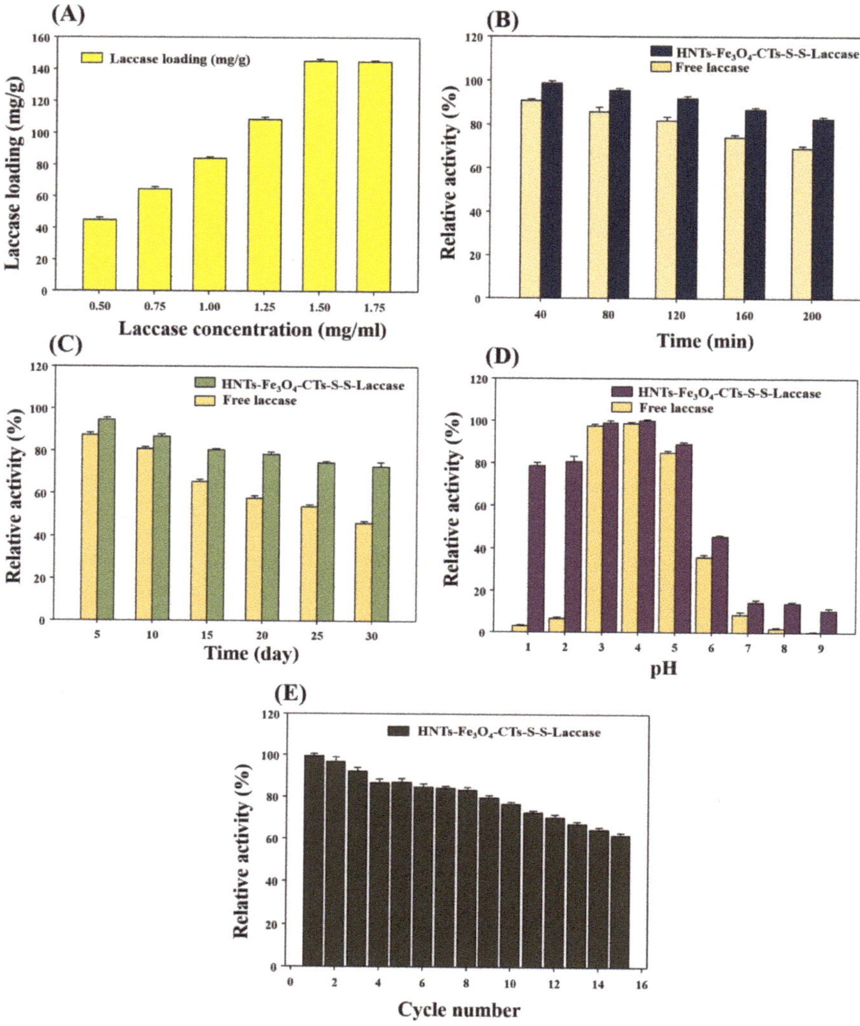

Figure 6. (**A**) Effect of initial laccase concentration on the immobilization on the HNTs-Fe$_3$O$_4$-CTs-SH, (**B**) temperature stability at 60 °C for 200 min by free laccase and HNTs-Fe$_3$O$_4$-CTs-S-S-Laccase, (**C**) storage stability for 30 days by free laccase and HNTs-Fe$_3$O$_4$-CTs-S-S-Laccase, (**D**) pH stabilities free laccase and HNTs-Fe$_3$O$_4$-CTs-S-S-Laccase, and (**E**) repeated cycle activities of the HNTs-Fe$_3$O$_4$-CTs-S-S-Laccase.

3.4. Biocatalytic Performance of the HNTs-Fe$_3$O$_4$-CTs-S-S-Laccase

After assessment of the immobilization activity recoveries and understanding of the laccase loading pattern on HNTs-Fe$_3$O$_4$-CTs-SH, the biocatalytic performance of the HNTs-Fe$_3$O$_4$-CTs-S-S-Laccase was evaluated by using the temperature stability, storage stability, pH stability, and readability potential studies. First, the data obtained for temperature stability is presented in Figure 6B. Enzyme being proteinaceous is extremely prone to denaturation due to elevated temperatures [23]. Excellent enzyme immobilization strategies were found to be effective to overcome this problem [44]. Hence, in this study free laccase and the immobilized enzyme system were assessed for temperature stability in acetate buffer (100 mM, pH 4.2) for 200 min at 60 °C. After 200 min, free laccase lost 31% of the initial activity (Figure 6B). However, HNTs-Fe$_3$O$_4$-CTs-S-S-Laccase lost 17% of the initial activity (Figure 6B). Hence, the obtained results in Figure 6B suggest significantly improved thermal stability of the laccase after the immobilization protocol. There are many applications of the enzyme laccase that can be enhanced by improved thermal stability [35]. Further, the storage stability of the immobilized and free laccase was tested (Figure 6C). The long-term storage of the enzyme in solution is a challenging task [35]. However, after immobilization, the enzyme native structure can be protected for a longer time [10]. This helps the utilization of enzyme-based catalysis and makes it economically applicable [9]. In this regard, the immobilized laccase and free laccase were tested for storage stabilities. After 30 days incubation of both the systems, free laccase retained 46% of the initial activity. However, HNTs-Fe$_3$O$_4$-CTs-S-S-Laccase retained 73% of the activity. A significant increase in activity stability was found after the immobilization. This can be helpful to the many laccase-based applications. Moreover, the pH of the system is a major key for the biocatalysts. The effect of pH on the immobilized laccase and free laccase with their incubation for 1 h was tested (Figure 6D). The HNTs-Fe$_3$O$_4$-CTs-S-S-Laccase exhibited excellent biocatalysis over the range of pH of 1–9, compared to the free laccase (Figure 6D). This will elicit the laccase application profile at various pH ranges [13]. Overall all the stability experiments proved HNTs-Fe$_3$O$_4$-CTs-S-S-Laccase as an excellent biocatalyst and HNTs-Fe$_3$O$_4$-CTs-SH can be used for the immobilization of other enzymes.

The free enzyme offers only a single use, as it is a very difficult as well as costly affair to separate enzyme and products after the reaction process [10]. Hence, nanosupport with immobilized enzyme provides leverage to overcome this limitation [1]. Besides, the magnetized nanosupports would be an excellent choice in this regard [44]. The HNTs-Fe$_3$O$_4$-CTs-S-S-Laccase is a capable biocatalyst and super-magnetic in nature. Hence, its potential was tested for repeated use. The data for repeated use application of the HNTs-Fe$_3$O$_4$-CTs-S-S-Laccase is given in Figure 6E. Upon 15 cycles of repeated use, HNTs-Fe$_3$O$_4$-CTs-S-S-Laccase possessed 61% of the initial activity. The HNTs-Fe$_3$O$_4$-CTs-S-S-Laccase exhibited significant potential of reusability. This can be highly applicable in many laccase-based applications. Being a support made from clay mineral (HNTs), super-magnetic Fe$_3$O$_4$ NPs, and thiol functionalized biopolymer (CTs), this nanosupport can be highly beneficial in many desired applications of the laccase. Furthermore, the HNTs-Fe$_3$O$_4$-CTs for laccase immobilization with glutaraldehyde (GTA) cross-linking route was reported in our previous study [13]. The same backbone nanosupport HNTs-Fe$_3$O$_4$-CTs was used in this study for laccase immobilization through the thiolation route. Hence, it is imperative to discuss and compare both the routes for laccase immobilization. The thiolated support gave a laccase loading capacity of 144 mg/g; however, GTA cross-linking exhibited 100 mg/g of laccase loading capacity. Thiolation enhanced the laccase-loading capacity, this might be due to the enhanced disulfide linkage formation in the immobilization process. Regarding storage and pH stability, thiolated HNTs-Fe$_3$O$_4$-CTs gave a better biocatalytic performance than GTA cross-linked HNTs-Fe$_3$O$_4$-CTs [13]. Furthermore, in a comparison of the temperature stabilities the GTA cross-linked possessed slightly higher temperature stability than thiolated HNTs-Fe$_3$O$_4$-CTs [13]. The thiolated HNTs-Fe$_3$O$_4$-CTs possessed remarkable enhancement in the repeated cycle studies [13]. These comparisons indicated significantly improved nanosupport HNTs-Fe$_3$O$_4$-CTs-SH was provided for the laccase immobilization. This upgraded nanosupport can enable the exploration of many laccase-based applications, and it can also be applied to other biocatalyst immobilization processes, to enhance their biocatalytic performances.

3.5. Application of HNTs-Fe$_3$O$_4$-CTs-S-S-Laccase in Environmental Pollutants Removal

Environmental pollutants such as textile dyes and pharmaceutical compounds cause significant damage to the natural ecosystem and human health. Synthetic reactive dyes released from the textile industry have very serious implications for water ecosystems and humans [3]. Similarly, pharmaceutical compounds released from domestic and industrial wastewater in small concentrations pose a significant challenge to the environment due to their bioactive nature [4]. Laccase catalyzes the redox-mediated degradation of various pollutants. Still, free laccase has many limitations to mitigate this challenge, however, immobilized laccase with improved properties is most suitable for redox-mediated removal of the various pollutants from water [1]. The enhanced biocatalytic performance by the immobilized laccase plays a crucial role in the degradation of environmental pollutants. Hence, in this study, we applied the developed immobilization system of HNTs-Fe$_3$O$_4$-CTs-S-S-Laccase for the removal of textile dye DR80 and the pharmaceutical compound ampicillin. The structures of environmental pollutants taken for study, DR80 and ampicillin, and the redox mediator compounds, such as p-Coumaric acid (CA), 1-Hydroxybenzotriazole hydrate (HBT), syringaldehyde (SA), guaiacol (GUA), and 2,2′-Azino-bis (3-ethylbenzothiazoline-6-sulfonic acid) (ABTS) are shown in Figure 7.

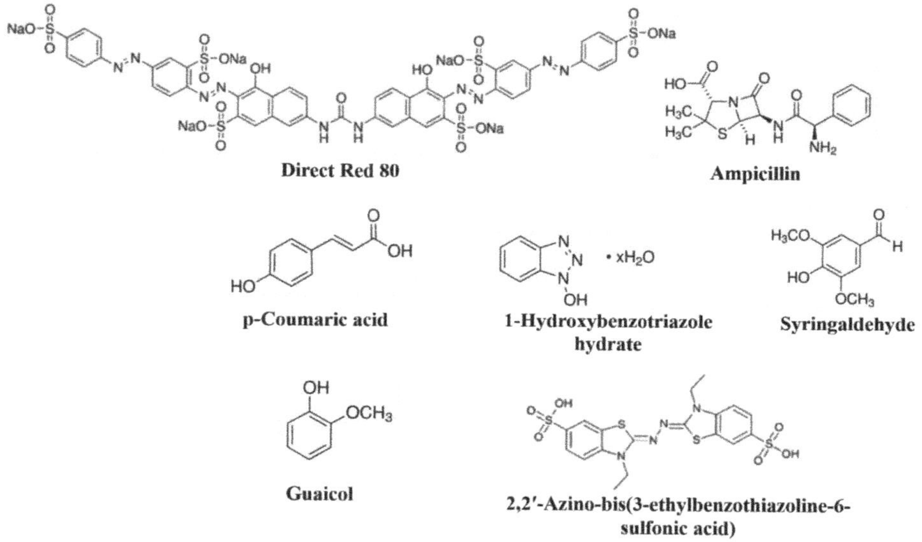

Figure 7. Structures of the environmental pollutants Direct Red 80 and ampicillin, and redox mediators, p-coumaric acid (CA), 1-1-hydroxy benzotriazole hydrate (HBT), syringaldehyde (SA), guaiacol (GUA), and 2,2′-Azino-bis(3-ethylbenzothiazoline-6-sulfonic acid) (ABTS).

The HNTs-Fe$_3$O$_4$-CTs-S-S-Laccase and free laccase were applied for the degradation of DR80. The obtained results are shown in Figure 8. The free laccase does not show the decolorization of DR80 without a redox mediator. Laccase requires the redox mediator system to degrade the environmental pollutants [16]. The HNTs-Fe$_3$O$_4$-CTs-S-S-Laccase without redox mediator gave 45% removal of DR80 Figure 8. Consequently, without a mediator immobilized laccase cannot attack the DR80 directly, and obtained 45% DR80 removal in the absence of the mediator was assigned to the adsorption process. Next, HNTs-Fe$_3$O$_4$-CTs-S-S-Laccase with redox mediators CA, HBT, SA, GUA, and ABTS, gave 63, 76, 60, 67, and 92% decolorization of DR80 Figure 8. However, free laccase exhibited 27, 39, 27, 30, and 74% decolorization of DR80 in the presence of the CA, HBT, SA, GUA, and ABTS (Figure 8). Thus, HNTs-Fe$_3$O$_4$-CTs-S-S-Laccase enhanced the decolorization of DR80 compared to the free laccase. This enhancement corresponded to the enhanced catalytic performance and possible adsorption effect of the HNTs-Fe$_3$O$_4$-CTs-S-S-Laccase. The immobilized laccase or laccase first oxidized the mediator, and the oxidized mediator carried out the oxidative degradation of the DR80. Among all redox mediators studied, the ABTS was found to be the best for the DR80 removal potential of the HNTs-Fe$_3$O$_4$-CTs-S-S-Laccase. This obtained result might be due to the better oxidation−reduction potential of the ABTS by HNTs-Fe$_3$O$_4$-CTs-S-S-Laccase. Hence, the developed laccase immobilization system was found to be successful in the removal of a toxic textile dye like DR80.

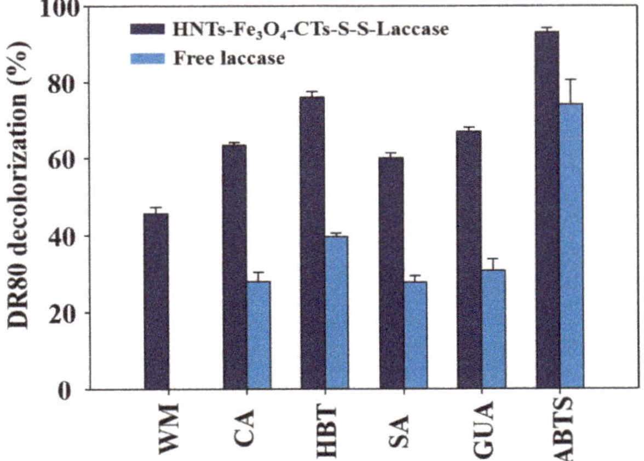

Figure 8. Removal of the DR80 by free laccase and HNTs-Fe$_3$O$_4$-CTs-S-S-Laccase; without a mediator (WM), and with redox mediators of CA, HBT, SA, GUA, and ABTS.

After investigating the decolorization potential of HNTs-Fe$_3$O$_4$-CTs-S-S-Laccase, the effect of pH on the decolorization of DR80 in the presence of redox mediator ABTS was assessed. pH played a crucial role in biocatalytic reactions of the laccase. The HNTs-Fe$_3$O$_4$-CTs-S-S-Laccase exhibited higher activities over all the pH range (Figure 6D). The decolorization performance of DR80 at various pH values of 1, 2, 3, 4, 5, 6, 7, 8, and 9 was observed as follows: 88, 89, 92, 92, 89, 46, 28, 20, and 16%, respectively (Figure 9). The pH range 1–5 displayed higher decolorization. The highest decolorization was displayed at pH 4, this obtained result was in agreement with previous observations from Figure 6D.

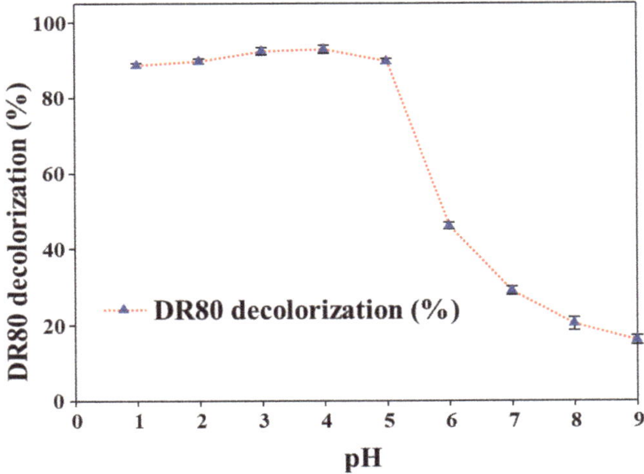

Figure 9. Effect of pH on removal of the DR80 by HNTs-Fe$_3$O$_4$-CTs-S-S-Laccase in the presence of the redox mediator ABTS.

Furthermore, the repeated cycle decolorization of the DR80 was assessed. It was very important to apply laccase in multiple cycles to enhance the remediation potential. The HNTs-Fe$_3$O$_4$-CTs-S-S-Laccase was assessed for 10 decolorization cycles of DR80 in the presence of redox mediator ABTS (Figure 10). The HNTs-Fe$_3$O$_4$-CTs-S-S-Laccase displayed 92, 91, 91, 85, 80, 76, 71, 71, 67, and 60% decolorization of the DR80. At the end of the 10th cycle, 60% decolorization was observed. This marks higher remediation potential of the developed immobilized laccase system. However, in the case of free laccase, it was very difficult to separate laccase from the first reaction mixture. However, the HNTs-Fe$_3$O$_4$-CTs-S-S-Laccase is super-magnetic, can be easily retrieved from the water, and applied in the next cycle of pollutant removal. The overall DR80 removal studies by HNTs-Fe$_3$O$_4$-CTs-S-S-Laccase gave an idea about its potential for the treatment of textile dyes.

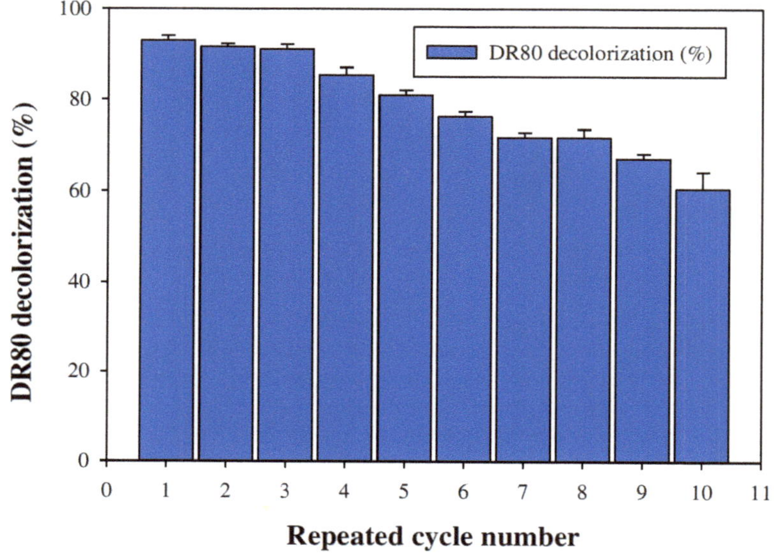

Figure 10. Repeated cycle removal of the DR80 by HNTs-Fe$_3$O$_4$-CTs-S-S-Laccase in the presence of the redox mediator ABTS.

Finally, the HNTs-Fe$_3$O$_4$-CTs-S-S-Laccase was tested for its potential in the degradation of the ampicillin. The fate of antibiotics like ampicillin in water bodies is causing serious environmental concern mainly due to the possible spread of antibiotic resistance [45]. Among all classes of antibiotics, the β-lactam class of antibiotics has captured over 65% of the world antibiotic market. Ampicillin is a widely used β-lactam antibiotics [46]. The role of laccase for degradation of the β-lactam antibiotics has been reported [47]. Hence, in this study, we assessed the potential of the HNTs-Fe$_3$O$_4$-CTs-S-S-Laccase for ampicillin degradation. The degradation of ampicillin by HNTs-Fe$_3$O$_4$-CTs-S-S-Laccase was observed by the HPLC analysis as mentioned in the report [48]. The obtained results for redox-mediated degradation of ampicillin by HNTs-Fe$_3$O$_4$-CTs-S-S-Laccase are shown in Figure 11. The HPLC of control ampicillin showed a very sharp peak at the retention time of the 3.9 min (Figure 11A). A similar kind of ampicillin peak was reported in earlier reports [49–51]. The ampicillin in the presence of the HNTs-Fe$_3$O$_4$-CTs-S-S-Laccase and with no redox mediator showed a very sharp peak at the retention time of the 3.9 min (Figure 11B). This obtained result corroborated that, without a mediator, no degradation was observed. The intensity of the peak also remained consistent suggesting no adsorption of ampicillin. Further, HNTs-Fe$_3$O$_4$-CTs-S-S-Laccase with redox mediator ABTS exhibited a changed profile of the peaks at different retention times; such as 2.4, 2.6, 2.79, and 7.07 min, respectively, see Figure 11C. This observation confirmed the complete degradation of the ampicillin. The main ampicillin peak of 3.9 min disappeared and new peaks arrived due to degradation by the HNTs-Fe$_3$O$_4$-CTs-S-S-Laccase. Similarly, the HNTs-Fe$_3$O$_4$-CTs-S-S-Laccase with mediators GUA, SA, and HBT gave degradation of ampicillin with different retention time peaks (Figure 11D–F). Thus, the overall ampicillin degradation study corroborated that various redox mediators can be used for ampicillin degradation by HNTs-Fe$_3$O$_4$-CTs-S-S-Laccase. Therefore, this system could be used for a wide range of pollutant removal. As in this study, we assessed the potential of immobilized laccase for textile dye DR80 and pharmaceutical compound ampicillin.

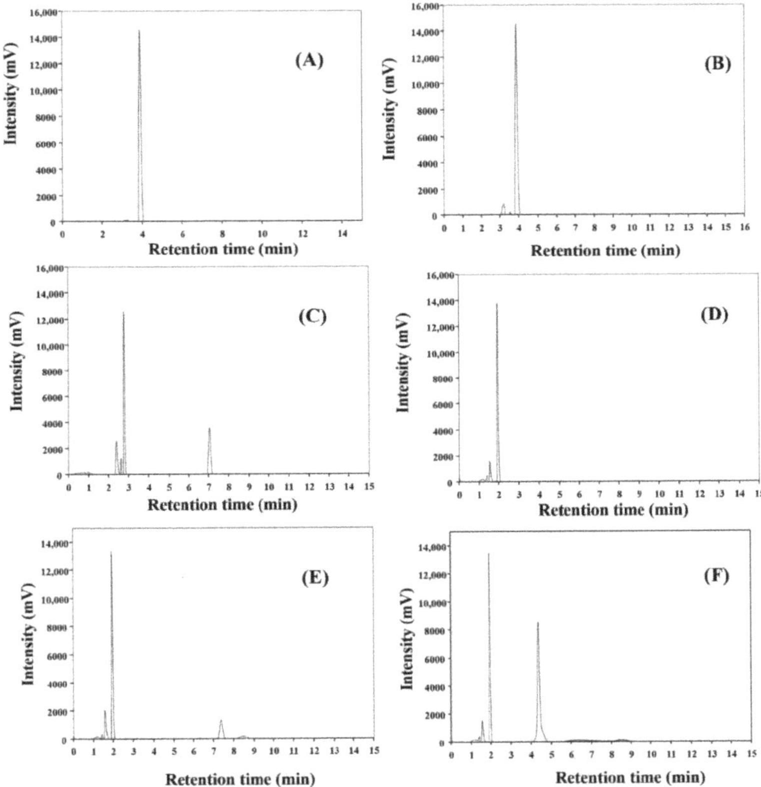

Figure 11. HPLC analysis of the ampicillin degradation by HNTs-Fe$_3$O$_4$-CTs-S-S-Laccase (**A**) control ampicillin, (**B**) ampicillin + HNTs-Fe$_3$O$_4$-CTs-S-S-Laccase (in absence of the redox mediator), (**C**) ampicillin + HNTs-Fe$_3$O$_4$-CTs-S-S-Laccase + ABTS, (**D**) ampicillin + HNTs-Fe$_3$O$_4$-CTs-S-S-Laccase + GUA, (**E**) ampicillin + HNTs-Fe$_3$O$_4$-CTs-S-S-Laccase + SA, and (**F**) ampicillin + HNTs-Fe$_3$O$_4$-CTs-S-S-Laccase + HBT.

4. Conclusions

In conclusion, this study investigated the new nanocomposite containing HNTs, Fe$_3$O$_4$ NPs, and thiolated CTs for laccase immobilization. The detailed characterizations, FE-SEM, HR-TEM, XPS, XRD, FT-IR, and VSM analyses of the nanocomposite corroborated successful synthesis. The immobilized laccase displayed outstanding biocatalytic performance with improved thermal, storage and pH stability. The immobilized laccase also gave redox-mediated degradation of environmental pollutants such as DR80 and ampicillin. This indicated that immobilized laccase can be applied for wastewater treatment. The super-magnetic nature can easily retrieve the nanosupport from the solution after the decontamination of the pollutant. Thus, the novel nanosupport developed in this study "HNTs-Fe$_3$O$_4$-CTs-SH" is highly efficient for laccase immobilization, and also can be applied for other enzyme-immobilization processes.

Author Contributions: Data curation, A.A.K.; Formal analysis, A.A.K., G.D.S., and R.G.S.; Investigation, A.A.K.; Methodology, A.A.K., B.S., S.K.S., and G.S.G.; Project administration, A.A.K., D.-Y.K., and J.-S.S.; Supervision, D.-Y.K., and J.-S.S.; Validation, A.A.K.; Writing – original draft, A.A.K.; Writing – review & editing, A.A.K., D.-Y.K., and J.-S.S. All authors have read and agreed to the published version of the manuscript.

Funding: This work was supported by the National Research Foundation of Korea (NRF) grant funded by the Korea government (MSIT) (NRF-2019R1G1A1009363) and the grant (2017001970003) from the Ministry of Environment of Korea.

Conflicts of Interest: There is no conflict of interest.

References

1. Datta, S.; Veena, R.; Samuel, M.S.; Selvarajan, E. Immobilization of laccases and applications for the detection and remediation of pollutants: A review. *Environ. Chem. Lett.* **2020**, 1–18. [CrossRef]

2. Datta, S.; Rajnish, K.N.; Samuel, M.S.; Pugazlendhi, A.; Selvarajan, E. Metagenomic applications in microbial diversity, bioremediation, pollution monitoring, enzyme and drug discovery. A review. *Environ. Chem. Lett.* **2020**, *18*, 1229–1241. [CrossRef]

3. Liu, Y.Q.; Maulidiany, N.; Zeng, P.; Heo, S. Decolourization of azo, anthraquinone and triphenylmethane dyes using aerobic granules: Acclimatization and long-term stability. *Chemosphere* **2020**, *263*, 128312. [CrossRef] [PubMed]

4. Taoufik, N.; Boumya, W.; Janani, F.Z.; Elhalil, A.; Mahjoubi, F.Z.; Barka, N. Removal of emerging pharmaceutical pollutants: A systematic mapping study review. *J. Environ. Chem. Eng.* **2020**, *8*, 104251. [CrossRef]

5. Antibiotics, N.; Family, H.; Sanchez, H.M.; Whitener, V.A.; Thulsiraj, V.; Amundson, A.; Collins, C.; Duran-gonzalez, M.; Giragossian, E.; Hornstra, A.; et al. Antibiotic Resistance of Escherichia coli Isolated from Owned Retail Broiler Chicken Meat. *Animals* **2020**, *10*, 2217.

6. Tabla-Hernández, J.; Rodríguez-Espinosa, P.F.; Hernandez-Ramirez, A.G.; Mendoza-Pérez, J.A.; Cano-Aznar, E.R.; Martínez-Tavera, E. Treatment of Eutrophic water and wastewater from Valsequillo Reservoir, Puebla, Mexico by Means of Ozonation: A multiparameter approach. *Water* **2018**, *10*, 1790. [CrossRef]

7. Ulu, A.; Birhanli, E.; Boran, F.; Köytepe, S.; Yesilada, O.; Ateş, B. Laccase-conjugated thiolated chitosan-Fe$_3$O$_4$ hybrid composite for biocatalytic degradation of organic dyes. *Int. J. Biol. Macromol.* **2020**, *150*, 871–884. [CrossRef]

8. Yavaşer, R.; Karagözler, A.A. Laccase immobilized polyacrylamide-alginate cryogel: A candidate for treatment of effluent. *Process Biochem.* **2010**, *101*, 137–146.

9. Ghodake, G.S.; Yang, J.; Shinde, S.S.; Mistry, B.M.; Kim, D.Y.; Sung, J.S.; Kadam, A.A. Paper waste extracted A-cellulose fibers super-magnetized and chitosan-functionalized for covalent laccase immobilization. *Bioresour. Technol.* **2018**, *261*, 420–427. [CrossRef]

10. Ba, S.; Arsenault, A.; Hassani, T.; Jones, J.P.; Cabana, H. Laccase immobilization and insolubilization: From fundamentals to applications for the elimination of emerging contaminants in wastewater treatment. *Crit. Rev. Biotechnol.* **2013**, *33*, 404–418. [CrossRef]

11. Tully, J.; Yendluri, R.; Lvov, Y. Halloysite Clay Nanotubes for Enzyme Immobilization. *Biomacromolecules* **2016**, *17*, 615–621. [CrossRef] [PubMed]

12. Cheng, C.; Song, W.; Zhao, Q.; Zhang, H. Halloysite nanotubes in polymer science: Purification, characterization, modification and applications. *Nanotechnol. Rev.* **2020**, *9*, 323–344. [CrossRef]

13. Kadam, A.A.; Jang, J.; Jee, S.C.; Sung, J.S.; Lee, D.S. Chitosan-functionalized supermagnetic halloysite nanotubes for covalent laccase immobilization. *Carbohydr. Polym.* **2018**, *194*, 208–216. [CrossRef] [PubMed]

14. Kadam, A.A.; Shinde, S.K.; Ghodake, G.S.; Saratale, G.D.; Saratale, R.G.; Sharma, B.; Hyun, S.; Sung, J.-S. Chitosan-Grafted Halloysite Nanotubes-Fe$_3$O$_4$ Composite for Laccase-Immobilization and Sulfamethoxazole-Degradation. *Polymers* **2020**, *12*, 2221. [CrossRef] [PubMed]

15. Hajizadeh, Z.; Maleki, A.; Rahimi, J.; Eivazzadeh-Keihan, R. Halloysite Nanotubes Modified by Fe_3O_4 Nanoparticles and Applied as a Natural and Efficient Nanocatalyst for the SymmetricalHantzsch Reaction. *Silicon* **2020**, *12*, 1247–1256. [CrossRef]
16. Kadam, A.A.; Jang, J.; Lee, D.S. Supermagnetically Tuned Halloysite Nanotubes Functionalized with Aminosilane for Covalent Laccase Immobilization. *ACS Appl. Mater. Interfaces* **2017**, *9*, 15492–15501. [CrossRef]
17. Liu, M.; Zhang, Y.; Wu, C.; Xiong, S.; Zhou, C. Chitosan/halloysite nanotubes bionanocomposites: Structure, mechanical properties and biocompatibility. *Int. J. Biol. Macromol.* **2012**, *51*, 566–575. [CrossRef]
18. Kim, M.; Jee, S.C.; Sung, J.-S.; Kadam, A.A. Anti-proliferative applications of laccase immobilized on super-magnetic chitosan-functionalized halloysite nanotubes. *Int. J. Biol. Macromol.* **2018**, *118*, 228–237. [CrossRef]
19. Federer, C.; Kurpiers, M.; Bernkop-Schnürch, A. Thiolated Chitosans: A Multi-talented Class of Polymers for Various Applications. *Biomacromolecules* **2020**. [CrossRef]
20. Leichner, C.; Jelkmann, M.; Bernkop-Schnürch, A. Thiolated polymers: Bioinspired polymers utilizing one of the most important bridging structures in nature. *Adv. Drug Deliv. Rev.* **2019**, *151–152*, 191–221. [CrossRef]
21. Hanif, M.; Zaman, M.; Qureshi, S. Thiomers: A Blessing to Evaluating Era of Pharmaceuticals. *Int. J. Polym. Sci.* **2015**, *2015*, 1–9. [CrossRef]
22. Zhu, X.; Su, M.; Tang, S.; Wang, L.; Liang, X.; Meng, F.; Hong, Y. *Zhu-2012-Synthesis of Thiolat.pdf*; 2012; pp. 1973–1982.
23. Verma, M.L.; Kumar, S.; Das, A.; Randhawa, J.S.; Chamundeeswari, M. Chitin and chitosan-based support materials for enzyme immobilization and biotechnological applications. *Environ. Chem. Lett.* **2020**, *18*, 315–323. [CrossRef]
24. Teymourian, H.; Salimi, A.; Khezrian, S. Biosensors and Bioelectronics Fe_3O_4 magnetic nanoparticles/reduced graphene oxide nanosheets as a novel electrochemical and bioeletrochemical sensing platform. *Biosens. Bioelectron.* **2013**, *49*, 1–8. [CrossRef] [PubMed]
25. Chauhan, K.; Singh, P.; Singhal, R.K. New Chitosan-Thiomer: An Efficient Colorimetric Sensor and Effective Sorbent for Mercury at Ultralow Concentration. *ACS Appl. Mater. Interfaces* **2015**, *7*, 26069–26078. [CrossRef] [PubMed]
26. Nandiyanto, A.B.D.; Oktiani, R.; Ragadhita, R. How to read and interpret ftir spectroscope of organic material. *Indones. J. Sci. Technol.* **2019**, *4*, 97–118. [CrossRef]
27. Taylor, P.; Trofimov, B.A.; Sinegovskaya, L.M.; Gusarova, N.K. Vibrations of the S–S bond in elemental sulfur and organic polysulfides: A structural guide. *J. Sulfur Chem.* **2009**, *30*, 518–554.
28. Ji, Y.; Yang, X.; Ji, Z.; Zhu, L.; Ma, N.; Chen, D.; Jia, X.; Tang, J.; Cao, Y. DFT-Calculated IR Spectrum Amide I, II, and III Band Contributions of N-Methylacetamide Fine Components. *ACS Omega* **2020**, *5*, 8572–8578. [CrossRef]
29. Sharma, B.; Kadam, A.A.; Sung, J.S.; Myung, J. ha Surface tuning of halloysite nanotubes with Fe_3O_4 and 3-D MnO2 nanoflakes for highly selective and sensitive acetone gas sensing. *Ceram. Int.* **2020**, *46*, 21292–21303. [CrossRef]
30. Wan, L.; Yan, D.; Xu, X.; Li, J.; Lu, T.; Gao, Y.; Yao, Y.; Pan, L. Self-assembled 3D flower-like Fe_3O_4/C architecture with superior lithium ion storage performance. *J. Mater. Chem. A* **2018**, *6*, 24940–24948. [CrossRef]
31. Chandra, S.; Das, R.; Kalappattil, V.; Eggers, T.; Harnagea, C.; Nechache, R.; Phan, M.H.; Rosei, F.; Srikanth, H. Epitaxial magnetite nanorods with enhanced room temperature magnetic anisotropy. *Nanoscale* **2017**, *9*, 7858–7867. [CrossRef]
32. Zhao, Q.; Liu, J.; Wang, Y.; Tian, W.; Liu, J.; Zang, J.; Ning, H.; Yang, C.; Wu, M. Novel in-situ redox synthesis of Fe_3O_4/rGO composites with superior electrochemical performance for lithium-ion batteries. *Electrochim. Acta* **2018**, *262*, 233–240. [CrossRef]
33. Stefanov, I.; Pérez-Rafael, S.; Hoyo, J.; Cailloux, J.; Santana Pérez, O.O.; Hinojosa-Caballero, D.; Tzanov, T. Multifunctional Enzymatically Generated Hydrogels for Chronic Wound Application. *Biomacromolecules* **2017**, *18*, 1544–1555. [CrossRef] [PubMed]
34. Kalkan, N.A.; Aksoy, S.; Aksoy, E.A.; Hasirci, N. Preparation of chitosan-coated magnetite nanoparticles and application for immobilization of laccase. *J. Appl. Polym. Sci.* **2012**, *123*, 707–716. [CrossRef]

35. Zhang, Y.; Piao, M.; He, L.; Yao, L.; Piao, T.; Liu, Z.; Piao, Y. Immobilization of laccase on magnetically separable biochar for highly efficient removal of bisphenol A in water. *RSC Adv.* **2020**, *10*, 4795–4804. [CrossRef]

36. Qiu, X.; Wang, S.; Miao, S.; Suo, H.; Xu, H.; Hu, Y. Co-immobilization of laccase and ABTS onto amino-functionalized ionic liquid-modified magnetic chitosan nanoparticles for pollutants removal. *J. Hazard. Mater.* **2020**, *401*, 123353. [CrossRef]

37. Cao, P.; Liu, H.; Wu, D.; Wang, X. Immobilization of laccase on phase-change microcapsules as self-thermoregulatory enzyme carrier for biocatalytic enhancement. *Chem. Eng. J.* **2020**, *405*, 126695. [CrossRef]

38. Jankowska, K.; Ciesielczyk, F.; Bachosz, K.; Zdarta, J.; Kaczorek, E.; Jesionowski, T. Laccase immobilized onto zirconia-silica hybrid doped with Cu2+ as an effective biocatalytic system for decolorization of dyes. *Materials* **2019**, *12*, 1252. [CrossRef]

39. Zhang, K.; Yang, W.; Liu, Y.; Zhang, K.; Chen, Y.; Yin, X. Laccase immobilized on chitosan-coated Fe_3O_4 nanoparticles as reusable biocatalyst for degradation of chlorophenol. *J. Mol. Struct.* **2020**, *1220*, 128769. [CrossRef]

40. Qiu, X.; Wang, Y.; Xue, Y.; Li, W.; Hu, Y. Laccase immobilized on magnetic nanoparticles modified by amino-functionalized ionic liquid via dialdehyde starch for phenolic compounds biodegradation. *Chem. Eng. J.* **2020**, *391*, 123564. [CrossRef]

41. Xue, P.; Liu, X.; Gu, Y.; Zhang, W.; Ma, L.; Li, R. Laccase-mediator system assembling co-immobilized onto functionalized calcium alginate beads and its high-efficiency catalytic degradation for acridine. *Colloids Surfaces B Biointerfaces* **2020**, *196*, 111348. [CrossRef]

42. Aricov, L.; Leonties, A.R.; Gîfu, I.C.; Preda, D.; Raducan, A.; Anghel, D.F. Enhancement of laccase immobilization onto wet chitosan microspheres using an iterative protocol and its potential to remove micropollutants. *J. Environ. Manage.* **2020**, *276*, 111326. [CrossRef] [PubMed]

43. Zofair, S.F.F.; Arsalan, A.; Khan, M.A.; Alhumaydhi, F.A.; Younus, H. Immobilization of laccase on Sepharose-linked antibody support for decolourization of phenol red. *Int. J. Biol. Macromol.* **2020**, *161*, 78–87. [CrossRef] [PubMed]

44. Daronch, N.A.; Kelbert, M.; Pereira, C.S.; de Araújo, P.H.H.; de Oliveira, D. Elucidating the choice for a precise matrix for laccase immobilization: A review. *Chem. Eng. J.* **2020**, *397*, 125506. [CrossRef]

45. Shen, L.; Liu, Y.; Xu, H. Lou Treatment of ampicillin-loaded wastewater by combined adsorption and biodegradation. *J. Chem. Technol. Biotechnol.* **2010**, *85*, 814–820. [CrossRef]

46. Elander, R.P. Industrial production of β-lactam antibiotics. *Appl. Microbiol. Biotechnol.* **2003**, *61*, 385–392. [CrossRef]

47. Zhang, C.; You, S.; Zhang, J.; Qi, W.; Su, R.; He, Z. An effective in-situ method for laccase immobilization: Excellent activity, effective antibiotic removal rate and low potential ecological risk for degradation products. *Bioresour. Technol.* **2020**, *308*, 123271. [CrossRef]

48. Kumar, R.; Park, B.J.; Jeong, H.R.; Lee, J.T.; Cho, J.Y. Biodegradation of β-lactam antibiotic "ampicillin" by white rot fungi from aqueous solutions. *J. Pure Appl. Microbiol.* **2013**, *4*, 3163.

49. Pajchel, G.; Pawłowski, K.; Tyski, S. CE versus LC for simultaneous determination of amoxicillin/clavulanic acid and ampicillin/sulbactam in pharmaceutical formulations for injections. *J. Pharm. Biomed. Anal.* **2002**, *29*, 75–81. [CrossRef]

50. Mem, H.; Ma, M. Development and validation of RP-HPLC method for determination of amoxicillin residues and application to NICOMAC coating machine. *J. Anal. Pharm. Res.* **2018**, *7*, 586–594. [CrossRef]

51. Yang, J.; Lin, Y.; Yang, X.; Ng, T.B.; Ye, X.; Lin, J. Degradation of tetracycline by immobilized laccase and the proposed transformation pathway. *J. Hazard. Mater.* **2017**, *322*, 525–531. [CrossRef]

Publisher's Note: MDPI stays neutral with regard to jurisdictional claims in published maps and institutional affiliations.

Article

Magnetic, Electrical, and Mechanical Behavior of Fe-Al-MWCNT and Fe-Co-Al-MWCNT Magnetic Hybrid Nanocomposites Fabricated by Spark Plasma Sintering

Alexandre Tugirumubano [1], **Sun Ho Go** [1], **Hee Jae Shin** [1], **Lee Ku Kwac** [1,2] and **Hong Gun Kim** [1,2,*]

[1] Institute of Carbon Technology, Jeonju University, Jeonju-si 55069, Korea; alexat123@yahoo.com (A.T.); royal2588@naver.com (S.H.G.); ostrich@jj.ac.kr (H.J.S.); kwac29@jj.ac.kr (L.K.K.)
[2] Department of Mechanical and Automotive Engineering, Jeonju University, Jeonju-si 55069, Korea
* Correspondence: hkim@jj.ac.kr; Tel.: +82-63-220-2613

Received: 5 February 2020; Accepted: 25 February 2020; Published: 29 February 2020

Abstract: This paper aims to investigate different properties of the Fe-Al matrix reinforced with multi-walled carbon nanotube (MWCNT) nanocomposites with the Al volume content up to 65%, according to the Fe-Al combination. In addition, the effect of adding Co content on the improvement of the soft magnetic properties of the nanocomposites was carried out. The nanocomposites were fabricated using the powder metallurgy process. The iron-aluminum metal matrix reinforced multi-walled carbon nanotube (Fe-Al-MWCNT) nanocomposites showed a continuous increase of saturation magnetization from 90.70 A.m^2/kg to 167.22 A.m^2/kg and microhardness, whereas the electrical resistivity dropped as the Al content increased. The incorporation of Co nanoparticles in Fe-Al-MWCNT up to 35 vol% of Co considerably improved the soft magnetic properties of the nanocomposites by reducing the coercivity and retentivity up to 42% and 47%, respectively. The results showed that Al-based magnetic nanocomposites with a high Al volume content can be tailored using ferromagnetic particles. The composites with a volume content of magnetic particles (Fe+Co) greater than 60 vol% exhibited higher saturation magnetization, higher coercivity, and higher retentivity than the standard Sendust core. Moreover, the produced composites can be used for the lightweight magnetic core in electromagnetic devices due to their low density and good magnetic and mechanical properties.

Keywords: magnetic hybrid nanocomposites; nanoparticles; magnetic properties; mechanical properties; spark plasma sintering

---

1. Introduction

Recently, the requirements for lightweight and high performance in various engineering structures and technologies have tremendously attracted many researchers and industries to composite materials. Among these composites, aluminum metal matrix composites are interesting and promising materials to fulfill the need for lightweight structure because of their excellent strength-to-weight and stiffness-to-weight ratio [1–3]. The final properties of these composites are strongly influenced by the reinforcing materials and the fabrication process. The mechanical, electrical, and thermal properties and corrosion behavior of the Al-based composites have been studied by many investigators, and improvement of these properties was achieved depending on the reinforcement and process parameters [4–6]. However, currently, there are no significant studies on the improvements in magnetic properties of those composites. Furthermore, very few studies can be found on the Fe-Al magnetic materials with high Al content.

The use of iron oxide (α-Fe$_2$O$_3$) nanoparticles as reinforcement in aluminum matrix composites were reported to produce Al-based magnetic composites [7–9]. However, these composites were found to present high coercivity and low saturation magnetization due to the poor magnetic properties of iron oxide [7]. The aluminum metal matrix composites reinforced with 30% of iron oxide Fe$_3$O$_4$ have shown a saturation magnetization of 13.43 A.m^2/kg, and the Fe$_3$O$_4$ reinforcement was found to reduce the electrical conductivity of the composites [10,11]. Fathy et al. [12] investigated the magnetic properties of Al metal matrix reinforced with a 5% to 15 % content of Fe particles as magnetic reinforcement. Although they found an enhancement of magnetic properties, the saturation magnetization of their composites was substantially low (less than 1 A.m^2/kg). This was attributed to the formation of a diamagnetic Al$_{13}$Fe$_4$ intermetallic compound during the manufacturing process. These magnetic properties seem to be relatively low for most magnetic applications such as the magnetic core in transformers, inductors, or other electromagnetic devices. Therefore, in order to have metal matrix magnetic composites with lightweight and good magnetic properties, new approaches are advised.

In this paper, Al particles and Fe nanoparticles were combined to form a dual matrix and reinforced with 2 vol% of multi-walled carbon nanotubes (MWCNTs). It is well known that aluminum is a lightweight material, compared to iron, and is paramagnetic. On the other hand, iron is a heavy metal and is a ferromagnetic material with the best saturation induction compared to other ferromagnetic materials. The multi-walled carbon nanotubes (MWCNTs) are known to be advanced, light, and strong materials which have different applications, including the strengthening of materials in various composite materials [13–15]. The MWCNTs have good electric, mechanical, and thermal properties, and they are promising as reinforcements in composite materials. In this work, the Fe-Al-MWCNT hybrid nanocomposites were produced by spark plasma sintering with the aluminum content varying between 28 and 65 vol%. Their morphology and magnetic, electrical, and mechanical properties were tested and evaluated according to the Fe and Al volume-percentage combination. In addition, the effect of Fe and Co combination on various properties was studied by maintaining the Al and MWCNT contents in composites constant.

2. Materials and Methods

2.1. Sample Preparation

The powders of iron (Fe) nanoparticles with an average particle size of 90–110 nm and purity of 99.9%, aluminum (Al) powders with an average particle size of 30–40 μm and purity of 99.9%, and multi-walled carbon nanotubes (MWCNTs) with a 8–15 nm diameter and a 5–20 μm length were prepared with a volume ratio based on a sample of 50 mm in diameter and 30 mm in length. Furthermore, the cobalt (Co) nanoparticles with an average particle size of 80–120 nm and purity of 99.8% were prepared and used to evaluate the effect of Co on the magnetic properties of the composites with the above-mentioned materials. The Fe, Al, and Co powders were supplied by Ditto Technology Co. Ltd., Gunpo, Korea, whereas MWCNT powders were supplied by Nanosolution, Jeonju, Korea. Tables 1 and 2 show the prepared samples according to the volume fraction of iron, aluminum, and cobalt nanoparticle powders, respectively.

Table 1. Element volume ratio in Fe-Al-MWCNT hybrid nanocomposite materials. MWCNT: multi-walled carbon nanotube.

Composites	Materials ID	Fe (vol%)	Al (vol%)	MWCNT (vol%)
33Fe-65Al-2MWCNT	M1	33	65	2
38Fe-60Al-2MWCNT	M2	38	60	2
43Fe-55Al-2MWCNT	M3	43	55	2
48Fe-50Al-2MWCNT	M4	48	50	2
50Fe-48Al-2MWCNT	M5	50	48	2
55Fe-43Al-2MWCNT	M6	55	43	2
60Fe-38Al-2MWCNT	M7	60	38	2
65Fe-33Al-2MWCNT	M8	65	33	2
70Fe-28Al-2MWCNT	M9	70	28	2

Table 2. Element volume ratio in Fe-Co-Al-MWCNT hybrid nanocomposite materials.

Composites	ID	Fe (vol%)	Co (vol%)	Al (vol%)	MWCNT (vol%)
70Fe-28Al-2MWCNT	M9	70	0	28	2
60Fe-10Co-28Al-2MWCNT	M10	60	10	28	2
50Fe-20Co-28Al-2MWCNT	M11	50	20	28	2
40Fe-30Co-28Al-2MWCNT	M12	40	30	28	2
35Fe-35Co-28Al-2MWCNT	M13	35	35	28	2

The powders were mixed using the ball-milling technique on a horizontal roll ball mill. Before milling, 2 g of stearic acid were added as the process agent to reduce the cold welding of the powder and limit the powders sticking to the walls of the ball mill jar. Sixty alumina balls with a 10-mm diameter were used as the milling media. The powder milling was performed with a milling speed of 300 rpm for 48 h [9].

In order to prepare the powders for sintering, graphite die and two punches were prepared. Graphite sheets were used as die-wall and punch lubricant. The powder mixtures were filled in graphite die and then initially compacted with a hydraulic press to adjust and equalize the outer length of the punches (upper punch and bottom punch). The die containing the pre-compacted powders was mounted in a spark plasma sintering (SPS) machine (SPS-3.20MK-V, SPS Syntex Inc., Kawasaki, Japan) for simultaneous compaction and sintering. This process was performed by applying an initial pressure of 20 MPa while increasing the temperature from room temperature to 500°C. From 500°C to 600°C, a pressure of 40 MPa was applied, maintained for 10 min at 600°C, and then completely released during SPS-chamber cooling from 600°C to room temperature. The obtained cylindrical samples were machined into specimens using an electrical discharge machine (EDM) for characterization.

2.2. Characterization

Scanning electron microscopy (SEM, CX-200, COXEM Co.Ltd., Daejeon, Korea) and field-emission scanning electron microscopy (FE-SEM, JSM-7100F, Jeol, Tokyo, Japan), combined with an energy dispersive X-ray spectrometer (EDS; Aztec Energy, Oxford Instruments, Abingdon, UK), were used to investigate the morphology and elemental distribution of the composites. X-ray powder diffraction (XRD, CuKα: λ = 1.54 Å) was used to examine the crystallography and phase identification.

The cubic specimens of 2 mm × 2 mm × 2 mm were prepared and tested for magnetic property measurements using a vibrating sample magnetometer (VSM 7404-S, Lakeshore Cryotronics, Inc., Westerville, OH, USA) with an applied field varying form –1 T to +1 T. The bar sample preparation and the transverse-rupture strength (TRS) testing were performed according to the ASTM standard B528 [16], using a universal tensile-testing machine. The TRS testing was conducted by considering three specimens for each sample. The Vickers microhardness was also tested with an applied load of 0.1 kgf for 10 s using a hardness tester (HM-123, Code No 810-990K, Mitutoyo Corporation, Kawasaki,

Japan). For each sample, 10 measurements were taken and averaged for microhardness characterization. The electrical resistivity was measured using the four-probe method on the rectangular samples having the same size as the TRS specimens (i.e., 31.7 mm in length, 12.7 mm in width, and 6.35 in thickness). For each material, 20 measurements were taken and averaged.

3. Results and Discussion

3.1. Microstructure Analysis of the As-Received Powders and Fe-Al-MWCNT Nanocomposites

Figure 1 shows the SEM images of the starting materials. The starting Fe powders shown in Figure 1a revealed a mixture of very small Fe particles (at nanoscale) and relatively big Fe particles. The iron (Fe) particles were of spherical shape. The Al particles shown in Figure 1b seemed to be large even at low magnification because they were at microscale. Figure 1c shows that the MWCNT were in the form of fibers with an entangled network. Figure 1d–i shows that the core shells of nanoparticles were created on the surface of the large particles after ball milling. It can be seen that a larger amount of nanoparticles were bonded on the big particles when the nanocomposites had a higher volume content of Fe nanoparticles (i.e., 65Fe-33Al-2MWCNT; Figure 1g–i) than when they had a lower volume content of Fe nanoparticles (i.e., 38Fe-60Al-2MWCNT; Figure 1d–f). The dispersion of MWNCT could also be observed at a high magnification of the ball-milled powders, as seen in Figure 1f,h.

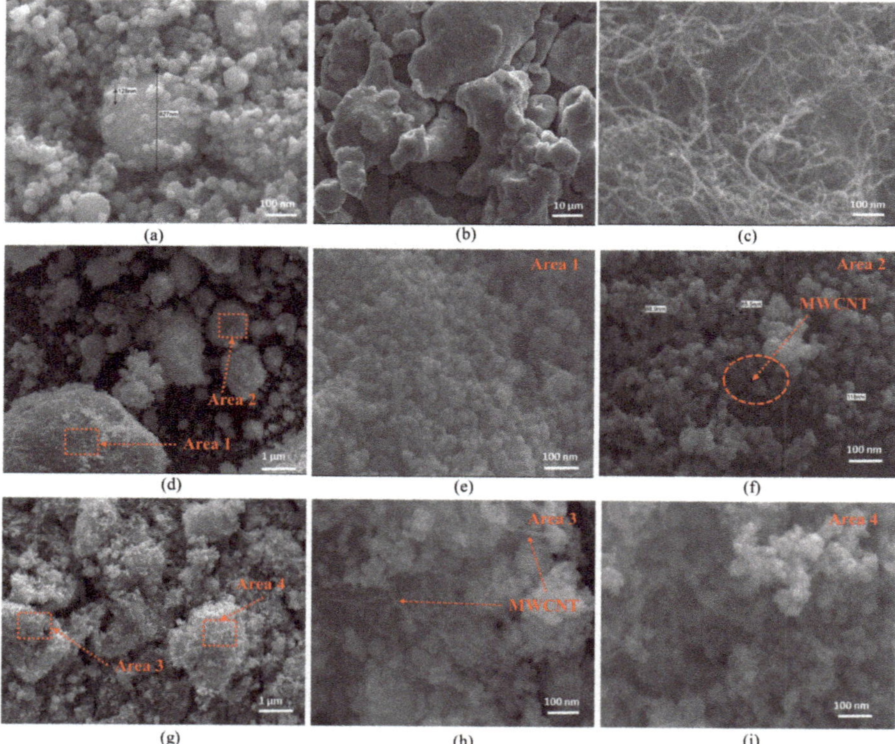

Figure 1. SEM images of powders: (**a**) As-received Fe nanopowders; (**b**) As-received Al powders; (**c**) As-received MWCNT; (**d**) Ball-milled (38Fe-60Al-2MWCNT) powders; (**e**) and (**f**) Higher Magnification of Selected Areas (1&2) in the images (**d**); (**g**) Ball-milled (65Fe-33Al-2MWCNT) powders; (**h**) and (**i**) Higher Magnification of Selected Areas (3&4) in the images (**g**).

Figure 2 shows SEM images of the composites. The Fe, Al, and MWCNTs are presented in the light gray, dark gray, and black colors, respectively, and are clearly indicated in Figure 2b. It is clear that Fe, Al, and MWCNTs were well distributed in the composites and the higher the volume content of Fe was, the larger its surface area was. The SEM images show that at a lower Fe content the Fe nanoparticles formed thin layers around the Al particles, and those layers got thicker as the volume fraction of Fe nanoparticles increased. The layers of Fe nanoparticles were the results of nanoparticles coating on the surface of large particles, using the mechanical ball-milling technique as discussed in previous research [17,18].

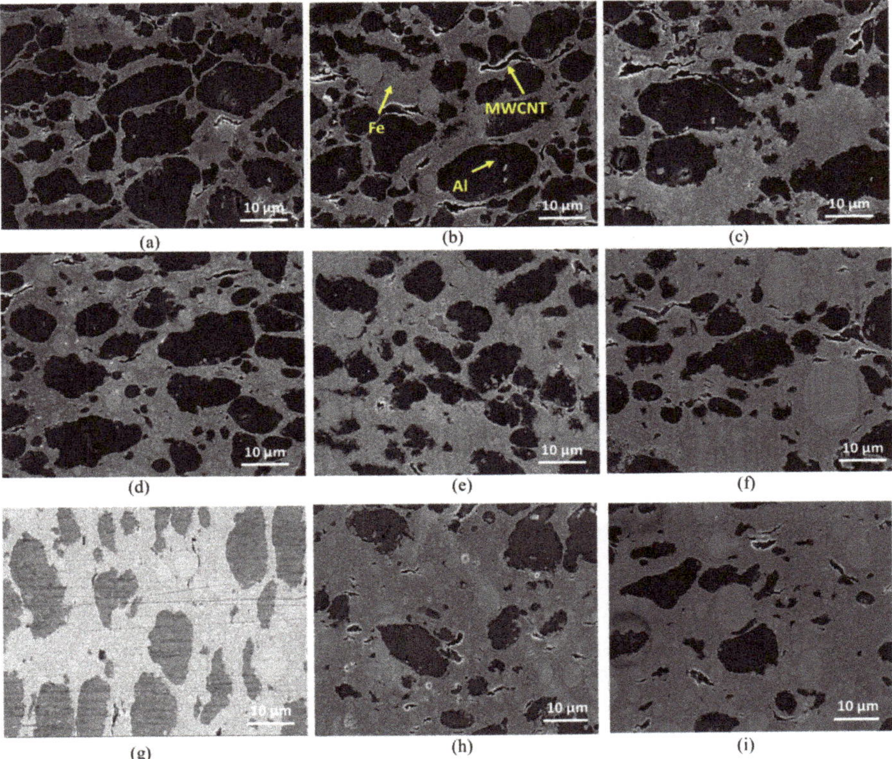

Figure 2. SEM images of Fe-Al-MWCNT hybrid nanocomposites: (**a**) 33Fe-65Al-2MWCNT (M1); (**b**) 38Fe-60Al-2MWCNT (M2); (**c**) 43Fe-55Al-2MWCNT (M3); (**d**) 48Fe-50Al-2MWCNT (M4); (**e**) 50Fe-48Al-2MWCNT (M5); (**f**) 55Fe-43Al-2MWCNT (M6); (**g**) 60Fe-38Al-2MWCNT (M7); (**h**) 65Fe-33Al-2MWCNT (M8); and (**i**) 70Fe-28Al-2MWCNT (M9).

At a Fe volume content greater than 50% (Figure 2e–i), the Al particles were distributed in Fe phases as reinforcing particles in the Fe matrix. It can be seen that most of MWCNTs were trapped inside the Fe areas. This was because a large number of iron nanoparticles had surrounded the Al particles and the clusters of MWCNTs as a result of ball milling. The SEM images of the composites revealed the presence of big spherical Fe particles in the Fe area. However, it can be seen that the total surface formed by the big spherical Fe particles seemed smaller than the rest of the Fe area in the SEM images. Therefore, it can be believed that the volume content of big Fe particles was smaller than the volume content of small particles (at nanoscale) which may explain the average particle size of 90–110 nm (as specified by the iron powder supplier Ditto Technology Co. Ltd.).

The elemental mappings performed using EDS are shown in Figure 3. The elements present in the materials can distinctively be seen in the areas of each element illustrated by their own colors. Figure 3a,d shows the distribution of elements in a composite with lower Fe volume content (33Fe-65Al-2MWCNT) and in a composite with higher Fe volume content (65Fe-33Al-2MWCNT), respectively. The EDS images show the presence of oxygen (O) which covers the surface of the iron (Fe) area. However, it can be seen that a high concentration of oxygen (O) is rather lower on the surface area of the large Fe particles than the surface area occupied by small Fe particles. This suggests that the area formed by small Fe particles were more oxidized than the area formed by big Fe particles. One possible reason for the presence of oxygen (O) in the composites may be the contamination of the sample during the ball milling due to the use of alumina balls as the ball-milling media. The second reason may be the oxidation of the surface of the samples after being polished for SEM and EDS analyses as the samples were exposed to the air. Figure 3b displays the selected area with a layer of Fe nanoparticles between Al particles and its EDS mapping at the high magnification of ×50,000 in the 33Fe-65Al-2MWCNT nanocomposite. Figure 3d shows the SEM image of 38Fe-60Al-2MWCNT nanocomposites, the high magnification of the selected area in the nanocomposites, and the MWCNT area that was mapped with the magnification of ×150,000. The image clearly shows the entangled MWCNTs.

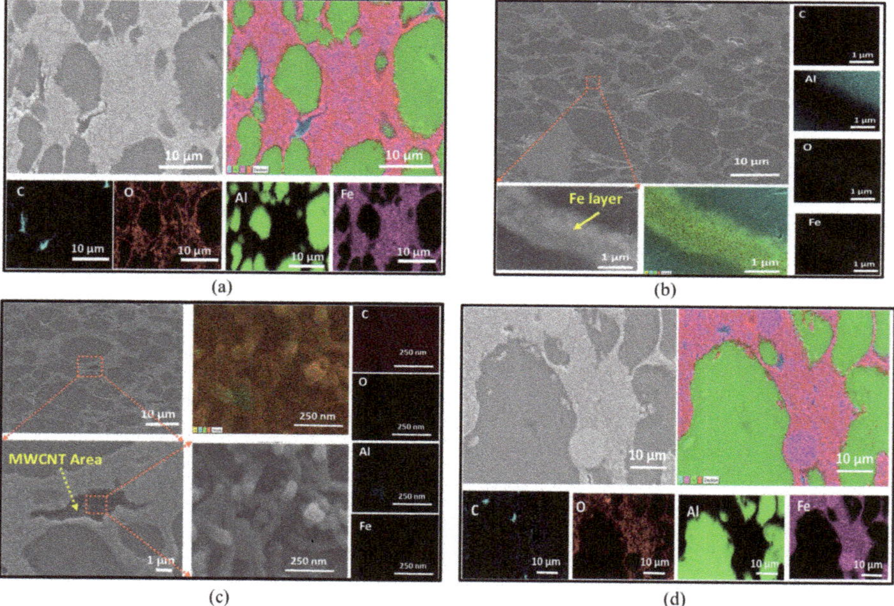

Figure 3. Energy Dispersive Spectroscopy (EDS) mapping: (**a**) 33Fe-65Al-2MWCNT (M1); (**b**) Higher magnification of selected region in 33Fe-65Al-2MWCNT (M1); (**c**) Higher magnification of selected region in 38Fe-60Al-2MWCNT (M2) with MWCNT area; and (**d**) 65Fe-33Al-2MWCNT (M8).

The X-ray diffraction (XRD) analysis was conducted on the polished surface of the composites to investigate the formation of phases in Fe-Al-MWCNT nanocomposites. The XRD patterns of the composites in comparison to that of MWCNTs, Al, and Fe are shown in Figure 4. The Al XRD peaks (111), (200), (220), and (311), and the Fe XRD peaks (110) and (200) were detected in the XRD patterns of the composites. The intensity of Al peaks (111) and (311) decreased as the Al content decreased from 65 vol% (M1) to 28 vol% (M9). It can be seen that the iron oxide peaks (220), (311), (440), and (511) appeared in the XRD patterns of the composites which revealed the formation of iron oxide. This might have resulted from either the exposition to the air of polished surface before testing or the oxygen contamination during the ball-milling process. The XRD patterns of the composites did not show peaks of the MWCNTs or carbide compound which suggests that no carbide formation may have occurred as it did in some previous studies on carbon-nanotube-reinforced aluminum composites [19–21].

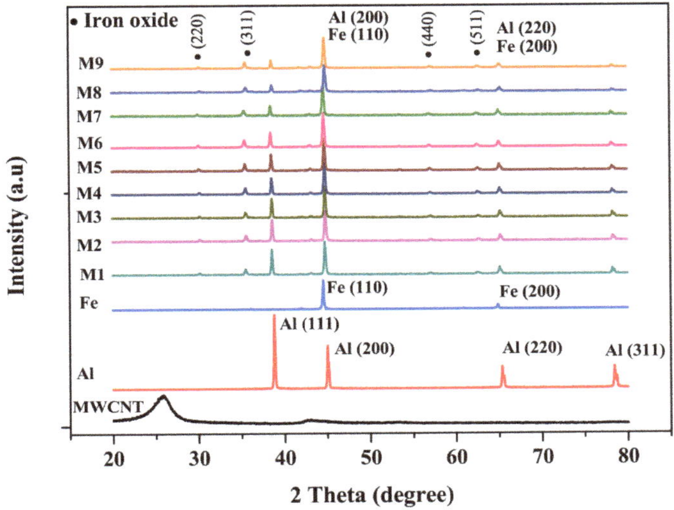

Figure 4. X-ray diffraction (XRD) patterns of Fe-Al-MWCNT hybrid nanocomposites.

3.2. Magnetic Properties of Fe-Al-MWCNT Hybrid Nanocomposites

The saturation magnetization, coercivity, and retentivity of the composites were recorded by VSM. Figure 5a,b and Figure 5c,d show the hysteresis loops of chosen powders after ball milling and the sintered composites, respectively. For comparison, the conventional Sendust core (CS610125, Fe-Si-Al alloy powder manufactured by Chang Sung Corporation, Korea) was purchased and tested with VSM measurements. The results of the magnetic properties of Sendust core were compared with the manufactured composites as shown in Figure 5c,d and Figure 6a,c. The Fe-Al-MWCNT composites had the hysteresis loops similar to that of standard Sendust core which is a ferromagnetic material. Thus, the Fe-Al-MWCNT nanocomposites can be considered ferromagnetic materials. It can be seen that the ball-milled powders and compacts exhibited ferromagnetic behavior.

Figure 6 compares the specific saturation magnetization, intrinsic coercivity, and retentivity of the powders and sintered composites. It was found that the saturation magnetizations of the compacts were, in general, higher than that of the ball-milled powders, and they increased as the content of Fe nanoparticles was increased. On the other hand, the coercivity and retentivity were comparatively reduced after sintering. A decrease of 35% to 51% in coercivity and 13% to 51% in retentivity was achieved as a result of the sintering of composites with Fe volume content varying from 33 vol% to 55 vol%. In fact, the coercivity and retentivity were high for the as-milled powders because of the presence of large gaps between particles when they were packed for measurement. However, the compact and sintering played an essential role in improving the densification of the nanocomposites by eliminating those gaps and therefore reducing the coercivity and retentivity and enhancing the saturation magnetization. Moreover, the stearic acid used as a process agent during ball milling stayed in its solid form after ball milling. But stearic acid was gradually decomposed while increasing the sintering temperature, leading to a lower volume of non-magnetic materials which contributed to the reduction of magnetic dilution and coercivity. Therefore, it can be inferred that it would be difficult to predict the trend of magnetic properties (especially coercivity and retentivity) based on the mixture of powders before the compaction and sintering process.

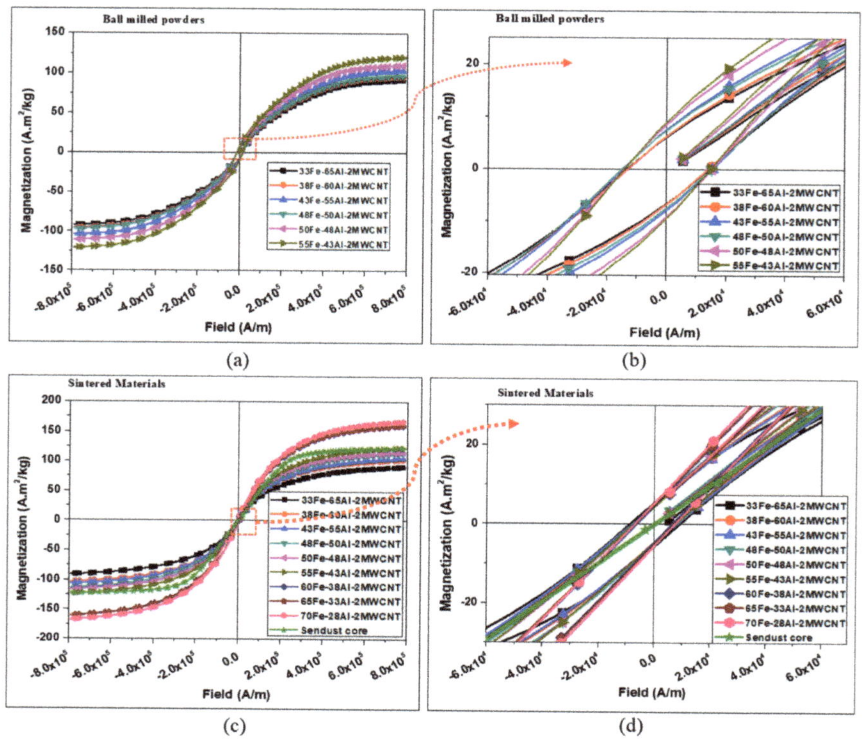

Figure 5. Magnetic hysteresis loops of the nanocomposites: (**a**) Full loops of ball-milled powder; (**b**) Magnification of selected region in (**a**); (**c**) Full loops of sintered composites; and (**d**) Magnification of selected region in (**c**).

The reduction of coercivity and retentivity narrowed the hysteresis loops and consequently reduced the hysteresis losses of materials because the hysteresis losses of magnetic material are proportional to the area of the hysteresis loop. For the compacted composites, the composite with the lowest Fe content (33 vol%) had the highest coercivity of 8875 A/m, whereas the coercivity of the other composites fluctuated between 15085.99 A/m and 6383.20 A/m (see Figure 6b). The saturation magnetization, coercivity, and retentivity of the Sendust core obtained by VSM measurements were 122.20 A.m^2/kg, 685.10 A/m, and 0.33 A.m^2/kg, respectively. Evidently, the Fe-Al-MWCNT composites with Fe content of 55 vol% and above exhibit greater saturation magnetization (123–167 A.m^2/kg) than the commercial Sendust core, but their coercivity and retentivity are extremely higher than that of the Sendust core (see Figures 5 and 6). Figure 5g shows that the Fe-Al-MWCNT composites have wider hysteresis loops than the conventional Sendust core, and Figure 6b shows that the Fe-Al-MWCNT composites have a coercivity greater than 5000 A/m. Therefore, since the Fe-Al-MWCNTs exhibit the coercivity in a range between 1000 and 100,000 A/m, they can be classified as semi-hard magnetic materials [22].

As shown in Figure 6d, a continuous increase in density from 3.94 g/cm^3 at low Fe content to 5.41 g/cm^3 at high Fe content was observed which was attributed to the high volume content of iron phase with high density (7.87 g/cm^3).

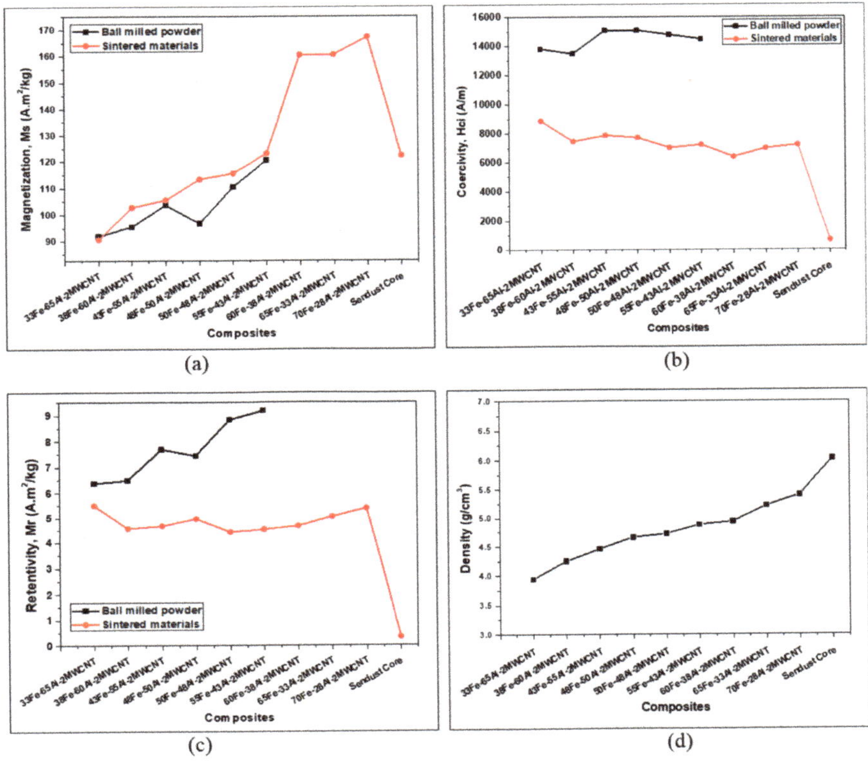

Figure 6. Magnetic properties and density of Fe-Al-MWCNT nanocomposites: (**a**) Saturation magnetization; (**b**) Coercivity; (**c**) Retentivity; and (**d**) Volume density.

3.3. Electrical and Mechanical Properties of Fe-Al-MWCNT Hybrid Nanocomposites

The electrical resistivity of Fe-Al-MWCNT nanocomposites was measured using the four-probe method. The results analyzed in Figure 7a are the average resistivity of 20 recorded data for each sample. The uninterrupted drop in electrical resistivity was observed as the iron content was increased while reducing the Al content. The decrease in electrical resistivity of Fe-Al-MWCNT composites due to the reduction of Al content was in agreement with the variation electrical properties of Fe-Al based alloys [23].

The transverse rupture strength of the Fe-Al-MWNCT nanocomposites was evaluated by using the three-point bending test method. The three-point bending method of evaluating the transverse rupture strength consisted of applying a load to the center of the bar specimen. Then the load at which the sample broke was used to calculate the transverse rupture strength (TRS; MPa), using the following Equation (1) [16]:

$$TRS = \frac{3PL}{2t^2w} \tag{1}$$

where P is the maximum load at rupture (N), L is the length specimen span relative to the fixture (25.4 mm, according to ASTM B528), t and w are the thickness and width of the specimens, respectively.

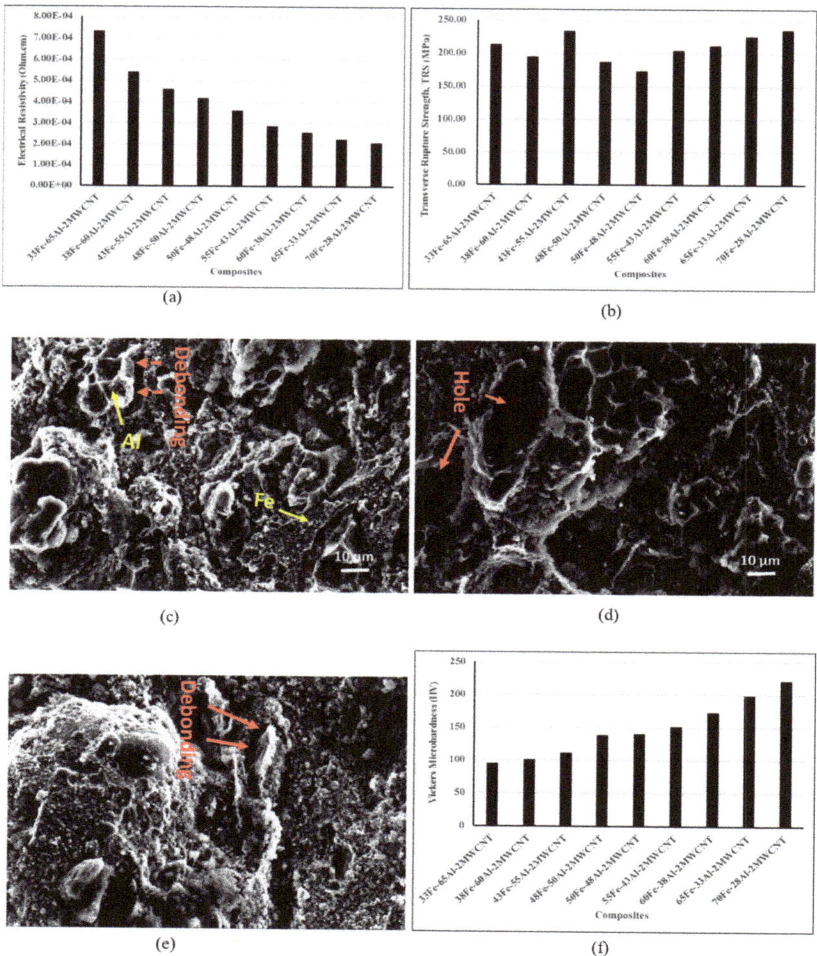

Figure 7. Electrical and mechanical properties of Fe-Al-MWCNT composites: (**a**) Electrical resistivity, (**b**) Mean transverse rupture strength (TRS); (**c**) Fractography SEM image of 33Fe-65Al-2MWCNT; (**d**) Fractography SEM image of 50Fe-48Al-2MWCNT; (**e**) Fractography SEM image of 70Fe-28Al-2MWCNT; and (**f**) Microhardness.

Figure 8b shows the average transverse rupture strength of three specimens tested for each material. The results showed that the lowest strength of 173.83 MPa was given by the 50Fe-48Al-2MWCNT (M5), whereas the highest strength of 234.39 MPa was given by 43Fe-55Al-2MWCNT (M3). It can be seen in Figure 7b that for the composites with Fe content lower than 50 vol%, the strengths were decreased as the Al content decreased, although this trend was interrupted by a sudden increase of strength for composites 43Fe-55Al-2MWCNT (M3). However, the composites with a Fe content greater than 50 vol%, the transverse rupture strength continuously increased up to a maximum of 235.99 MPa with an increase in iron volume content. In fact, at a high-volume content of aluminum, it was seen from SEM images (Figure 2a–d) that Fe nanoparticles formed brittle coating layers around ductile Al particles. Therefore, at this level, the transverse rupture strength of the composites was mainly handled by the elasticity and plasticity of aluminum particles. As the volume ratio of iron nanoparticles increased, the Fe layer thickness increased which led to an expansion of space between Al particles and an increased brittleness of the composites due to insufficient bonding between ductile Al particles, MWCNTs, and brittle Fe nanoparticles. Consequently, a decrease in strength was observed. However, when the Al content was decreased from 48 vol% to 28 vol% while increasing Fe content from 50 vol% to 70 vol%, it was seen in the SEM images (Figure 2e–i) that large areas of Fe phases were formed, and the Al particles were randomly distributed in the Fe matrix. It can be assessed for the nanocomposites with high Fe content that the resistance to the applied load and load transfer in materials were mainly ensured by strong bonding of Fe nanoparticles that formed the iron matrix, leading to the enhancement of mechanical strength as the Fe content was augmented. The surface fracture analysis of the selected composites 33Fe-65Al-2MWCNT, 50Fe-48Al-2MWCNT, and 70Fe-28Al-2MWCNT was carried out after the transverse rupture strength testing, as shown in Figure 7c–e. It can be seen in fractography images that the Fe domains underwent brittle failure, whereas the failure of Al domains was ductile. The fracture of nanocomposites was mainly due to the combined debonding of Al to Fe and brittleness of Fe areas. The debonding between Al particles and Fe domain created holes on the fracture surface as can be seen in Figure 7d.

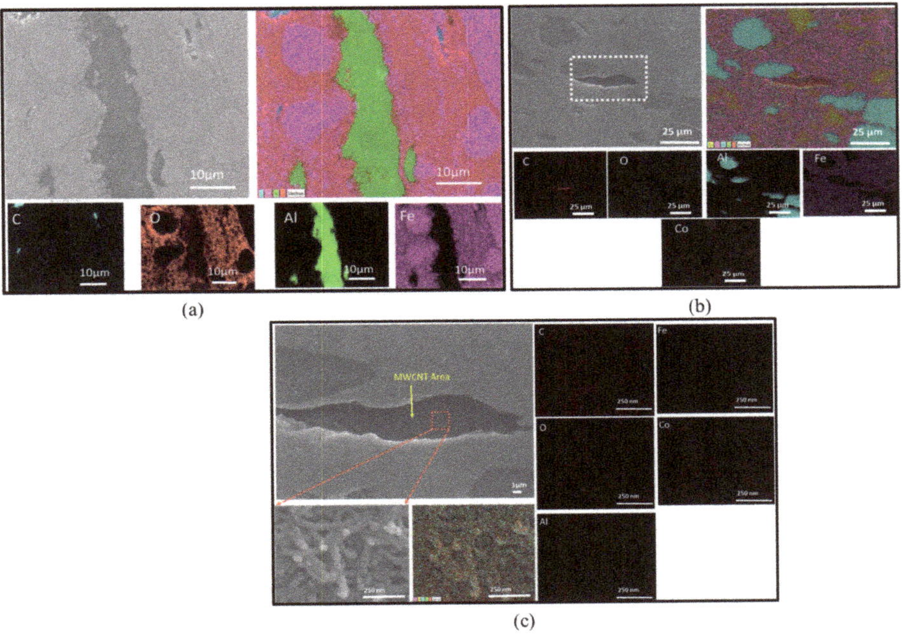

Figure 8. EDS mapping of the hybrid nanocomposites: (**a**) 70Fe-28Al-2MWCNT; (**b**) 60Fe-10Co-28Al-2MWCNT; and (**c**) High magnification of a selected region in (**b**) for 60Fe-10Co-28Al-2MWCNT.

Figure 7f shows the microhardness obtained by averaging 10 measurements for each specimen. The results showed a continuous enhancement of hardness as the content of iron nanoparticles augmented. The lowest hardness was HV94.92 and the highest was HV 221.49 for the higher Al content and lower Al content, respectively. Thus, an increase of 57.14% in hardness was achieved by augmenting the Fe content from 33 vol% to 70 vol% and reducing the Al content. This can evidently be attributed to the higher hardness of iron grains than the Al grains. The improvement of microhardness as a function of Fe content in Fe-Al based composites was in good agreement with the results reported by Fathy et al. [12].

3.4. Effect of Co Content on the Properties of (70-x)Fe-xCo-28Al-2MWCNT Hybrid Nanocomposites (x = 0 to 35 vol%)

3.4.1. Morphology

Figure 8a,b shows the EDS mapping of the selected composites with high Fe content (70Fe-28Al-2MWCNT and 60Fe-10Co-28Al-2MWCNT). Figure 8 distinctively shows the area of each element in the composites. The EDS mapping displayed oxygen (O) over the surface of iron. Similar to the previous samples (Fe-Al-MWCNT composites), in Figure 3, a higher concentration of oxygen (O) was detected on the surface occupied by small Fe particles than the area occupied by large Fe particles. This indicated the possible oxidation of iron that might have occurred after polishing the samples and being exposed to the air before being visualized with EDS. There may have also been contamination during the preparation of powder mixtures before sintering as a result of the exposure of powders to the air while weighing and the residues of alumina balls that were left during the ball milling. Figure 8c shows an EDS analysis image of a multi-walled carbon nanotube (MWCNT) area map at a magnification of ×150 000.

3.4.2. Magnetic and Electrical Properties

In order to improve the soft magnetic properties of Fe-Al-MWCNT nanocomposites, the Co nanoparticles were incorporated in the Fe-Al-MWCNT composite with best magnetic properties. The hysteresis curves of the as-received Fe and Co nanoparticles that were obtained using VSM are plotted in Figure 9. The iron nanoparticles had relative higher coercivity and higher saturation magnetization and consequently wider hysteresis loop than cobalt. The Fe and Co nanoparticles had the saturation magnetization of 175.79 A.m^2/kg and 152.91 A.m^2/kg, respectively. Their coercivities were 15309.71 A/m and 8425.96 A/m, respectively.

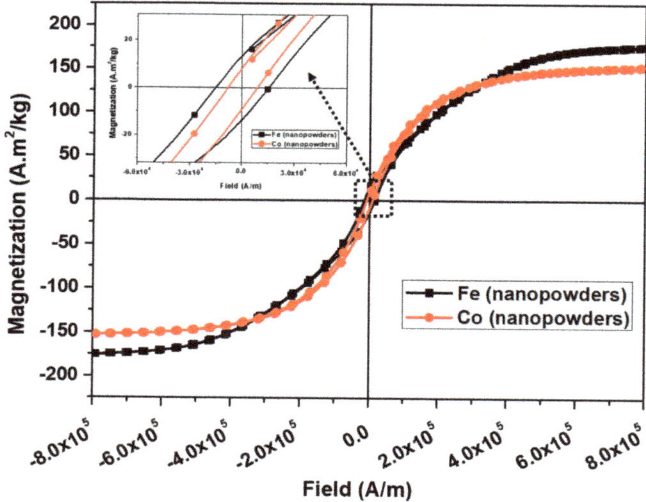

Figure 9. Hysteresis curves of as-received Fe and Co nanoparticles.

Figure 10 shows the hysteresis curves of Fe-Co-Al-MWCNT nanocomposites, according to the Fe-Co combination in comparison with the hysteresis curve of commercial Sendust core (CS610125, Fe-Si-Al alloy powder). The typical magnetic properties of the composites extracted from the hysteresis curves are plotted in Figure 11. The saturation magnetization, shown in Figure 11a, were found to decrease using Co content up to 20 vol% and then start to increase when the Co volume fraction was increased. This might be caused by the modification of magnetic ordering in the composite and crystallinity of materials as the composition ratio changed. However, all Fe-Co based nanocomposites had lower saturation magnetization than the 70Fe-28Al-2MWCNT composite (without Co, 167.22 A.m^2/kg). In Fe-Co based nanocomposites, 50Fe-20Co-28Al-2MWCNT exhibited the lowest saturation magnetization of 145.09 A.m^2/kg, whereas 35Fe-35Co-28Al-2MWCNT exhibited the highest saturation magnetization of 161.86 A.m^2/kg.

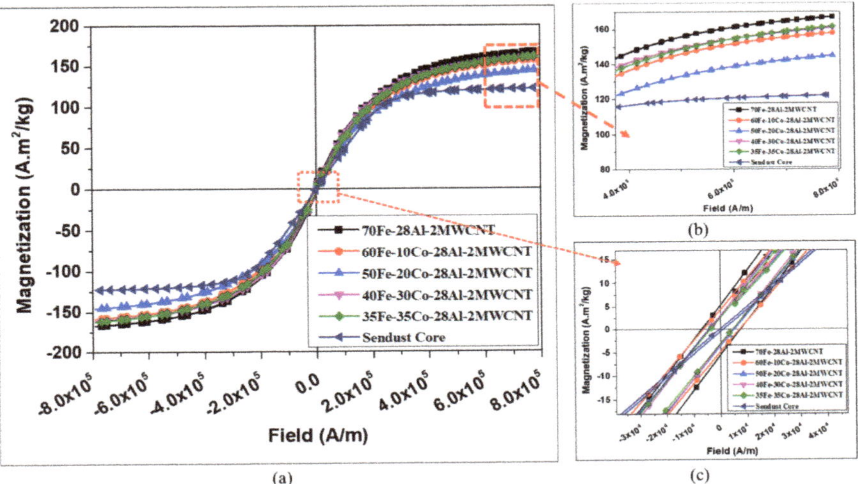

Figure 10. Comparison of hysteresis loops of the Fe-Al-MWCNT and Fe-Co-Al-MWCNT composites: (**a**) Full loops; (**b**) and (**c**) magnification of selected regions.

The coercivity (Figure 11b) and retentivity (Figure 11c) dropped by increasing the volume content of Co nanoparticles. The coercivity reduced from 7210.19 A/m to 4132.01 A/m when the volume content of Co increased from 0 vol% to 35 vol%. This shows a 42.7% drop in coercivity. The retentivity also decreased due to high Co content, and it continuously declined to 47.4% when 35 vol% of Co was used, in comparison to the 70Fe-28Al-2MWCNT nanocomposite. The composites exhibited higher saturation magnetization, higher coercivity, and higher retentivity than the standard Sendust core.

The electrical resistivity obtained from four-probe measurements for each composite is shown in Figure 11d. It is clear, in Figure 11d, that the use of cobalt nanoparticles led to a drop of electrical resistivity, and consequently, to an enhancement of electrical conductivity. The value of resistivity for the 70Fe-28Al-2MWCNT nanocomposite was 2.10×10^{-4} Ω.cm which continuously decreased as the Co content increased when the Fe-Co combination was used in composites. The lowest resistivity of 3.82×10^{-5} Ω.cm was given by the 35Fe-35Co-28Al-2MWCNT composite. This may be associated with lower resistivity of Co when compared to that of Fe.

3.4.3. Mechanical Properties and Volume Density

The material densities were calculated as the ratio of the mass-to-volume of each nanocomposite, and the results are shown in Figure 12a. As expected, the density of composites increased with the Co volume content and their values were between 5.41 g/cm3 and 5.97 g/cm3. This can be explained by the higher density of Co material than that of other materials present in composites.

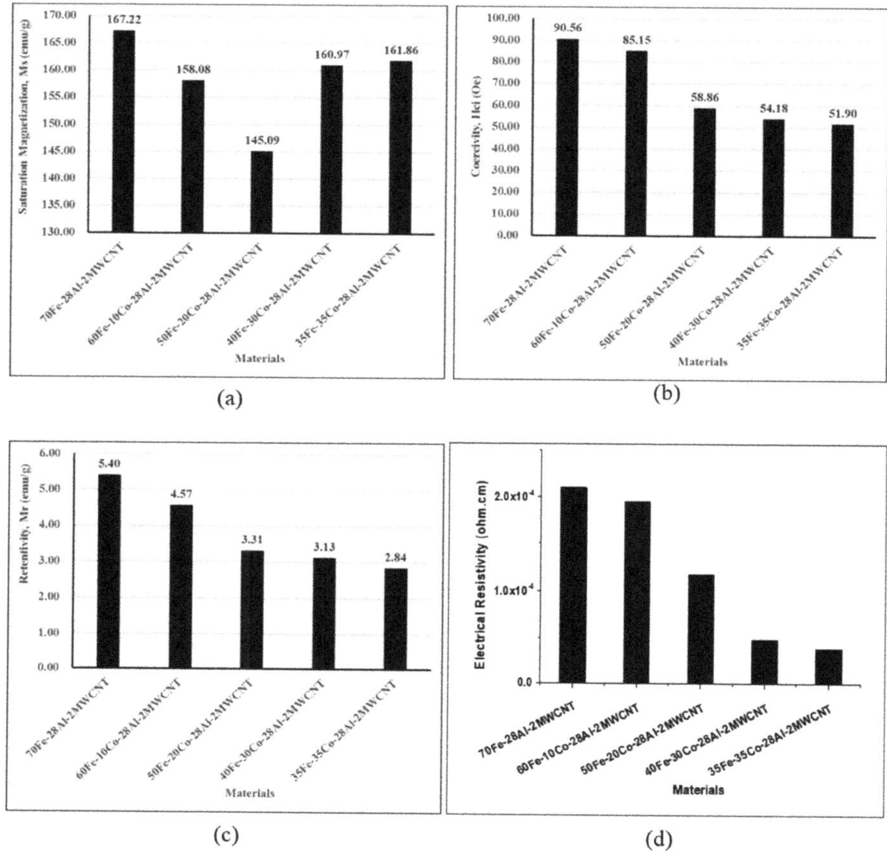

Figure 11. Magnetic and electrical properties of the Fe-Co-Al-MWCNT composites: (**a**) Saturation magnetization; (**b**) Coercivity; (**d**) Retentivity; and (**c**) Electrical resistivity.

Figure 12b summarizes the average transverse rupture strength (TRS). It can be seen that at first, the TRS has dropped when Co 10 vol% was used and then started to increase to the maximum for composite with Co 30 vol%, and then dropped again. In this work, the Fe-Co-based composites had generally lower TRS than the composite without Co nanoparticles except for 40Fe-30Co-28Al-2MWCNT, which had the TRS of 239.47 MPa against 235.99 MPa of 70Fe-28Al-2MWCNT. The composite flexural strengths (TRSs) were in the range of 197 MPa to 240 MPa.

The effect of Co content on microhardness showed that Co had improved the hardness of the composites, as shown in Figure 12c. The hardness was found to increase from HV 221.49 at Co 0 vol% to HV 393.65 at Co 20 vol% and then reduced to HV 351.63 at 35 vol%. Therefore, an increase of 43.73% in Vickers hardness was achieved by incorporating Co 20 vol% in the composite. The enhancement of composite hardness due to cobalt incorporation is evidently due to the higher hardness of cobalt than iron [24]. Moreover, the variation in microhardness of the composites with respect to Co content may be explained by the change in microstructure, dislocation, interstitial defects and particle arrangements as a result of the manufacturing process.

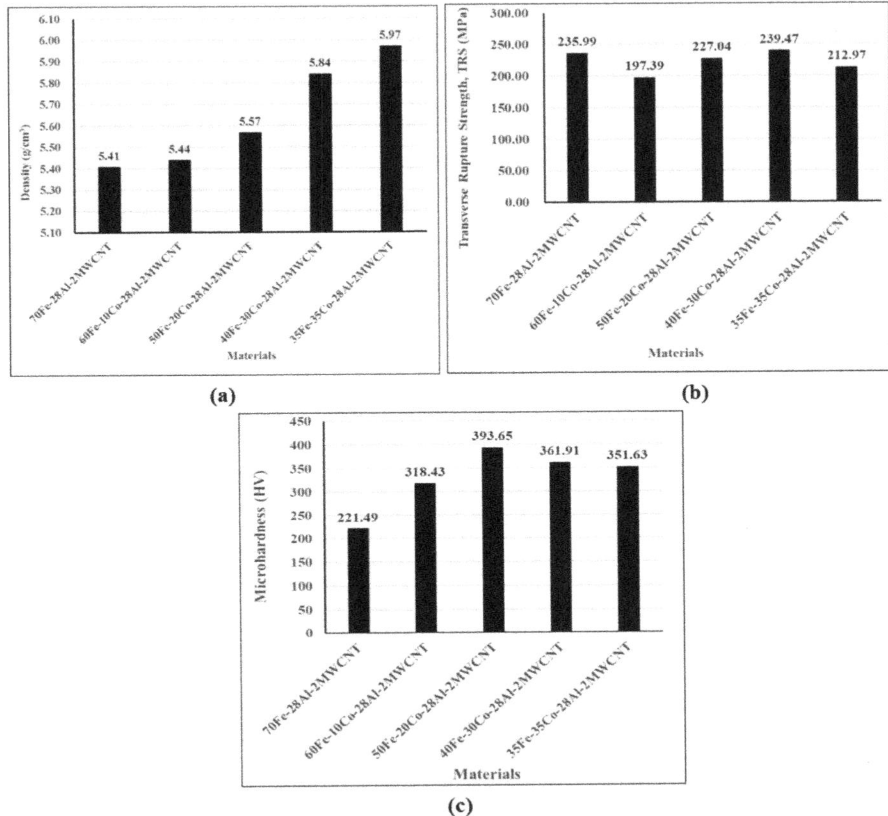

Figure 12. Effect of Co volume content on (**a**) volume density; (**b**) transverse rupture strength (TRS); and (**c**) Vickers microhardness of Fe-Co-Al-MWCNT nanocomposites.

4. Conclusions

In this work, the Fe-Al and Fe-Co-Al-based hybrid nanocomposites with MWCNT as reinforcements were fabricated by ball milling, followed by spark plasma sintering. The characterization of the composite morphology and magnetic, mechanical, and electrical properties was conducted according to the metallic particle combination. The morphology of composites distinctively showed the presence of all materials and their proper distribution. All composites showed a ferromagnetic behavior, and the magnetic properties were found to improve with an increase in iron content. The saturation magnetization continually increased to 167.22 A.m^2/kg when the Fe content increased up to Fe 70 vol%. Although the incorporation of Co in Fe-Co-Al-MWCNT slightly reduced the saturation magnetization, it remarkably reduced the coercivity and retentivity to 42.68 % and 47.4 %, respectively. This suggests a significant improvement in the soft magnetic properties and reduction in hysteresis losses of the magnetic composite materials due to the use of Co nanoparticles. The manufactured composites in this study can be added to the group of semi-hard magnetic materials because their coercivities fall in the range of 1000 A/m and 100,000A/m. Although the composites were magnetically harder than the conventional Sendust core, it was found that the composites with more than 60 vol% of ferromagnetic particles had better saturation magnetization than the Sendust core.

In general, the composites had good mechanical strength with values ranging between 173 MPa and 236 MPa for Fe-Al-MWCNT composites. The Co showed an enhancement of 1.25% in transverse rupture strength. The Fe content was found to be the leading element in hardness for the Fe-Al-based composites, where its increase led to an improvement of 57.14 % in Vickers microhardness with the largest value of HV 221.49 which was further enhanced to HV 393.65 by incorporating 20 vol% of Co nanoparticles. We believe that the fabricated nanocomposites can be used as magnetic cores for electromagnetic devices.

Author Contributions: Conceptualization, H.G.K., L.K.K., and H.J.G.; methodology, A.T., H.J.G., and S.H.G; formal analysis, A.T., S.H.G, and H.J.S; investigation, H.G.K., L.K.K, and A.T; resources, S.H.G., H.J.S., and H.G.K.; data curation, A.T. and S.H.G.; writing—original draft preparation, A.T.; writing—review and editing, H.G.K., L.K.K., and A.T.; visualization, S.H.G., H.J.S., and A.T.; supervision, H.G.K., H.J.S, and L.K.K.; project administration, H.G.K., H.J.S, and L.K.K.; funding acquisition, H.G.K., and H.J.S. All authors have read and agreed to the published version of the manuscript.

Funding: This research was supported by the Basic Science Research Program through the National Research Foundation of Korea (NRF), funded by the Ministry of Education (MOE) [No. 2016R1A6A1A03012069] and [No. 2017R1D1A1B03035690].

Acknowledgments: The authors would like to thank the staff of the Institute of Carbon Technology for their support.

Conflicts of Interest: The authors declare no conflict of interest.

References

1. Bayraktar, E.; Ayari, F.; Tan, M.J.; Tosun-Bayraktar, A.; Katundi, D. Manufacturing of aluminum matrix composites reinforced with iron oxide (Fe3O4) nanoparticles: Microstructural and mechanical properties. *Metall. Mater. Trans. B Process Metall. Mater. Process. Sci.* **2014**, *45*, 352–362. [CrossRef]
2. Panwar, N.; Chauhan, A. Fabrication methods of particulate reinforced Aluminium metal matrix composite-A review. *Mater. Today Proc.* **2018**, *5*, 5933–5939. [CrossRef]
3. Daoud, A. Microstructure and tensile properties of 2014 Al alloy reinforced with continuous carbon fibers manufactured by gas pressure infiltration. *Mater. Sci. Eng. A* **2005**, *391*, 114–120. [CrossRef]
4. Yashpal, K.; Suman, K.; Jawalkar, C.S.; Verma, A.S.; Suri, N.M. Fabrication of Aluminium Metal Matrix Composites with Particulate Reinforcement: A Review. *Mater. Today Proc.* **2017**, *4*, 2927–2936. [CrossRef]
5. Subba Rao, E.; Ramanaiah, N. Influence of Heat Treatment on Mechanical and Corrosion Properties of Aluminium Metal Matrix composites (AA 6061 reinforced with MoS2). *Mater. Today Proc.* **2017**, *4*, 11270–11278. [CrossRef]
6. Ezatpour, H.R.; Torabi Parizi, M.; Sajjadi, S.A.; Ebrahimi, G.R.; Chaichi, A. Microstructure, mechanical analysis and optimal selection of 7075 aluminum alloy based composite reinforced with alumina nanoparticles. *Mater. Chem. Phys.* **2016**, *178*, 119–127. [CrossRef]
7. Tugirumubano, A.; Sun Ho, G.; Vijay, S.; Jae Shin, H.; Ku Kwac, L.; Gun Kim, H. Magnetic properties of Aluminum Matrix Reinforced with Hematite Powders. In Proceedings of the Korea Society for Composite Materials, Jeju, Korea, 5–7 April 2018.
8. Vijay, S.J.; Tugirumubano, A.; Go, S.H.; Kwac, L.K.; Kim, H.G. Numerical simulation and experimental validation of electromagnetic properties for Al-MWCNT-Fe2O3 hybrid nano-composites. *J. Alloys Compd.* **2018**, *731*, 465–470. [CrossRef]
9. Tugirumubano, A. Development and Characterization of Magnetic Particulate Composites for Magnetic Core Applications. PhD. Thesis, Jeonju University, Jeonju-si, Korea, 2020.
10. Ferreira, L.M.P.; Bayraktar, E.; Robert, M.H. Magnetic and electrical properties of aluminium matrix composite reinforced with magnetic nano iron oxide (Fe3O4). *Adv. Mater. Process. Technol.* **2016**, *2*, 165–173.
11. Ferreira, L.M.P.; Bayraktar, E.; Robert, M.H.; Miskioglu, I. *Optimization of Magnetic and Electrical Properties of New Aluminium Matrix Composite Reinforced with Magnetic Nano Iron Oxide (Fe3O4)*; Conference Proceedings of the Society for Experimental Mechanics Series; Springer: New York, NY, USA, 2016; Volume 7, pp. 11–18.
12. Fathy, A.; El-Kady, O.; Mohammed, M.M.M. Effect of iron addition on microstructure, mechanical and magnetic properties of Al-matrix composite produced by powder metallurgy route. *Trans. Nonferrous Met. Soc. China Engl. Ed.* **2015**, *25*, 46–53. [CrossRef]
13. Dhandapani, S.; Rajmohan, T.; Palanikumar, K.; Charan, M. Synthesis and characterization of dual particle (MWCT+B4C) reinforced sintered hybrid aluminum matrix composites. *Part. Sci. Technol.* **2016**, *34*, 255–262. [CrossRef]

14. Yoo, S.; Wang, H.; Oh, S.; Kang, S.; Choa, Y. Synthesis and Densification of CNTs/Fe/Al2O3 Nanocomposite Powders by Chemical Vapor Deposition. *Key Eng. Mater.* **2006**, *317–318*, 665–668. [CrossRef]

15. Singla, D.; Amulya, K.; Murtaza, Q. CNT Reinforced Aluminium Matrix Composite—A Review. *Mater. Today Proc.* **2015**, *2*, 2886–2895. [CrossRef]

16. ASTM. *ASTM B528-99: Standard Test Methods for Transverse Rupture Strength of Metal Powder Specimens*; ASTM Standards; ASTM: West Conshohocken, PA, USA, 2004.

17. Geng, K.; Xie, Y.; Xu, L.; Yan, B. Structure and magnetic properties of ZrO2-coated Fe powders and Fe/ZrO2 soft magnetic composites. *Adv. Powder Technol.* **2017**, *28*, 2015–2022. [CrossRef]

18. Wu, Z.; Fan, X.; Wang, J.; Li, G.; Gan, Z.; Zhang, Z. Core loss reduction in Fe-6.5 wt.%Si/SiO2 core-shell composites by ball milling coating and spark plasma sintering. *J. Alloys Compd.* **2014**, *617*, 21–28. [CrossRef]

19. Esawi, A.M.K.; Morsi, K.; Sayed, A.; Taher, M.; Lanka, S. The influence of carbon nanotube (CNT) morphology and diameter on the processing and properties of CNT-reinforced aluminium composites. *Compos. Part A Appl. Sci. Manuf.* **2011**, *42*, 234–243. [CrossRef]

20. Chen, B.; Shen, J.; Ye, X.; Imai, H.; Umeda, J.; Takahashi, M.; Kondoh, K. Solid-state interfacial reaction and load transfer efficiency in carbon nanotubes (CNTs)-reinforced aluminum matrix composites. *Carbon N. Y.* **2017**, *114*, 198–208. [CrossRef]

21. Najimi, A.A.; Shahverdi, H.R. Microstructure and mechanical characterization of Al6061-CNT nanocomposites fabricated by spark plasma sintering. *Mater. Charact.* **2017**, *133*, 44–53. [CrossRef]

22. Tumanski, S. *Handbook of Magnetic Measurements*, 1st ed.; Taylor & Francis: Milton Park, UK, 2011.

23. Holladay, J.W. *Review of Developments in Iron-Aluminum-Base Alloys*; Battelle Memorial Institute, Defense Metals Information Center: Columbus, OH, USA, 1961.

24. Brinell Hardness of the Elements. Available online: https://periodictable.com/Properties/A/BrinellHardness. v.log.wt.html (accessed on 15 January 2020).

 *nanomaterials*

Article

An Ab Initio Study of Magnetism in Disordered Fe-Al Alloys with Thermal Antiphase Boundaries

Martin Friák [1,*] **, Miroslav Golian** [1]**, David Holec** [2]**, Nikola Koutná** [3]** and Mojmír Šob** [1]

[1]   Institute of Physics of Materials, v.v.i., Czech Academy of Sciences, Žižkova 22,
       CZ-616 62 Brno, Czech Republic; 473873@mail.muni.cz (M.G.); mojmir@ipm.cz (M.Š.)
[2]   Department of Materials Science, Montanuniversität Leoben, Franz-Josef-Strasse 18, A-8700 Leoben, Austria;
       david.holec@unileoben.ac.at
[3]   Institute of Materials Science and Technology, TU Wien, Getreidemarkt 9, A-1060 Vienna, Austria;
       nikola.koutna@mail.muni.cz
*   Correspondence: friak@ipm.cz

Received: 10 November 2019; Accepted: 18 December 2019; Published: 23 December 2019

Abstract: We have performed a quantum-mechanical study of a B2 phase of $Fe_{70}Al_{30}$ alloy with and without antiphase boundaries (APBs) with the {001} crystallographic orientation of APB interfaces. We used a supercell approach with the atoms distributed according to the special quasi-random structure (SQS) concept. Our study was motivated by experimental findings by Murakami et al. (Nature Comm. 5 (2014) 4133) who reported significantly higher magnetic flux density from A2-phase interlayers at the thermally-induced APBs in $Fe_{70}Al_{30}$ and suggested that the ferromagnetism is stabilized by the disorder in the A2 phase. Our computational study of sharp APBs (without any A2-phase interlayer) indicates that they have moderate APB energies (≈ 0.1 J/m^2) and cannot explain the experimentally detected increase in the ferromagnetism because they often induce a ferro-to-ferrimagnetic transition. When studying thermal APBs, we introduce a few atomic layers of A2 phase of $Fe_{70}Al_{30}$ into the interface of sharp APBs. The averaged computed magnetic moment of Fe atoms in the whole B2/A2 nanocomposite is then increased by 11.5% w.r.t. the B2 phase. The A2 phase itself (treated separately as a bulk) has the total magnetic moment even higher, by 17.5%, and this increase also applies if the A2 phase at APBs is sufficiently thick (the experimental value is 2–3 nm). We link the changes in the magnetism to the facts that (i) the Al atoms in the first nearest neighbor (1NN) shell of Fe atoms nonlinearly reduce their magnetic moments and (ii) there are on average less Al atoms in the 1NN shell of Fe atoms in the A2 phase. These effects synergically combine with the influence of APBs which provide local atomic configurations not existing in an APB-free bulk. The identified mechanism of increasing the magnetic properties by introducing APBs with disordered phases can be used as a designing principle when developing new magnetic materials.

Keywords: Fe-Al; antiphase boundaries; magnetism; ab initio; stability; disorder

---

1. Introduction

Antiphase boundaries (APBs) are very frequently occurring extended defects in crystals containing ordered sublattices. They are at the interface of two regions of the same ordered phase which are shifted one with respect to the other. The shift is formed, for example, during ordering processes when two grains crystallizing from the melt have the origin of their lattices in a distance which is not a multiple of translational vectors of the superlattice. As the formation of the above described interfaces occurs at elevated temperatures when diffusion processes are sufficiently active, an intermediate disordered phase can form (so-called thermal APBs). Dislocations with Burgers vectors that are not translation vectors of the ordered superlattice can also create APBs at any temperature, as they move through an ordered phase (so-called deformation APBs with sharp interfaces).

Our theoretical study is focused on APBs in $Fe_{70}Al_{30}$. This alloy belongs into a very promising family of Fe-Al-based materials possessing interesting properties including, e.g., remarkable resistance to oxidation, relatively low density, electrical resistivity, or low cost of raw materials [1–8]. Consequently, they have been very intensively studied both experimentally (see, e.g., Refs. [9–19]) and theoretically (see, e.g., Refs. [20–41]). Focusing on APBs, they have been observed in Fe-Al-based materials with sublattices by means of the transmission electron microscopy (TEM). For example, Marcinkowski and Brown in their classical works [42,43] observed APBs in thin foils of Fe-Al alloys by TEM and reported two types of APBs for the $D0_3$ superlattice. One of them is specific to the $D0_3$ superlattice but the other one, which is crucial for our study, can appear also in the B2 lattice. It is characterized by a shift of the interfacing grains by the $1/2\langle 111\rangle a$ where a is the lattice parameter of the 2-atom elementary cell of the body-centered cubic (bcc) lattice. It interrupts the chemical order of the first nearest neighbors (APB-NN or APB-B2 type). As deformation APBs, both types separate partials of superdislocations in Fe-Al materials [44–46]. Other studies may be found in Refs. [47–52].

Our research is focused on APB-NN (APB-B2) in a B2-phase $Fe_{70}Al_{30}$, and it was motivated by recent experiments by Murakami et al. [50]. Murakami and co-workers combined TEM characterization with direct magnetization measurements of thermally induced APBs. They were found to possess a finite width (2–3 nm) and a significant atomic disorder (an A2 phase). Importantly, electron holography studies of Murakami et al. revealed a magnetic flux density at the APBs higher than that of the matrix by approximately 60% (at 293 K). The authors concluded that the ferromagnetic state of APBs is stabilized by the disorder within APBs. To test this interpretation, we performed a theoretical study of APBs in the B2-phase of $Fe_{70}Al_{30}$. We first simulated sharp APBs and, after identifying important structure-property relations, we expanded the sharp interfaces of APBs by interlayers of disordered A2 phase. The Fe atoms in the A2 phase indeed exhibit (on average) a higher magnetic moment.

2. Methods

Our quantum-mechanical calculations were done with the help of the Vienna Ab initio Simulation Package (VASP) [53,54]. The software implements the density functional theory [55,56]. We have utilized projector augmented wave (PAW) pseudopotentials [57,58]. The generalized gradient approximation (GGA) for the exchange and correlation energy was employed in the parametrization according to Perdew and Wang [59] (PW91) in combination with the Vosko–Wilk–Nusair correction [60]. The plane-wave energy cut-off was equal to 400 eV and the product of the number of Monkhorst–Pack k-points and the number of atoms was equal to 27 648 (e.g., $8 \times 8 \times 8$ k-point mesh in the case of 54-atom supercell of the B2 phase in Figure 1a). All studied supercells were fully relaxed (with respect to their atomic positions, cell shape as well as the supercell volume) and the forces were reduced under 0.001 eV/Å. All calculated states were initially set up as ferromagnetic.

The B2 phase of $Fe_{70}Al_{30}$ is modeled by the special quasi-random structure (SQS) concept [61] and generated by the USPEX code [62–64] (see it in Figure 1a). It consists of two types of {001} atomic planes. One contains only Fe atoms and the other one both Fe and Al atoms distributed according to the above-mentioned SQS concept. As each Fe-Al plane has a different distribution of Fe and Al atoms, they are numbered 1–3 in Figure 1a. It is convenient to write the overall composition $Fe_{70}Al_{30}$ as $Fe_{50}(Fe_{20}Al_{30})$ distinguishing between the two sublattices (one occupied by solely Fe atoms and the other by both Fe and Al). It is worth mentioning that the Fe-Al sublattice is Al-rich while the overall alloy is Fe-rich (this aspect will be important for our analysis below). As the APB energy typically depends on the crystallographic orientation of the interface only very weakly, we believe that our choice of the {001} interface plane is sufficiently representative for a broader range of orientations.

Two of the B2-phase 54-atom cells shown in Figure 1a stacked along the [001] direction form an 108-atom supercell (Figure 1b) which was used for constructing the studied APBs. We applied a $\langle 111\rangle$ shift to all atoms in the upper half of the supercell in Figure 1b to form a supercell with two sharp APBs (dashed lines in Figure 1c). Other three atomic configurations of the B2 phase with sharp APBs and our model of the disordered A2 phase at the thermal APBs in $Fe_{70}Al_{30}$ are described below.

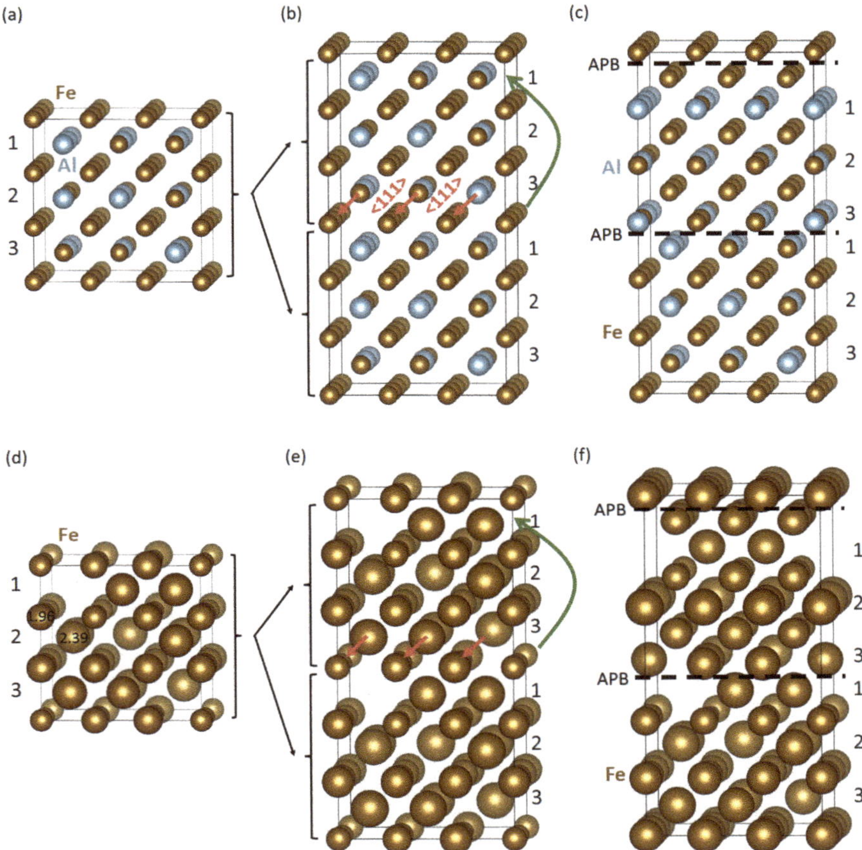

Figure 1. Schematic visualizations of our way of constructing sharp antiphase boundaries (APBs) in $Fe_{70}Al_{30}$ alloy. A 54-atom supercell in part (**a**) is a $3 \times 3 \times 3$ multiple of 2-atom body-centered cubic (bcc) elementary cell and it is our model of the B2 phase of $Fe_{70}Al_{30}$. It consists of atomic planes (i) containing only Fe atoms which are separated by planes (ii) containing both Fe and Al atoms (the latter with the atoms distributed according to the special quasi-random structure (SQS) concept). As each Fe-Al plane is different, they are numbered 1–3. Two of these 54-atom cells stacked along the [001] direction form a supercell (**b**) which was used for constructing the studied APBs. In particular, when applying a $\langle 111 \rangle$ shift to all atoms in the upper half of the supercell (**b**), a supercell (**c**) with two sharp APBs (see dashed lines) is formed. The $\langle 111 \rangle$ shift is indicated by the red vectors in part (**b**). In order to apply the $\langle 111 \rangle$ shift to all atoms in the upper half of (**b**), one Fe atomic plane is cyclically shifted as schematically visualized by a curved green arrow. The computed local magnetic moments of atoms in supercells (**a**–**c**) are shown in parts (**d**–**f**), respectively. The magnitude of local magnetic moments are visualized by the diameter of spheres representing atoms—two values in Bohr magnetons μ_B are given in (**d**) in order to show the scaling.

3. Results

Our results related to both APB-free B2 phase of $Fe_{70}Al_{30}$ and that containing sharp APBs are given in Figure 1 and Table 1. The APB-free B2 phase is disordered and, therefore, each Fe atom has a different local atomic environment, and these differences are sensitively reflected by the value of their local magnetic moment. The computed magnitude of local magnetic moments in both APB-free B2 phase (Figure 1b) and the APB-containing one (Figure 1c) are visualized in Figure 1e,f, respectively, by spheres with different radius. As the local magnetic moments of Fe atoms in Fe-rich Fe-Al binaries decrease with an increasing number of Al atoms in the first nearest neighbors shell (1NN) of Fe atoms [22,39], we analyze these relations also here. The computed values of local magnetic moments of Fe atoms are plotted as functions of the number of Al atoms in the 1NN in Figure 2. The visualized trends show the local magnetic moments of Fe atoms decreasing with the increasing number of Al atoms in the 1NN of Fe atoms in a qualitative agreement with our previous studies [22,39]. Regarding the Al atoms in the 2NN shell, we did not find any clear impact—see Appendix A.

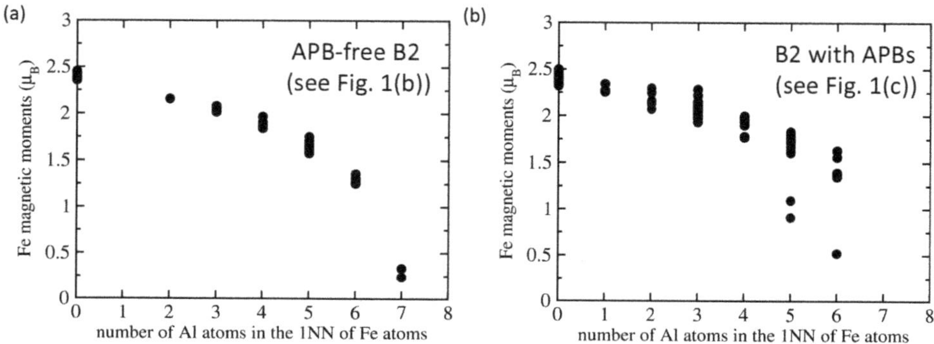

Figure 2. Computed local magnetic moments of Fe atoms as a function of the number of Al atoms in the first nearest neighbor (1NN) shell of Fe atoms. Part (**a**) shows them in the APB-free B2 phase (visualized in Figure 1b) and part (**b**) contains the values for the B2 phase with APBs shown in Figure 1c.

A statistical summary of the number of Al atoms in the 1NN shell of Fe atoms is given in Table 1. It contains percentages of Fe atoms with different numbers of Al atoms in their 1NN shells. In the APB-free B2 phase, the Fe atoms at the Fe-only sublattice have the atoms at the Fe-Al sublattice as their 1NN neighbors and vice versa. In contrast, the studied APBs introduce two types of APB-specific environments. In particular, one APB is characterized by two adjacent Fe-only atomic planes (see it close to the top of the supercell in Figure 1c). The other one has an interface formed by two adjacent Fe-Al planes containing both Fe and Al atoms (see it in the center of the supercell in Figure 1c).

When comparing the percentages of Fe atoms with different numbers of Al atoms in the APB-free B2 phase and that with the APBs, it is important to realize that the atoms are, due to the presence of APBs, re-distributed so that there are more Fe atoms with fewer Al atoms in the 1NN shells of these Fe atoms. Despite the fact that the number of Fe atoms with no Al atom in the 1NN shell is lower in the case of APB-containing B2 phase, the percentages of Fe atoms with 1, 2, and 3 Al atoms in the 1NN shell are significantly higher in the case of APBs and percentages of Fe atoms with 4 or more Al atoms in the 1NN shell are significantly lower. These two findings result in higher local magnetic moments of the Fe atoms in the APB atomic configurations in Figure 1c. This enhancement of local magnetic moments is also reflected by higher values of the average magnetic moment of Fe atoms listed in Table 1 (an increase from 1.83 to 2.00 μ_B). As far as different volumes per atom are concerned (see Table 1), they indicate a possibility of strains which can lead to incoherent APB interfaces, but such states are beyond the scope of our study. Lastly, the APB energy is equal to 0.103 J/m^2.

Table 1. Computed properties of the studied atomic configurations including the B2 phase without APBs as well as variants of APB-containing B2 phase. The table summarizes volumes, APB energies in the case of APB-containing configurations, percentages of Fe atoms with different number of Al atoms (from 0 to 8) in the 1NN shell and the average value of magnetic moments of Fe atoms $\langle \mu^{Fe} \rangle$.

	Volume	$\langle \gamma^{APB} \rangle$	% of Fe Atoms with # Al Atoms in 1NN									$\langle \mu^{Fe} \rangle$
	(Å^3/atom)	(J/m^2)	0 Al	1 Al	2 Al	3 Al	4 Al	5 Al	6 Al	7 Al	8 Al	(μ_B)
B2 phase—Figure 1a	11.80	–	29	0	3	11	13	26	13	5	0	1.83
B2 with APB—Figure 1c	11.98	0.103	20	8	12	20	9	20	9	3	0	2.00
B2 with APB—Figure 3a	11.89	0.099	20	8	12	20	9	20	9	3	0	1.86
B2 with APB—Figure 3b	11.92	0.019	16	11	11	20	12	17	9	5	0	1.85
B2 with APB—Figure 3c	11.98	0.165	24	8	13	18	11	16	8	3	0	1.89

3.1. Compositional Changes at Sharp APB Interfaces

Next, we check the sensitivity of our computed properties of sharp APBs to compositional changes at the APB interfaces. Using the supercell of the B2 phase (Figure 1a) we have applied three different cyclic shifts to the atomic planes in the two interfacing grains in Figure 1c—see these atomic configurations and their corresponding local magnetic moments in Figure 3. As the cyclic shifts were applied to the same supercell of the B2 phase (Figure 1a), we can use its energy as the same reference as before when analyzing the properties of the APB shown in Figure 1c. The cyclic shifts only changed local atomic configurations at the APB interfaces. These changes are well visible on the order of different {001} planes containing both Fe and Al atoms (see the numbers 1–3 assigned to them). Their properties are given in Table 1.

Regarding APB energies, the atomic configuration in Figure 3a has it equal to 0.099 J/m^2, i.e., very close to that which we obtained for the APB configuration in Figure 1c, 0.103 J/m^2. In contrast, the APB energy of the configuration shown in Figure 3b is significantly lower, only 0.019 J/m^2, while that of the configuration in Figure 3c is significantly higher, 0.165 J/m^2. In order to explain the above discussed changes, we suggest to focus on differences in the Al concentration in the two adjacent {001} Fe-Al planes above and below the APB interface. In particular, the two Fe-Al planes above and below the APB in the middle of the supercell in Figure 1c contain 5 + 7 = 12 Al atoms (out of 18 atoms in total, i.e., 66.7 at.% Al). This is similar to 4 + 7 = 11 Al atoms (61.1 at.%) in the case of atomic configuration in Figure 3a. Both of these values are close to the average 60 at.% Al concentration in the Fe-Al planes in the B2 phase. In contrast, the Al concentration is significantly lower (4 + 4 = 8 Al atoms, 44.4 at.%) in the case of configuration visualized in Figure 3b and higher (7 + 7 = 14 Al atoms, 77.8 at.%) in Figure 3c. They represent models for fluctuations in the Al concentration at the APB interfaces.

The APB energies which increase with increasing average concentration of Al atoms in the two Fe-Al atomic planes adjacent to the APB interface can be approximated by a linear trend. The two quantities are thus correlated. The linear fitting function (for APB energies in J/m^2) has the form $\langle \gamma^{APB} \rangle = 0.0043 c^{Al} - 0.1695$, where c^{Al} is the concentration of Al in at.%. The level of correlation could be judged from the value of the R^2, which is equal to 0.9789. Regarding the other pair of {001} planes adjacent to APBs, which are formed by Fe-only planes, they are the same in all configurations.

As far as magnetic properties are concerned, the magnitude of local magnetic moments corresponding to the atomic configurations shown in Figure 3a–c are visualized in Figure 3d–f, respectively, by the diameter of spheres representing the atoms. The magnitudes of these moments are also summarized as functions of the number of Al atoms in the 1NN shells of Fe atoms in Figure 4. The obtained computed trends confirm the decrease of local magnetic moments of Fe atoms with the increasing number of Al atoms in the 1NN shell.

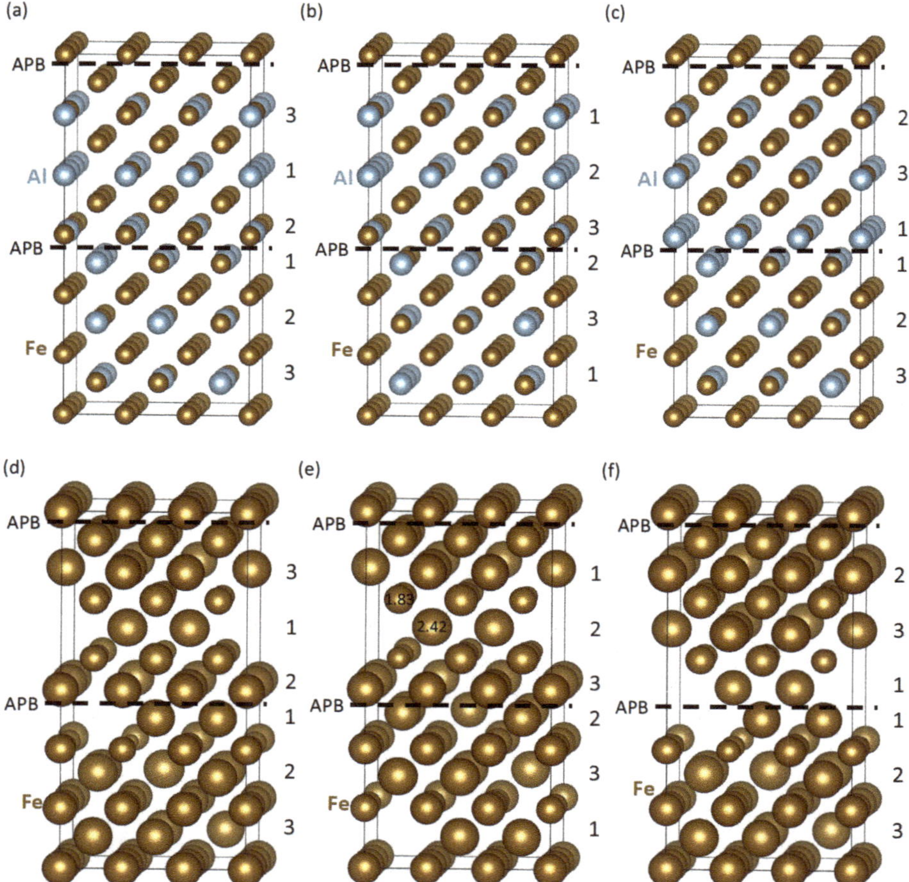

Figure 3. Schematic visualizations of three additional variants of sharp APBs with different compositions of two atomic planes containing both Fe and Al atoms. In particular, two Fe-Al planes above and below the APB in the middle of the supercell in Figure 1c contain 5 and 7 Al atoms, respectively, while those shown here contain 4 and 7 Al atoms in part (**a**), 4 and 4 Al atoms in part (**b**), and 7 and 7 Al atoms in part (**c**). The local magnetic moments of atoms corresponding to supercells (**a**–**c**) are shown in parts (**d**–**f**), respectively. The magnitude of local magnetic moments are visualized by the diameter of spheres representing atoms—two values in Bohr magnetons μ_B are given in part (**e**) in order to show the scaling.

Statistical information for each configuration is given in Table 1. The configurations in Figure 3 have higher percentages of Fe atoms with lower number of Al atoms (1–4 Al atoms) in the 1NN shell and lower percentages of Fe atoms with 5 and more Al atoms in the 1NN shell (compared with the APB-free B2 phase). The increase of magnetism is smaller because some of the Fe atoms with higher number of Al atoms in the 1NN shell have their moments antiparallel to the moments of other Fe atoms. There is thus an APB-induced change from a ferromagnetic state of the B2 phase of $Fe_{70}Al_{30}$ to a ferrimagnetic one.

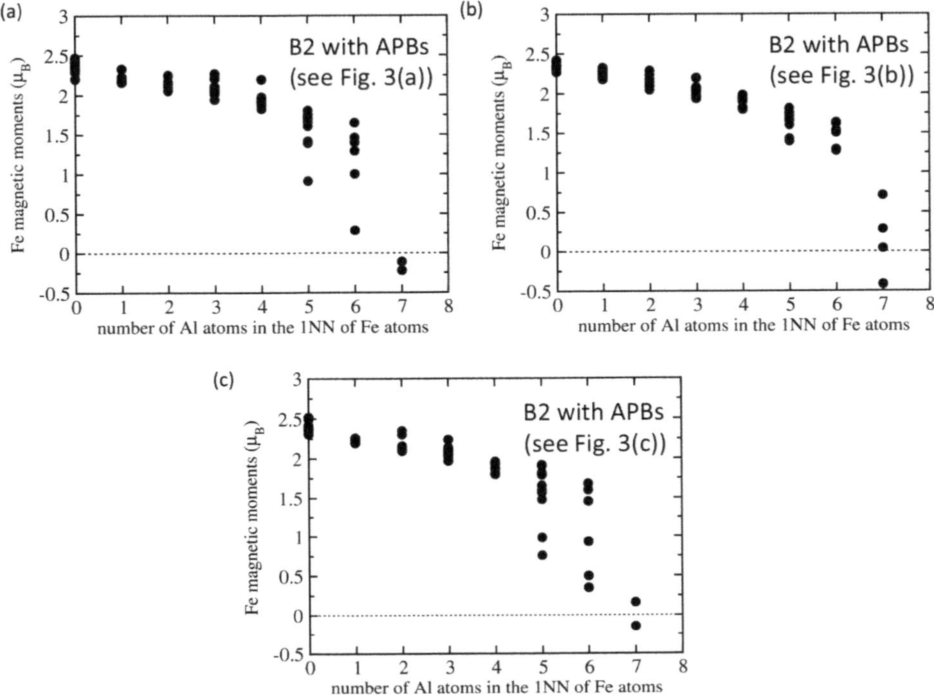

Figure 4. Computed local magnetic moments of Fe atoms in the supercells visualized in Figure 3a–c are summarized in parts (**a**–**c**), respectively, as a function of the number of Al atoms in the 1NN shell of Fe atoms.

3.2. Calculations of Thermally-Induced APBs

After studying APBs with sharp interfaces, we make an attempt to simulate thermally-induced ones which were experimentally probed by Murakami et al. [50]. In order to do so, we introduce an interlayer of disordered A2 phase to each of the two APB interfaces visualized in Figure 1c. We performed our simulations for this particular APB atomic configuration because it has a moderate value of the APB energy (0.103 J/m^2) close to the average of the extreme values obtained for APB atomic configurations shown in Figure 3b,c. We model a disordered A2 phase by a 54-atom supercell visualized in Figure 5a where the atomic positions are generated according to the SQS concept.

There are six {001} atomic planes in our A2-phase supercell and its size in the $\langle 001 \rangle$ direction is about 0.9 nm. While it is less than the reported experimental values (2–3 nm), we consider properties of the A2 phase computed separately as a bulk as our model for the experimental thick layers. If we take the bulk APB-free B2 phase and bulk A2 phase as references and handle the atomic configuration in Figure 5b as a nanocomposite with four interfaces between the two phases (B2 and A2), the averaged interface energy $\langle \gamma \rangle$ is:

$$\langle \gamma \rangle = \{ E_{216}(B2/A2/B2/A2) - (2 \times E_{54}(B2)) - (2 \times E_{54}(A2)) \} / (4 \times A),$$

where $E_{216}(B2/A2/B2/A2)$ is the energy of the atomic configuration in Figure 5b, $E_{54}(B2)$ is the energy of the supercell in Figure 1a, $E_{54}(A2)$ is the energy of the supercell in Figure 5a and A is the interface area.

The calculated value equals to 0.083 J/m^2. This averaged interface energy is listed in Table 2 and, importantly, it is lower than that of the sharp APB in Figure 1c. This indicates that the A2 layers would form if permitted by diffusional processes in the case of thermally-induced APBs.

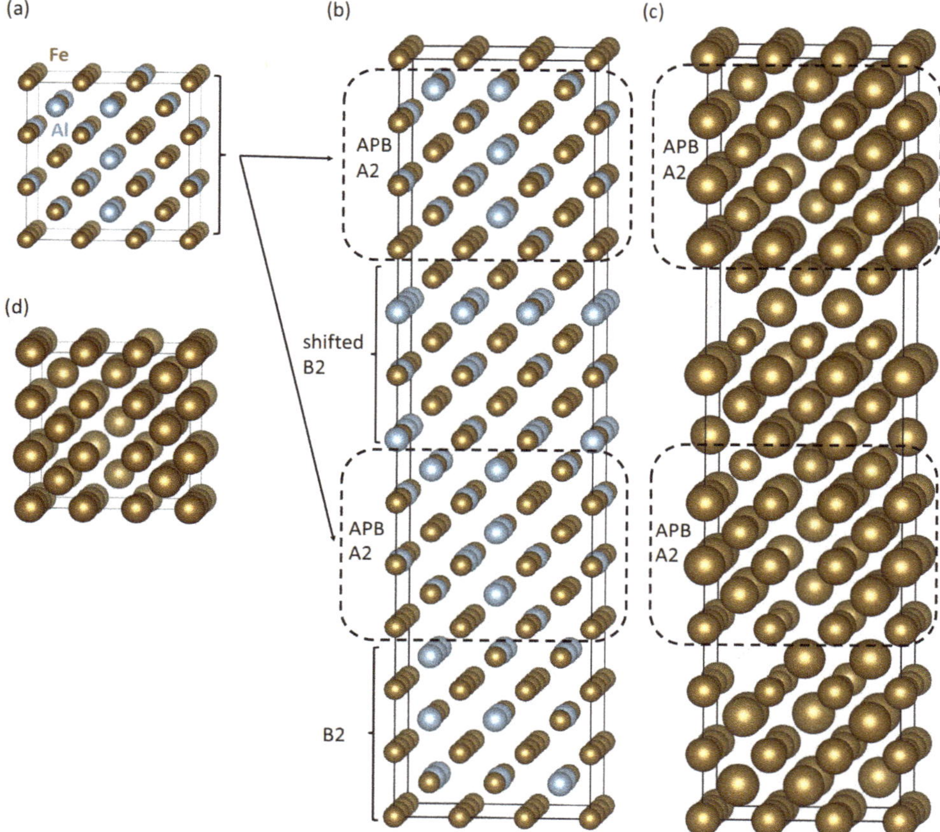

Figure 5. Schematic visualizations of thermal APBs in Fe$_{70}$Al$_{30}$ alloy. A 54-atom supercell in part (**a**) is a $3 \times 3 \times 3$ multiple of 2-atom body-centered cubic (bcc) elementary cell and represents our model for a perfectly disordered A2 phase of Fe$_{70}$Al$_{30}$. It consists of atomic planes containing both Fe and Al atoms distributed according to the SQS concept. Two of these 54-atom A2-phase supercells are introduced at each of the two sharp APBs in the supercell shown in Figure 1b to form the calculated thermal APBs. The computed local magnetic moments of atoms in supercells (**a,b**) are shown in parts (**c,d**), respectively. The magnitude of local magnetic moments are visualized by the diameter of spheres representing atoms—the scaling is the same as in Figures 1 and 3.

Regarding magnetic properties of the atomic configuration shown in Figure 5b, the magnitudes of local magnetic moments are visualized by the diameter of the spheres representing the atoms in Figure 5c. Furthermore, similarly as above, we also analyze the relation between the magnitude of local magnetic moments of Fe atoms and the number of Al atoms in their 1NN shell. The trends for the A2 phase (Figure 5a) and the thermally-induced APB (Figure 5c) are summarized in Figure 6a,b, respectively.

Comparing the plots shown in Figure 6a, which is related to the A2 phase, with that in Figure 6b with local magnetic moments of Fe atoms obtained for the thermally-induced APB, it is obvious that the disordered A2 phase has a higher number of Fe atoms with fewer Al atoms in the 1NN sphere. The percentages are listed in Table 2. While only 5% of Fe atoms have no Al atom in the 1NN shell, nearly one half (45%) of Fe atoms in the A2 phase has only 2 Al atoms as the first nearest neighbors and no iron atom exhibits more than 5 Al atoms in the 1NN shell. As all analyzed trends between the local magnetic moment of Fe atoms and the number of Al atoms in their 1NN shell show decreasing magnetic moments with increasing concentration of Al (see Figures 2, 4 and 6), the above discussed percentages found in the A2 phase mean that the Fe atoms in this phase would be more magnetically polarized than those in the B2 phase. A higher averaged magnetic moment of Fe atoms in the A2 phase (see Table 2) illustrates these findings.

Table 2. Calculated properties of the studied atomic configurations of the A2 phase as a bulk and at the APB interface as a model for the thermally-induced APBs. The table summarizes volumes, APB energies, percentages of Fe atoms with different number of Al atoms (from 0 to 8) in the 1NN shell of the Fe atoms and the average value of magnetic moments Fe atoms $\langle \mu^{Fe} \rangle$.

	Volume	$\langle \gamma^{APB} \rangle$	% of Fe Atoms with # Al Atoms in 1NN									$\langle \mu^{Fe} \rangle$
	(Å^3/atom)	(J/m^2)	0 Al	1 Al	2 Al	3 Al	4 Al	5 Al	6 Al	7 Al	8 Al	(μ_B)
A2 phase	12.14	–	5	13	45	18	13	5	0	0	0	2.15
B2 APB with A2	12.02	0.083	15	10	23	15	17	14	5	0	0	2.04

4. Discussion

Our theoretical results qualitatively confirm the interpretation of experimental findings published by Murakami et al. [50] that the ferromagnetic state of APBs is stabilized by structural disorder within APBs. In particular, the thermally-induced disordered A2 phase at the APB interfaces contains Fe atoms with higher magnetic moments when compared with those in the B2 phase and, importantly, the A2 phase is in a ferromagnetic state. Both of these aspects are in agreement with experiments. However, our theoretical study provides also an atomic-scale type of information not available in the study [50].

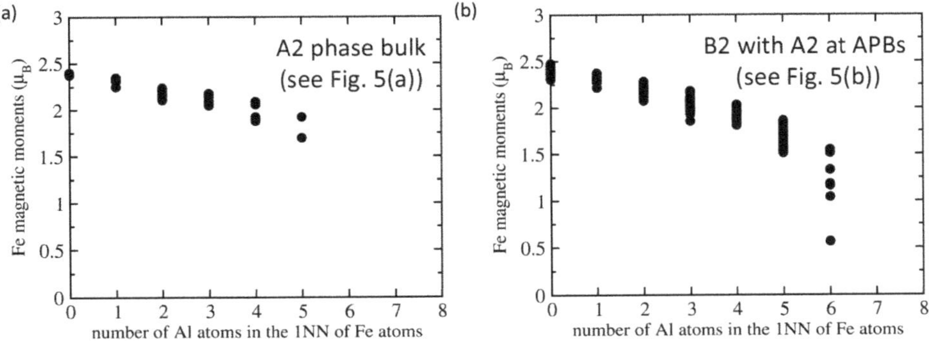

Figure 6. Calculated local magnetic moments of Fe atoms in the supercells visualized in Figure 5a,b as a function of the number of Al atoms in the 1NN. The local magnetic moments of Fe atoms in atomic configurations shown in Figure 5a,b are displayed in parts (**a**,**b**), respectively.

Our results allow for obtaining a deeper insight into the actual mechanisms behind the observed phenomena and a better understanding of them. In particular, we clearly show that the local magnetic moments of Fe atoms decrease with an increasing concentration of Al atoms in the 1NN shell of the Fe atoms. This trend is in the case of the B2 phase clearly nonlinear when considering the whole concentration range of Al atoms (see Figure 2a). The decrease is weaker and quite linear for Fe atoms with up to 4 Al atoms in the 1NN shell (i.e., up to 50% of the atoms) but becomes much steeper for higher concentrations of Al atoms (Fe atoms with 7 Al atoms in the 1NN shell are nearly nonmagnetic—see Figure 2a). This nonlinear dependence of the magnetic moment of Fe atoms as function of the concentration of Al atoms in their 1NN shell is crucial for our understanding of differences between the B2 and A2 phases of $Fe_{70}Al_{30}$ alloy as discussed below.

Analyzing the structure of the B2 phase first, one half of its {001} atomic planes contains only Fe atoms and, consequently, all Al atoms are located in the other half of {001} atomic planes where the concentration of Al becomes much higher (60%) than the overall value of 30 at.% in the $Fe_{70}Al_{30}$. As the Fe-only and Fe-Al planes alternate in the B2 phase along the $\langle 001 \rangle$ direction, all those Fe atoms from the Fe-only atomic planes have the 1NN shell formed by atoms from the Fe-Al planes (and vice versa). Due to the fact that the average Al concentration in the Fe-Al planes is 60%, the Fe atoms from the Fe-only planes have their magnetic moments significantly reduced. Two aspects are important for the overall reduction of the magnetization. First, the reduction of the local magnetic moments is more significant because the above discussed stronger decrease has onset for the Al concentration equal to $\approx$50 at.%. Second, the Fe atoms from the Fe-only planes, which have their magnetic moments nonlinearly reduced, represent 5/7 of all Fe atoms. The remaining 2/7 of Fe atoms in the Fe-Al planes have their 1NN shell full of Fe atoms (from the Fe-only planes) and their magnetic moment reaches the maximum values, but they represent only minority of all Fe atoms.

The situation in the A2 phase is quantitatively very different. All of the atomic planes have on average the same Al concentration and it is only 30 at.%. Leaving aside local fluctuations, 30% is then also the average concentration of Al atoms in the 1NN shells of all Fe atoms. Considering the fact that the decrease of the magnetic moment is weaker for concentration of Al atoms below 50% (prior the onset of nonlinearly stronger reduction), the magnetic moments of Fe atoms in the A2 phase will be reduced less (see Figure 6a) than those in the B2 phase (see Figure 2a).

4.1. Linear Relation between the Al Concentration and the Energy of Sharp APBs

Another insight obtained from our simulations of sharp APBs in the B2 phase is the theoretically identified relation between the average APB energy and the concentration of Al atoms in the two atomic planes adjacent to the APB interface. The simulated APB shift leads to the situation when these planes are formed by either two Fe-only planes or two Fe-Al planes (each containing both Fe and Al atoms). None of these APB-related atomic environments exists in the APB-free B2 phase. We have performed calculations of four different sharp APBs in the B2 phase (Figures 1c and 3a–c) which all contained one APB interface formed by two Fe-only planes (identical in all four atomic distributions) but differ in the concentration of Al in the pair of APB-adjacent Fe-Al planes. The averaged APB energy turns out to decrease with a decreasing concentration of Al atoms in these two Fe-Al planes (within the range of Al concentrations between 8/18, i.e., 44.4%, and 14/18, i.e., 77.8%). Despite the fact that this relation is deduced from only a few computed cases and concentrations of Al, it can help us to explain the formation of the A2 phase at APBs. In particular, sharp APBs in the B2 phase with two interfacing Fe-Al {001} planes would have the average concentration of Al atoms close to that in these planes, i.e., 60 at.% Al. When an A2 phase forms at the APB interface and separates the pair of Fe-Al planes of the B2 phase by atomic planes of the A2 phase, the concentration of Al in the pair of planes adjacent to the newly formed two interfaces is lower (only $(60 + 30)/2 = 45\%$) because the average concentration of Al in the A2 atomic planes is on average only 30%. According to the above discussed relation between the average APB energy and the Al concentration, the energy of the newly formed B2/A2 interfaces would be lower than the original sharp APBs in the B2 phase.

4.2. Thermodynamic Stability of the APB Interface States

The change of the Al concentration at the APB interfaces by the formation of the A2 phase to 45 at.% deserves further attention as it can be partly justified in the context of thermodynamic stability of the binary Fe-Al system. According to the thermodynamic assessment by Sundman and co-workers [65], the compositional dependence of the enthalpy of formation has the minimum close to 50 at.% of Al in a phase, which is close to ordered stoichiometric FeAl with the CsCl structure. The experimental data also show that the crystallographic structures for Al-rich compositions do not have atoms in positions related to a bcc lattice. The pair of adjacent disordered Fe-Al planes with the Al concentrations on average equal to 60 at.% is therefore very likely to have a high energy. The reasons are related to the differences from the stoichiometric ordered FeAl phase in the minimum of the enthalphy curve: the Al concentration is much too high, the atoms are disordered, and the structure is not the one minimizing the enthalpy for this Al concentration. The above discussed reduction of the Al concentration of the two APB-adjacent Fe-Al layers from 60 at.% to 45 at.% (A2/B2 interface) changes the Al concentration closer to the 50 at.% for which the enthalpy has the minimum.

The thermodynamic perspective can help us to explain why the insertion of A2 phase between the two Fe-only atomic planes of the originally sharp APB in the B2 phase of $Fe_{70}Al_{30}$ would result in a decrease of the energy. The two Fe-only adjacent planes form an environment that is similar to that in the elemental ferromagnetic bcc Fe. However, the elemental Fe is, from the thermodynamic point of view, not preferred over Fe-Al states with less than 50 at.% Al (the above discussed enthalpy minimum is found for the FeAl compound [65]). Therefore, the energy can be expected to decrease due to locally increasing Al concentration. This happens exactly at the sharp APB interface formed by two Fe-only planes when one of them is replaced by an atomic plane of the A2 phase containing on average 30 at.% of Al. The average Al concentration of the pair formed by one Fe-only plane and one A2-phase plane would be 15 at.%. In the equilibrium phase diagram, this concentration corresponds to a disordered solid solution of Al atoms in a bcc Fe matrix. Therefore, a local atomic distribution in the pair of those two APB-related adjacent planes would be quite similar to the equilibrium one.

4.3. A Comparison of Thermodynamic Stability of the B2 and A2 Phase

However, the studied systems are not formed only by the two atomic planes adjacent to the APB interface. Regarding the formation of the A2 phase at the APBs, it should be noted that, according to our calculations, the energy of the A2 phase of $Fe_{70}Al_{30}$ is by 18.5 meV per atom higher than that of the B2 phase. Therefore, the above described process which reduces the APB energy of sharp APBs by formation of the A2 phase is, in fact, a complex competition among several different mechanisms. The energy of the atomic planes at the APBs is, on one hand, reduced by changing from sharp APBs in the B2 phase to energetically less costly B2/A2 interfaces, but, on the other hand, the number of the A2/B2 interfaces is twice as high, and the A2 phase itself has a higher energy. Another fact, which can be important at elevated temperatures, is that the configurational entropy of the A2 phase is different from that of the B2 phase. We therefore evaluate the ideal molar configurational entropy S^{conf} below.

As the B2 and A2 phases exhibit different numbers of ordered and disordered atomic sites (sublattices), we use a generalized formula (see, e.g., Ref. [66]) derived for the sublattice model [67] $S^{conf} = -R \sum_\alpha a_\alpha \sum_i f_i^\alpha \ln f_i^\alpha$ where R is the universal gas constant, i runs over different chemical species, α over different sublattices, a_α is the ratio of lattice sites of a sublattice α with respect to the total number of all lattice sites, and f_i^α is the concentration of a chemical species i on a sublattice α. The B2 phase has only one half of planes disordered, and the Al concentration in these disordered planes equals 60 at.%. The A2 phase has all lattice sites fully disordered, and the Al concentration is equal to 30 at.%. The molar configurational entropy (in the units of R) of the B2 phase amounts to 0.3365 and that of the A2 phase is equal to 0.6109. If the energy difference of 18.5 meV per atom is to be compensated solely by the difference in the configurational entropy, it would happen at the temperature of 784 K. The experimental B2–A2 second-order transition temperature is significantly higher, 1287 K [68], but there are several good reasons for this discrepancy. First, our calculations for static lattices did not include any phonons or magnons and, second, more importantly, the computed energy difference is between two ferromagnetic states while the experimental transition occurs above the Curie temperature between two paramagnetic states. The two above-mentioned temperatures cannot be, therefore, directly compared. It is worth noting that the above discussed competition of different phenomena would likely limit the width of interlayers formed at the APBs by the A2 phase and make the width of A2 layers rather sensitive to the temperature as well as to other conditions, such as a thermo-mechanical history of the samples.

4.4. Magnetism of Both Sharp and Thermally-Induced APBs

The existence of the thermally-induced A2 phase at the APBs is really crucial for the increase of magnetism detected in experiments by Murakami et al. [50]. The sharp APBs in the B2 phase do not often increase the magnetization enough (see the average magnetic moments of Fe atoms in Table 1). This is due to the fact that three of the four computed atomic configurations of sharp APBs induce a transition from a ferromagnetic state to a ferrimagnetic one and the magnetic moments with antiparallel orientation reduce the total magnetic moment. We expect that these ferro-to-ferrimagnetic transitions would be rather common close to the sharp APBs in the B2 phase because they are induced by APBs with the APB energies from the whole range of computed values (see them in Table 1 for the atomic configurations shown in Figure 3). Despite the fact that the antiparallel orientation is obtained only in the case of one or two atoms (out of 76 Fe atoms in our 108-atom supercells) and the magnitudes of these antiparallel magnetic moments are rather small (under 0.5 μ_B), the phenomenon can be possibly enhanced by temperature effects or other conditions. In fact, all four studied sharp APBs in the B2 phase exhibit slightly higher values of the average magnetic moment of Fe atoms than the APB-free B2 phase (in particular by 9.3% in the case of the atomic configuration in Figure 1c). However, it is the inception of the A2 phase at the APB interface which increases the average magnetic moment further more. The whole B2/A2 nanocomposite in Figure 5b shows a 11.5% higher averaged magnetic moment of Fe atoms than that in the APB-free B2 phase. This increase is still moderate, but we should keep in mind the fact that the A2 phase layers at the APBs in experiments are much thicker (2–3 nm) than our simulated ones (about 0.9 nm in Figure 5b). Such a thick A2 phase would have magnetic properties similar to those which we obtained for the bulk A2 phase (see Figure 5a,d). The average magnetic moment in the A2 phase would then be significantly higher, by 17.5%, than that in the APB-free bulk B2 phase.

The increase of the averaged magnetic moment of Fe atoms in the A2 phase by 17.5% (w.r.t to the APB-free B2 phase of $Fe_{70}Al_{30}$) is still not directly comparable with the experimental increase by 60% reported by Murakami et al. [50]. When searching for reasons for this discrepancy, it is worth mentioning that Murakami et al. detected the magnetic flux density at the APBs at 293 K while our quantum-mechanical study performed for static lattices (corresponding to very low temperatures close to 0 K) was focused on changes in the magnetic moments of individual atoms. The experimental change of the magnetic flux density is thus not directly comparable with the theoretical increase of the average magnetic moment of Fe atoms. However, our study provides a very valuable insight into thestructure–property relations connecting (i) the local atomic (dis)order and details of atomic configurations (including chemical composition) on one hand and (ii) the values of local magnetic moments of individual Fe atoms on the other hand. We therefore hope that the above identified and analyzed mechanisms, which increase the average magnetic moments of Fe atoms, are among the decisive ones when interpreting the experimental data reported by Murakami et al. [50].

Finally, the identified mechanism of increasing the magnetic properties in materials by introducing thermally-induced APBs with disordered phases can possibly be used as a designing principle when developing new magnetic materials. It should be applicable when magnetic species co-exist with some other (non-magnetic) chemical species which decrease the magnetic moment of the magnetic elements. If this reduction of magnetism is enhanced by thermodynamically-driven formation of ordered sublattices, then the APBs offer a way of decreasing the level of order in the system and that results in a statistically higher probability of magnetic species to be in magnetically more favorable environment (see, e.g., our recent study of impact of APBs in Fe-Al-Ti [69]).

5. Conclusions

We have performed an ab initio study of B2 phase of $Fe_{70}Al_{30}$ alloy with and without antiphase boundaries (APBs). Our study was motivated by experimental findings by Murakami et al. [50] who reported higher magnetic flux density from A2-phase interlayers at the thermally-induced APBs in $Fe_{70}Al_{30}$. They suggested to connect the enhancement of the ferromagnetism with the disorder in the A2 phase. We show that the averaged magnetic moment of Fe atoms in the A2 phase is by 17.5% higher than that in the B2 phase. While we can not treat the A2 layers of the experimental thickness (2–3 nm [50]), our simulations of thinner (about 0.9 nm) A2 layers within a B2/A2 nanocomposite resulted in the average magnetic moment of Fe atoms by 11.5% higher than that in the APB-free B2 bulk. We explain the changes in magnetism by (dis)order-dependent reduction of local magnetic moments of Fe atoms by Al atoms in the 1NN shell of Fe atoms (see also Refs. [22,39]). This effect is synergically combined with the influence of APBs, which provide local atomic configurations not existing in a APB-free bulk. The studied sharp APBs can increase the local magnetic moments of Fe atoms, but they more often lead to an APB-induced ferromagnetic-to-ferrimagnetic transition.

Regarding the formation of the A2 phase at the APBs, we link it to the energetics of atomic configurations occurring at both the sharp and A2-containing thermal APBs in the B2 phase of $Fe_{70}Al_{30}$. The studied sharp APBs have rather low APB energies (between 0.019 and 0.165 J/m^2), and these were found to be increasing with increasing Al concentration in the two atomic planes adjacent to the APB interface. These two atomic planes represent local atomic configurations which are APB-specific and have either much too high or much too low concentration of Al. The insertion of A2-phase atomic planes leads to the change of Al concentration accompanied by lowering of the energy. This mechanism can be understood in terms of equilibrium thermodynamic of the Fe-Al binary system (the enthalpy has the minimum for the Al concentration close to 50 at.%). The studied mechanism of increasing the magnetic properties by introducing thermally-induced APBs with disordered phases can possibly be used as a designing principle when developing new magnetic materials.

Author Contributions: Writing—Original Draft Preparation, M.F.; Conceptualization, M.G. and M.F.; Methodology, N.K., D.H. and M.F.; Writing—Review and Editing, M.G., N.K., D.H. and M.Š.; Visualization, M.F., M.G. and D.H.; Validation, M.G. and D.H.; Resources, M.F.; Project Administration, M.F.; Funding Acquisition, M.F. All authors have read and agreed to the published version of the manuscript.

Funding: The authors acknowledge the Czech Science Foundation for the financial support received under Project No. 17-22139S (M.F., M.G.).

Acknowledgments: M.F., M.G., and M.Š. also acknowledge the support from the Academy of Sciences of the Czech Republic for its institutional support (Institutional Project No. RVO:68081723) and from the Ministry of Education, Youth and Sports of the Czech Republic via the research infrastructure IPMINFRA, LM2015069. Computational resources were made available by the Ministry of Education, Youth and Sports of the Czech Republic under, in particular, the Project of the IT4Innovations National Supercomputer Center (Project No. LM2015070) within the program Projects of Large Research, Development and Innovations Infrastructures and partly also via the CESNET (Project No. LM2015042) and CERIT-Scientific Cloud (Project No. LM2015085). Figures 1, 3, and 5 were visualized using the VESTA [70].

Conflicts of Interest: The authors declare no conflict of interest.

Appendix A

Figure A1 shows computed dependences of local magnetic moments of Fe atoms as functions of the number of Al atoms in the second nearest neighbor shell (2NN) of the Fe atoms for the APB-free bulk B2 phase (see Figure A1a) and the APB-free bulk A2 phase (see Figure A1b). As far as the values for the former are concerned, the Fe atoms located in the Fe-only planes in the B2 phase have their second nearest neighbors formed by the other Fe atoms on this sublattice. As it is Fe-only sublattice, they have no Al atoms in the 2NN shell and their magnetic moments (which vary significantly) seem to be determined by atoms in other shells, apparently by those in their 1NN shell (see Figure 2a).

The Fe atoms on the mixed Fe-Al sublattice in the B2 bulk phase have their 2NN atoms similarly represented by other atoms on this mixed sublattice and can have some Al atoms in their 2NN shell. However, the computed values show very weak dependence on the number of 2NN Al atoms (see Figure A1a). Regarding the bulk A2 phase, there is only one type of mixed Fe-Al sites, but, again, the calculated values of local magnetic moments of the Fe atoms seem to depend on the number of Al atoms in the 2NN shell of the Fe atoms only very weakly (see Figure A1b). A weaker impact of the second coordination sphere found in this study is in line with our recent results related to the Fe-Al system [39,41].

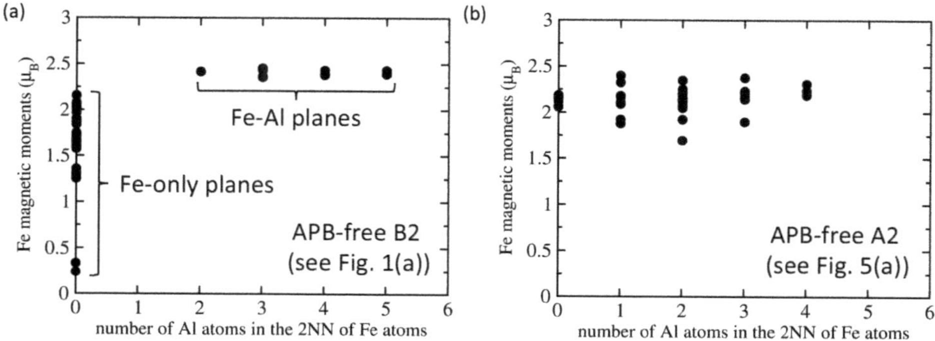

Figure A1. Calculated local magnetic moments of Fe atoms as a function of the number of Al atoms in their second nearest neighbor (2NN) shell of the Fe atoms. Part (**a**) summarizes them in the APB-free bulk B2 phase (visualized in Figure 1a) and part (**b**) for the APB-free bulk A2 phase shown in Figure 5a.

References

1. Sauthoff, G. *Intermetallics*; VCH Verlagsgesellschaft: Weinheim, Germany, 1995.

2. Liu, C.T.; Stringer, J.; Mundy, J.N.; Horton, L.L.; Angelini, P. Ordered intermetallic alloys: An assessment. *Intermetallics* **1997**, *5*, 579–596. doi:10.1016/S0966-9795(97)00045-9.

3. Stoloff, N.S. Iron aluminides: Present status and future prospects. *Mater. Sci. Eng. A* **1998**, *258*, 1–14. doi:10.1016/S0921-5093(98)00909-5.

4. Liu, C.T.; Lee, E.H.; McKamey, C.G. An environmental-effect as the major cause for room-temperature embrittlement in FeAl. *Scr. Metall. Mater.* **1989**, *23*, 875–880. doi:10.1016/0036-9748(89)90263-9.

5. Lynch, R.J.; Heldt, L.A.; Milligan, W.W. Effects of alloy composition on environmental embrittlement of B2 ordered iron aluminides. *Scr. Metall. Mater.* **1991**, *25*, 2147–2151. doi:10.1016/0956-716X(91)90290-H.

6. Liu, C.T.; McKamey, C.G.; Lee, E.H. Environmental-effects on room-temperature ductility and fracture in Fe₃Al. *Scr. Metall. Mater.* **1990**, *24*, 385–389. doi:10.1016/0956-716X(90)90275-L.

7. Lynch, R.J.; Gee, K.A.; Heldt, L.A. Environmental embrittlement of single-crystal and thermomechanically processed B2-ordered iron aluminides. *Scr. Metall. Mater.* **1994**, *30*, 945–950. doi:10.1016/0956-716X(94)90420-0.

8. Zamanzade, M.; Barnoush, A.; Motz, C. A Review on the Properties of Iron Aluminide Intermetallics. *Crystals* **2016**, *6*, 10. doi:10.3390/cryst6010010.

9. Kattner, U.; Burton, B. Al-Fe (Aluminium-Iron). In *Phase Diagrams of Binary Iron Alloys*; Okamoto, H., Ed.; ASM International: Materials Park, OH, USA, 1993; pp. 12–28.

10. Palm, M.; Inden, G.; Thomas, N. The Fe-Al-Ti system. *J. Phase Equilibria* **1995**, *16*, 209–222. doi:10.1007/BF02667305.

11. Palm, M.; Lacaze, J. Assessment of the Al-Fe-Ti system. *Intermetallics* **2006**, *14*, 1291–1303. doi:10.1016/j.intermet.2005.11.026.

12. Palm, M.; Sauthoff, G. Deformation behaviour and oxidation resistance of single-phase and two-phase L2₁-ordered Fe-Al-Ti alloys. *Intermetallics* **2004**, *12*, 1345–1359. doi:10.1016/j.intermet.2004.03.017.

13. Dobes, F.; Vodickova, V.; Vesely, J.; Kratochvil, P. The effect of carbon additions on the creep resistance of Fe-25Al-5Zr alloy. *Metall. Mater. Trans. A-Phys. Metall. Mater. Sci.* **2016**, *47A*, 6070–6076. doi:10.1007/s11661-016-3770-6.

14. Vernieres, J.; Benelmekki, M.; Kim, J.H.; Grammatikopoulos, P.; Bobo, J.F.; Diaz, R.E.; Sowwan, M. Single-step gas phase synthesis of stable iron aluminide nanoparticles with soft magnetic properties. *APL Mater.* **2014**, *2*, 116105. doi:10.1063/1.4901345.

15. Jirásková, Y.; Pizúrová, N.; Titov, A.; Janičkovič, D.; Friák, M. Phase separation in Fe-Ti-Al alloy—Structural, magnetic, and Mössbauer study. *J. Magn. Magn. Mater.* **2018**, *468*, 91–99. doi:10.1016/j.jmmm.2018.07.065.

16. Dobeš, F.; Dymáček, P.; Friák, M. Force-to-Stress Conversion Methods in Small Punch Testing Exemplified by Creep Results of Fe-Al Alloy with Chromium and Cerium Additions. *IOP Conf. Ser. Mater. Sci. Eng.* **2018**, *461*, 012017. doi:10.1088/1757-899x/461/1/012017.

17. Dobeš, F.; Dymáček, P.; Friák, M. Small punch creep of Fe-Al-Cr alloy with Ce addition and its relation to uniaxial creep tests. *Kovové Mater. Met. Mater.* **2018**, *56*, 205. doi:10.4149/km 2018 4 205.

18. Dymáček, P.; Dobeš, F.; Jirásková, Y.; Pizúrová, N.; Friák, M. Tensile, creep and fracture testing of prospective Fe-Al-based alloys using miniature specimens. *Theor. Appl. Fract. Mech.* **2019**, *99*, 18–26. doi:10.1016/j.tafmec.2018.11.005.

19. Dobeš, F.; Dymáček, P.; Friák, M. The Influence of Niobium Additions on Creep Resistance of Fe-27 at. % Al Alloys. *Metals* **2019**, *9*, 739. doi:10.3390/met9070739.

20. Watson, R.E.; Weinert, M. Transition-metal aluminide formation: Ti, V, Fe, and Ni aluminides. *Phys. Rev. B* **1998**, *58*, 5981–5988. doi:10.1103/PhysRevB.58.5981.

21. Gonzales-Ormeno, P.; Petrilli, H.; Schon, C. Ab-initio calculations of the formation energies of BCC-based superlattices in the Fe-Al system. *Calphad* **2002**, *26*, 573–582. doi:10.1016/S0364-5916(02)80009-8.

22. Friák, M.; Neugebauer, J. Ab initio study of the anomalous volume-composition dependence in Fe-Al alloys. *Intermetallics* **2010**, *18*, 1316–1321. doi:10.1016/j.intermet.2010.03.014.

23. Amara, H.; Fu, C.C.; Soisson, F.; Maugis, P. Aluminum and vacancies in α-iron: Dissolution, diffusion, and clustering. *Phys. Rev. B* **2010**, *81*, 174101. doi:10.1103/PhysRevB.81.174101.

24. Liu, S.; Duan, S.; Ma, B. First-principles calculation of vibrational entropy for Fe-Al compounds. *Phys. Rev. B* **1998**, *58*, 9705–9709.

25. Kulikov, N.I.; Postnikov, A.V.; Borstel, G.; Braun, J. Onset of magnetism in B2 transition-metal aluminides. *Phys. Rev. B* **1999**, *59*, 6824–6833. doi:10.1103/PhysRevB.59.6824.

26. Fähnle, M.; Drautz, R.; Lechermann, F.; Singer, R.; Diaz-Ortiz, A.; Dosch, H. Thermodynamic properties from ab-initio calculations: New theoretical developments, and applications to various materials systems. *Phys. Status Solidi B Basic Solid State Phys.* **2005**, *242*, 1159–1173. doi:10.1002/pssb.200440010.

27. Friák, M.; Deges, J.; Krein, R.; Frommeyer, G.; Neugebauer, J. Combined ab initio and experimental study of structural and elastic properties of Fe$_3$Al-based ternaries. *Intermetallics* **2010**, *18*, 1310. doi:10.1016/j.intermet.2010.02.025.

28. Kirklin, S.; Saal, J.E.; Hegde, V.I.; Wolverton, C. High-throughput computational search for strengthening precipitates in alloys. *Acta Mater.* **2016**, *102*, 125–135. doi:10.1016/j.actamat.2015.09.016.

29. Airiskallio, E.; Nurmi, E.; Heinonen, M.H.; Vayrynen, I.J.; Kokko, K.; Ropo, M.; Punkkinen, M.P.J.; Pitkanen, H.; Alatalo, M.; Kollar, J.; et al. High temperature oxidation of Fe-Al and Fe-Cr-Al alloys: The role of Cr as a chemically active element. *Corros. Sci.* **2010**, *52*, 3394–3404. doi:10.1016/j.corsci.2010.06.019.

30. Medvedeva, N.I.; Park, M.S.; Van Aken, D.C.; Medvedeva, J.E. First-principles study of Mn, Al and C distribution and their effect on stacking fault energies in fcc Fe. *J. Alloy. Compd.* **2014**, *582*, 475–482. doi:10.1016/j.jallcom.2013.08.089.

31. Čížek, J.; Lukáč, F.; Procházka, I.; Kužel, R.; Jirásková, Y.; Janičkovič, D.; Anwand, W.; Brauer, G. Characterization of quenched-in vacancies in Fe-Al alloys. *Phys. B* **2012**, *407*, 2659–2664. doi:10.1016/j.physb.2011.12.122.

32. Ipser, H.; Semenova, O.; Krachler, R. Intermetallic phases with D0(3)-structure: A statistical-thermodynamic model. *J. Alloy. Compd.* **2002**, *338*, 20–25. doi:10.1016/S0925-8388(02)00177-9.

33. Lechermann, F.; Welsch, F.; Elsässer, C.; Ederer, C.; Fähnle, M.; Sanchez, J.; Meyer, B. Density-functional study of Fe$_3$Al: LSDA versus GGA. *Phys. Rev. B* **2002**, *65*, 132104. doi:10.1103/PhysRevB.132104.

34. Connetable, D.; Maugis, P. First, principle calculations of the kappa-Fe$_3$AlC perovskite and iron-aluminium intermetallics. *Intermetallics* **2008**, *16*, 345–352. doi:10.1016/j.intermet.2007.09.011.

35. Lechermann, F.; Fähnle, M.; Meyer, B.; Elsässer, C. Electronic correlations, magnetism, and structure of Fe-Al subsystems: An LDA+U study. *Phys. Rev. B* **2004**, *69*, 165116. doi:10.1103/PhysRevB.69.165116.

36. Kellou, A.; Grosdidier, T.; Raulot, J.M.; Aourag, H. Atomistic study of magnetism effect on structural stability in Fe$_3$Al and Fe$_3$AlX (X = H, B, C, N, O) alloys. *Phys. Status Solidi B Basic Solid State Phys.* **2008**, *245*, 750–755. doi:10.1002/pssb.200743301.

37. Šesták, P.; Friák, M.; Holec, D.; Všianská, M.; Šob, M. Strength and brittleness of interfaces in Fe-Al superalloy nanocomposites under multiaxial loading: An ab initio and atomistic study. *Nanomaterials* **2018**, *8*, 873. doi:10.3390/nano8110873.

38. Friák, M.; Slávik, A.; Miháliková, I.; Holec, D.; Všianská, M.; Šob, M.; Palm, M.; Neugebauer, J. Origin of the low magnetic moment in Fe$_2$AlTi: An Ab initio study. *Materials* **2018**, *11*, 1732. doi:10.3390/ma11091732.

39. Miháliková, I.; Friák, M.; Jirásková, Y.; Holec, D.; Koutná, N.; Šob, M. Impact of Nano-Scale Distribution of Atoms on Electronic and Magnetic Properties of Phases in Fe-Al Nanocomposites: An Ab Initio Study. *Nanomaterials* **2018**, *8*, 1059. doi:10.3390/nano8121059.

40. Friák, M.; Holec, D.; Šob, M. Quantum-Mechanical Study of Nanocomposites with Low and Ultra-Low Interface Energies. *Nanomaterials* **2018**, *8*, 1057. doi:10.3390/nano8121057.

41. Miháliková, I.; Friák, M.; Koutná, N.; Holec, D.; Šob, M. An Ab Initio Study of Vacancies in Disordered Magnetic Systems: A Case Study of Fe-Rich Fe-Al Phases. *Materials* **2019**, *12*, 1430. doi:10.3390/ma12091430.

42. Marcinkowski, M.; Brown, N. Theory and direct observation of dislocations in the Fe3Al superlattices. *Acta Metall.* **1961**, *9*, 764–786. doi:10.1016/0001-6160(61)90107-9.

43. Marcinkowski, M.J.; Brown, N. Direct Observation of Antiphase Boundaries in the Fe$_3$Al Superlattice. *J. Appl. Phys.* **1962**, *33*, 537–552. doi:10.1063/1.1702463.

44. McKamey, C.G.; Horton, J.A.; Liu, C.T. Effect of chromium on properties of Fe$_3$Al. *J. Mater. Res.* **1989**, *4*, 1156–1163. doi:10.1557/JMR.1989.1156.

45. Morris, D.; Dadras, M.; Morris, M. The influence of Cr addition on the ordered microstructure and deformation and fracture-behavior of a Fe-28-%-Al intermetallic. *Acta Metall. Mater.* **1993**, *41*, 97–111. doi:10.1016/0956-7151(93)90342-P.

46. Kral, F.; Schwander, P.; Kostorz, G. Superdislocations and antiphase boundary energies in deformed Fe$_3$Al single crystals with chromium. *Acta Mater.* **1997**, *45*, 675–682. doi:10.1016/S1359-6454(96)00181-4.

47. Allen, S.; Cahn, J. Microscopic theory for antiphase boundary motion and its application to antiphase domain coarsening. *Acta Metall.* **1979**, *27*, 1085–1095. doi:10.1016/0001-6160(79)90196-2.

48. Wang, K.; Wang, Y.; Cheng, Y. The Formation and Dynamic Evolution of Antiphase Domain Boundary in FeAl Alloy: Computational Simulation in Atomic Scale. *Mater. Res. Ibero-Am. J. Mater.* **2018**, *21*. doi:10.1590/1980-5373-MR-2017-1048.

49. Balagurov, A.M.; Bobrikov, I.A.; Sumnikov, V.S.; Golovin, I.S. Antiphase domains or dispersed clusters? Neutron diffraction study of coherent atomic ordering in Fe$_3$Al-type alloys. *Acta Mater.* **2018**, *153*, 45–52. doi:10.1016/j.actamat.2018.04.015.

50. Murakami, Y.; Niitsu, K.; Tanigaki, T.; Kainuma, R.; Park, H.S.; Shindo, D. Magnetization amplified by structural disorder within nanometre-scale interface region. *Nat. Commun.* **2014**, *5*, 4133. doi:0.1038/ncomms5133.

51. Oguma, R.; Matsumura, S.; Eguchi, T. Kinetics of B2-and D0$_3$ type ordering and formation of domain structures in Fe-Al alloys. *J. Phys. Condens. Matter* **2008**, *20*, 275225. doi:10.1088/0953-8984/20/27/275225.

52. Friák, M.; Všianská, M.; Šob, M. A Quantum–Mechanical Study of Clean and Cr–Segregated Antiphase Boundaries in Fe$_3$Al. *Materials* **2019**, *12*, 3954. doi:10.3390/ma12233954.

53. Kresse, G.; Hafner, J. Ab initio molecular dynamics for liquid metals. *Phys. Rev. B* **1993**, *47*, 558–561. doi:10.1103/PhysRevB.47.558.

54. Kresse, G.; Furthmüller, J. Efficient iterative schemes for ab initio total-energy calculations using a plane-wave basis set. *Phys. Rev. B* **1996**, *54*, 11169–11186. doi:10.1103/PhysRevB.54.11169.

55. Hohenberg, P.; Kohn, W. Inhomogeneous electron gas. *Phys. Rev. B* **1964**, *136*, B864–B871. doi:10.1103/PhysRev.136.B864.

56. Kohn, W.; Sham, L.J. Self-consistent equations including exchange and correlation effects. *Phys. Rev. A* **1965**, *140*, A1133–A1138. doi:10.1103/PhysRev.140.A1133.

57. Blöchl, P.E. Projector augmented-wave method. *Phys. Rev. B* **1994**, *50*, 17953–17979. doi:10.1103/PhysRevB.50.17953.

58. Kresse, G.; Joubert, D. From ultrasoft pseudopotentials to the projector augmented-wave method. *Phys. Rev. B* **1999**, *59*, 1758–1775. doi:10.1103/PhysRevB.59.1758.

59. Perdew, J.P.; Wang, Y. Accurate and simple analytic representation of the electron-gas correlation energy. *Phys. Rev. B* **1992**, *45*, 13244–13249. doi:10.1103/PhysRevB.45.13244.

60. Vosko, S.H.; Wilk, L.; Nusair, M. Accurate spin-dependent electron liquid correlation energies for local spin density calculations: A critical analysis. *Can. J. Phys.* **1980**, *58*, 1200. doi:10.1139/p80-159.

61. Zunger, A.; Wei, S.; Ferreira, L.; Bernard, J. Special quasirandom structures. *Phys. Rev. Lett.* **1990**, *65*, 353–356. doi:10.1103/PhysRevLett.65.353.

62. Oganov, A.R.; Glass, C.W. Crystal structure prediction using ab initio evolutionary techniques: Principles and applications. *J. Chem. Phys.* **2006**, *124*, 244704. doi:10.1063/1.2210932.

63. Lyakhov, A.O.; Oganov, A.R.; Stokes, H.T.; Zhu, Q. New developments in evolutionary structure prediction algorithm USPEX. *Comput. Phys. Commun.* **2013**, *184*, 1172–1182. doi:10.1016/j.cpc.2012.12.009.

64. Oganov, A.R.; Lyakhov, A.O.; Valle, M. How Evolutionary Crystal Structure Prediction Works—In addition, Why. *Acc. Chem. Res.* **2011**, *44*, 227–237. doi:10.1021/ar1001318.

65. Sundman, B.; Ohnuma, I.; Dupin, N.; Kattner, U.R.; Fries, S.G. An assessment of the entire Al–Fe system including D0$_3$ ordering. *Acta Mater.* **2009**, *57*, 2896–2908. doi:10.1016/j.actamat.2009.02.046.

66. Miracle, D.; Senkov, O. A critical review of high entropy alloys and related concepts. *Acta Mater.* **2017**, *122*, 448–511. doi:10.1016/j.actamat.2016.08.081.

67. Hillert, M. *Phase Equilibria, Phase Diagrams and Phase Transformations: Their Thermodynamic Basis*, 2nd ed.; Cambridge University Press: Cambridge, UK, 2008.

68. Stein, F.; Palm, M. Re-determination of transition temperatures in the Fe–Al system by differential thermal analysis. *Int. J. Mater. Res.* **2007**, *98*, 580–588. doi:10.3139/146.101512.

69. Friák, M.; Buršíková, V.; Pizúrová, N.; Pavlů, J.; Jirásková, Y.; Homola, V.; Miháliková, I.; Slávik, A.; Holec, D.; Všianská, M.; et al. Elasticity of Phases in Fe-Al-Ti Superalloys: Impact of Atomic Order and Anti-Phase Boundaries. *Crystals* **2019**, *9*, 299. doi:10.3390/cryst9060299.

70. Momma, K.; Izumi, F. VESTA 3 for three-dimensional visualization of crystal, volumetric and morphology data. *J. Appl. Crystallogr.* **2011**, *44*, 1272–1276. doi:10.1107/S0021889811038970.

MDPI
St. Alban-Anlage 66
4052 Basel
Switzerland
Tel. +41 61 683 77 34
Fax +41 61 302 89 18
www.mdpi.com

Nanomaterials Editorial Office
E-mail: nanomaterials@mdpi.com
www.mdpi.com/journal/nanomaterials